LONDON MATHEMATICAL SOCIETY LECTURE NOTE SERIES

Managing Editor: Professor J.W.S. Cassels, Department of Pure Mathematics and Mathematical Statistics, University of Cambridge, 16 Mill Lane, Cambridge CB2 1SB, England

The titles below are available from booksellers, or, in case of difficulty, from Cambridge University Press.

L'impression qui se dégageait de cette tête était celle d'une force juvénile, joyeuse, qui semblait jaillir... L'expression était celle d'un garnement espiègle, ravi de quelque coup qu'il viendrait ou méditerait de faire...Il s'en dégageait surtout une joie de vivre intense, contenue, fusant en jeu.

Alexandre Grothendieck, *Récoltes et Semailles*

London Mathematical Society Lecture Note Series. 242

Geometric Galois Actions

1. Around Grothendieck's Esquisse d'un Programme

Edited by

Leila Schneps
CNRS

and

Pierre Lochak
CNRS

CAMBRIDGE
UNIVERSITY PRESS

CAMBRIDGE UNIVERSITY PRESS
Cambridge, New York, Melbourne, Madrid, Cape Town,
Singapore, São Paulo, Delhi, Tokyo, Mexico City

Cambridge University Press
The Edinburgh Building, Cambridge CB2 8RU, UK

Published in the United States of America by
Cambridge University Press, New York

www.cambridge.org
Information on this title: www.cambridge.org/9780521596428

First published 1997

A catalogue record for this publication is available from the British Library

Library of Congress Cataloguing in Publication data

ISBN 978-0-521-59642-8 Paperback

Table of Contents

Preface

The idea of this book germinated during the Luminy conference on "Geometry and Arithmetic of Moduli Spaces", which took place in late August 1995. Our goal in organizing that conference was to encourage the natural generation of ideas which occurs when specialists in a domain are confined weeklong at close quarters with non-specialists who urgently need to learn about that domain for the purposes of their particular research. Moduli spaces are of course ubiquitous, but the research we particularly had in mind was any concerning topics in the early part of Grothendieck's *Esquisse d'un Programme*.

The *Esquisse* was not published earlier because Alexandre Grothendieck could not be found, much less his permission requested. It was during the Luminy conference that we learned from Jean Malgoire that he had obtained permission from Grothendieck to publish all and any of Grothendieck's mathematical manuscripts which lay in his possession; he told us of this and suggested that we include the *Esquisse* in a book of proceedings of the conference, a suggestion which we welcomed with enthusiasm. We thought of producing a book of proceedings which could appear as a kind of sequel to *The Grothendieck Theory of Dessins d'Enfants**, itself a volume of proceedings of a previous Luminy conference.

The inclusion of the *Esquisse* among the "contributions" generated enthusiasm among the participants certainly most unusual in such situations, and we quickly gathered more contributions than could be included in a single volume. Furthermore, we were offered – and in some cases, we solicited – manuscripts from a number of people close to the subject and deeply interested by the *Esquisse*, who were not present at the conference. In particular, the "Grothendieck day" in Utrecht organized by F. Oort was a fruitful source of relevant texts, most notably the famous unpublished letter from Grothendieck to G. Faltings, a copy of which the latter brought to Utrecht. He kindly appended his signature to the letter we presented to him authorizing ourselves to include Grothendieck's letter in this book.

The plethora of contributions led us finally to the following division: a first volume (the present one) containing the letter and the *Esquisse* together with a series of easily accessible texts chosen either to clarify points

---

* this series, n° 200

in the *Esquisse*, such as F. Oort's general introduction to Grothendieck's theory of the algebraic fundamental group (actually written for the Utrecht "Grothendieck day", as was the very short introduction to the published portion of Grothendieck's *Longue Marche à travers la théorie de Galois*) and the article on "π_1 at infinity" defining and explaining the meaning and context of this phrase used by Grothendieck. The contribution by J. Wolfart gives an accessible proof of the "easy" part of Belyi's theorem and also discusses the question of when compact Riemann surfaces have many automorphisms. The article by T. Oda gives a firm foundation to the theory of algebraic fundamental groups of stacks, notably the moduli spaces of curves; this article is a slightly emended version of a manuscript which has circulated in unpublished form since 1990, and we are particularly grateful for the opportunity to publish it here.

Other texts survey progress in areas covered in the letter and in the *Esquisse*, for instance the two on anabelian geometry by F. Pop and by Y. Ihara and H. Nakamura which explore the panorama of the anabelian theory conceived by Grothendieck in his letter to Faltings, and the articles on $\widehat{GT}$ and on Galois orbits of dessins d'enfants, which represent an exploration in the direction of Grothendieck's suggestion of characterizing the absolute Galois group in a combinatorial way via its action on profinite fundamental groups of moduli spaces, and explicitly determining its action on dessins. H. Nakamura's contribution represents an advance in this direction, explicitly generalizing the known Galois action on the fundamental (braid) groups of the genus zero moduli spaces to an action on the mapping class groups of all genera. We are grateful to J-P. Serre for his permission to publish here two letters containing unpublished results on non-abelian cohomology for good groups, beautiful in themselves and crucial to the proofs in some of the articles above. Finally, we are pleased to include the contribution by B. Teissier, containing a short critical and prospective survey on a part of the *Esquisse* which has hitherto received comparatively little attention, namely that concerning the foundations of "tame topology".

The remaining contributions received in the months following the Luminy conference, mostly research articles dealing with more specialized aspects of the topics discussed above, are collected in the companion volume to this book, *Geometric Galois Actions II*, subtitled *Inverse Galois Problem, Moduli and Mapping Class Groups*. As for the "unpublishable" and anonymous contributions of time, editorial help and moral support in the face of all the technical difficulties inherent in such a project, we insert here a special word of thanks to Xavier Buff and Leonardo Zapponi, as well as to all the referees.

Before closing this preface, we add a word on the translations of the letter to Faltings and the *Esquisse d'un Programme* which we have included here. The original texts were typed on Grothendieck's portable typewriter; we preserved the original numbering of the pages in the margin. First and foremost, these translations, with all their inadequacies and defects, are entirely the responsibility of the editors. We tried to render Grothendieck's extraordinarily rich and original usage of his two mother tongues, but (as any translator knows) it is not possible to succeed in such a project satisfactorily. In particular, conceiving of the translations principally as an aid to readers who, we hope, will attempt as much as possible to read the two texts in the original, we chose to adhere to an accurate respect of Grothendieck's style and sentence structure, sometimes producing a rather curious English.

We dedicate all our efforts to Alexandre Grothendieck, with warmest sincerity and affection.

Leila Schneps and Pierre Lochak

Paris, October 15th, 1996

ESQUISSE D'UN PROGRAMME

par Alexandre Grothendieck

N.B. Les astérisques (*) renvoient aux notes figurant au bas de la même page, les renvois numérotés de (^{1}) à (^{7}) aux notes (rajoutées ultérieurement) réunies à la fin du rapport.

ESQUISSE D'UN PROGRAMME

par Alexandre Grothendieck

1. Comme la conjoncture actuelle rend de plus en plus illusoire pour moi les perspectives d'un enseignement de recherche à l'Université, je me suis résolu à demander mon admission au CNRS, pour pouvoir consacrer mon énergie à développer des travaux et perspectives dont il devient clair qu'il ne se trouvera aucun élève (ni même, semble-t-il, aucun congénère mathématicien) pour les développer à ma place.

En guise de document "Titres et Travaux", on trouvera à la suite de ce texte la reproduction intégrale d'une esquisse, par thèmes, de ce que je considérais comme mes principales contributions mathématiques au moment d'écrire ce rapport, en 1972. Il contient également une liste d'articles publiés à cette date. J'ai cessé toute publication d'articles scientifiques depuis 1970. Dans les lignes qui suivent, je me propose de donner un aperçu au moins sur quelques thèmes principaux de mes réflexions mathématiques depuis lors. Ces réflexions se sont matérialisées au cours des années en deux volumineux cartons de notes manuscrites, difficilement déchiffrables sans doute à tout autre qu'à moi-même, et qui, après des stades de décantations successives, attendent leur heure peut-être pour une rédaction d'ensemble tout au moins provisoire, à l'intention de la communauté mathématique. Le terme "rédaction" ici est quelque peu impropre, alors qu'il s'agit bien plus de développer des idées et visions multiples amorcées au cours de ces douze dernières années, en les précisant et les approfondissant, avec tous les rebondissements imprévus qui constamment accompagnent ce genre de travail – un travail de découverte donc, et non de compilation de notes pieusement accumulées. Et je compte bien, dans l'écriture des "Réflexions Mathématiques" commencée depuis février 1983, laisser apparaître clairement au fil des pages la démarche de la pensée qui sonde et qui découvre, en tâtonnant dans la pénombre bien souvent, avec des trouées de lumière subites quand quelque tenace image fausse, ou simplement inadéquate, se trouve enfin débusquée et mise à jour, et que les choses qui semblaient de guingois se mettent en place, dans l'harmonie mutuelle qui leur est propre.

Quoi qu'il en soit, l'esquisse qui suit de quelques thèmes de réflexions des dernières dix ou douze années, tiendra lieu en même temps d'esquisse de programme de travail pour les années qui viennent, que je compte consacrer au développement de ces thèmes, ou au moins de certains d'entre eux. Elle est destinée, d'une part aux collègues du Comité National appelés à statuer sur ma demande, d'autre part à quelques autres collègues, anciens élèves, amis, dans l'éventualité où certaines des idées esquissées ici pourraient intéresser l'un d'entre eux.

2. Les exigences d'un enseignement universitaire, s'adressant donc à des étudiants (y compris les étudiants dits "avancés") au bagage mathématique modeste (et souvent moins que modeste), m'ont amené à renouveler de façon draconienne les thèmes de réflexion à proposer à mes élèves, et de fil en aiguille et de plus en plus, à moi-même également. Il m'avait semblé important de partir d'un bagage intuitif commun, indépendant de tout langage technique censé l'exprimer, bien antérieur à tout tel langage – il s'est avéré que l'intuition géométrique et topologique des formes, et plus particulièrement des formes bidimensionnelles, était un tel terrain commun. Il s'agit donc de thèmes qu'on peut grouper sous l'appellation de "topologie des surfaces" ou "géométrie des surfaces", étant entendu dans cette dernière appellation que l'accent principal se trouve sur les propriétés topologiques des surfaces, ou sur les aspects combinatoires qui en constituent l'expression technique la plus terre-à-terre, et non sur les aspects différentiels, voire conformes, riemaniens, holomorphes et (de là) l'aspect "courbes algébriques complexes". Une fois ce dernier pas franchi cependant, voici soudain la géométrie algébrique (mes anciennes amours!) qui fait irruption à nouveau, et ce par les objets qu'on peut considérer comme les pierres de construction ultimes de toutes les autres variétés algébriques. Alors que dans mes recherches d'avant 1970, mon attention systématiquement était dirigée vers les objets de généralité maximale, afin de dégager un langage d'ensemble adéquat pour le monde de la géométrie algébrique, et que je ne m'attardais sur les courbes algébriques que dans la stricte mesure où cela s'avérait indispensable (notamment en cohomologie étale) pour développer des techniques et énoncés "passe-partout" valables en toute dimension et en tous lieux (j'entends, sur tous schémas de base, voire tous topos annelés de base...), me voici donc ramené, par le truchement d'objets si simples qu'un enfant peut les connaître en jouant, aux débuts et origines de la géométrie algébrique, familiers à Riemann et à ses émules!

Depuis environ 1975, c'est donc la géométrie des surfaces (réelles), et à partir de 1977 les liens entre les questions de géométrie des surfaces et la géométrie algébrique des courbes algébriques définies sur des corps tels que C, R ou des extensions de type fini de Q, qui ont été ma principale source d'inspiration, ainsi que mon fil conducteur constant. C'est avec surprise et avec émerveillement qu'au fil des ans je découvrais (ou plutôt, sans doute, redécouvrais) la richesse prodigieuse, réellement inépuisable, la profondeur insoupçonnée de ce thème, d'apparence si anodine. Je crois y sentir un point névralgique entre tous, un point de convergence privilégié des principaux courants d'idées mathématiques, comme aussi des principales structures et des visions des choses qu'elles expriment, depuis les plus spécifiques, (tels les anneaux $\mathbf{Z}, \mathbf{Q}, \overline{\mathbf{Q}}, \mathbf{R}, \mathbf{C}$ ou le groupe Sl(2) sur l'un de ces anneaux, ou les groupes algébriques réductifs généraux) aux plus "abstraits", telles les "multiplicités" algébriques, analytiques complexes ou analytiques réelles. (Celles-ci s'introduisent naturellement quand il s'agit d'étudier systématiquement des "variétés de modules" pour les objets géométriques envisagés, si on veut dépasser le point de vue notoirement insuffisant des "modules grossiers", qui revient à "tuer" bien malencontreusement les groupes d'automorphismes de ces objets.) Parmi ces multiplicités modulaires, ce sont celles de Mumford-Deligne pour les courbes algébriques "stables" de genre g, à ν points marqués, que je note $\widehat{M}_{g,\nu}$ (compactification de la multiplicité "ouverte" $M_{g,\nu}$ correspondant aux courbes lisses), qui depuis quelques deux ou trois années ont exercé sur moi une fascination particulière, plus forte peut-être qu'aucun autre objet mathématique à ce jour. A vrai dire, il s'agit plutôt du système de toutes les multiplicités $M_{g,\nu}$ pour g, ν variables, liées entre elles par un certain nombre d'opérations fondamentales (telles les opérations de "bouchage de trous" i.e. de "gommage" de points marqués, celle de "recollement", et les opérations inverses), qui sont le reflet en géométrie algébrique absolue de caractéristique zéro (pour le moment) d'opérations géométriques familières du point de vue de la "chirurgie" topologique ou conforme des surfaces. La principale raison sans doute de cette fascination, c'est que cette structure géométrique très riche sur le système des multiplicités modulaires "ouvertes" $M_{g,\nu}$ se réflète par une structure analogue sur les groupoïdes fondamentaux correspondants, les "groupoïdes de Teichmüller" $\widehat{T}_{g,\nu}$, et que ces opérations au niveau des $\widehat{T}_{g,\nu}$ ont un caractère suffisamment intrinsèque pour que le groupe de Galois Γ de $\overline{\mathbf{Q}}/\mathbf{Q}$ opère sur toute cette "tour" de groupoïdes de Teichmüller, en respectant toutes ces structures. Chose plus extraordinaire encore, cette opération est fidèle – à vrai dire, elle est fidèle déjà sur le premier "étage" non trivial de cette tour, à savoir $\widehat{T}_{0,4}$ – ce qui signifie aussi, essentiellement, que l'action extérieure de Γ sur le groupe fondamental $\hat{\pi}_{0,3}$ de la droite projective standard $\mathbf{P}^1$ sur $\mathbf{Q}$, privée des trois points 0, 1, ∞, est déjà fidèle. Ainsi le groupe de Galois Γ se réalise comme un groupe d'automorphismes d'un groupe profini des plus concrets, respectant d'ailleurs certaines structures essentielles de ce groupe. Il s'ensuit qu'un élément de Γ peut être "paramétré" (de diverses façons équivalentes d'ailleurs) par un élément convenable de ce groupe profini $\hat{\pi}_{0,3}$ (un groupe profini libre à deux générateurs), ou par un système de tels éléments, ce ou ces éléments étant d'ailleurs soumis à certaines conditions simples, nécessaires (et sans doute non suffisantes) pour que ce ou ces éléments correspond(nt) bien à un élément de Γ. Une des tâches les plus fascinantes ici, est justement d'appréhender des conditions nécessaires et suffisantes sur un automorphisme extérieur de $\widehat{\pi}_{0,3}$ i.e. sur le ou les paramètres correspondants, pour qu'il provienne d'un élément de Γ – ce qui fournirait une description "purement algébrique", en termes de groupes profinis et sans référence à la théorie de Galois des corps de nombres, du groupe de Galois $\Gamma = \mathrm{Gal}(\overline{\mathbf{Q}}/\mathbf{Q})$!

Peut-être une caractérisation même conjecturale de Γ comme sous-groupe de $\mathrm{Autext}(\widehat{\pi}_{0,3})$ est-elle pour le moment hors de portée (^{1}); je n'ai pas de conjecture à proposer encore. Une autre tâche par contre est abordable immédiatement, c'est celle de décrire l'action de Γ sur toute la tour de Teichmüller, en termes de son action sur le "premier étage" $\widehat{\pi}_{0,3}$, i.e. exprimer un automorphisme de cette tour, en termes du "paramètre" dans $\widehat{\pi}_{0,3}$, qui repère l'élément courant γ de Γ. Ceci est lié à une représentation de la tour de Teichmüller (en tant que groupoïde muni d'une opération de "recollement") par générateurs et relations, qui donnera en particulier une présentation par générateurs et relations, au sens ordinaire, de chacun des $\widehat{T}_{g,\nu}$ (en tant que groupoïde profini). Ici, même pour $g = 0$ (donc quand les groupes de Teichmüller correspondants sont des groupes de tresses "bien connus"), les générateurs et relations connus à ce jour dont j'ai eu connaissance, me semblent inutilisables tels quels, car ils ne présentent pas les caractères d'invariance et de symétrie indispensables pour que l'action de Γ soit directement lisible sur cette présentation. Ceci est lié notamment au fait que les gens s'obstinent encore, en calculant avec des groupes fondamentaux, à fixer un seul point base, plutôt que d'en choisir astucieusement tout un paquet qui soit invariant par les symétries de la situation, lesquelles sont donc perdues en route. Dans certaines situations (comme des théorèmes de descente à la Van Kampen pour groupes fondamentaux) il est bien plus élégant, voire indispensable pour y comprendre quelque chose, de travailler avec des groupoïdes fondamentaux par rapport à un paquet de points base convenable, et il en est certainement ainsi pour la tour de Teichmüller. Il semblerait (incroyable, mais vrai!) que la géométrie même du premier étage de la tour de Teichmüller (correspondant donc aux "modules" soit pour des droites projectives avec quatre points marqués, soit pour des courbes elliptiques (!)) n'ait jamais été bien explicitée, par exemple la relation entre le cas de genre 0 avec la géométrie de l'octaèdre, et celle du tétraèdre. A fortiori les multiplicités modulaires $M_{0,5}$ (pour les droites projectives avec cinq points marqués) et $M_{1,2}$ (pour les courbes de genre 1 avec deux points marqués), d'ailleurs quasiment isomorphes entre elles, semblent-elles terre vierge – les groupes de tresses ne vont pas nous éclairer à leur sujet! J'ai commencé à regarder $M_{0,5}$ à des moments perdus, c'est un véritable joyau, d'une géométrie très riche étroitement liée à celle de l'icosaèdre.

L'intérêt a priori d'une connaissance complète des deux premiers étages de la tour (savoir, les cas où la dimension modulaire $N = 3g - 3 + \nu$ est ≤ 2) réside dans ce principe, que la tour entière se reconstitue à partir des deux premiers étages, en ce sens que via l'opération fondamentale de "recollement", l'étage 1 fournit un système complet de générateurs, et l'étage 2 un système complet de relations. Il y a une analogie frappante, et j'en suis persuadé, pas seulement formelle, entre ce principe, et le principe analogue de Demazure pour la structure des groupes algébriques réductifs, si on remplace le terme "étage" ou "dimension modulaire" par "rang semi-simple du groupe réductif". Le lien devient plus frappant encore, si on se rappelle que le groupe de Teichmüller $T_{1,1}$ (dans le contexte discret transcendant maintenant, et non dans le contexte algébrique profini, où on trouve les complétions profinies des premiers) n'est autre que $\mathrm{Sl}(2, \mathbf{Z})$, i.e. le groupe des points entiers du schéma en groupes simple de rang 1 "absolu" $\mathrm{Sl}(2)_{\mathbf{Z}}$. Ainsi, la pierre de construction fondamentale pour la tour de Teichmüller, est essentiellement la même que celle pour "la tour" des groupes réductifs de tous rangs – un groupe d'ailleurs dont on peut dire sans doute qu'il est présent dans toutes les disciplines essentielles des mathématiques.

Ce principe de construction de la tour de Teichmüller n'est pas démontré à l'heure actuelle – mais je n'ai aucun doute qu'il ne soit valable. Il résulterait (via une théorie de dévissage des structures stratifiées – en l'occurrence les $\widehat{M}_{g,\nu}$ – qui resterait à écrire, cf. par. 5) d'une propriété extrêmement plausible des multiplicités modulaires ouvertes $M_{g,\nu}$ dans le contexte analytique complexe, à savoir que pour une dimension modulaire $N \geq 3$, le groupe fondamental de $M_{g,\nu}$ (i.e. le groupe de Teichmüller habituel $T_{g,\nu}$) est isomorphe au "groupe fondamental à l'infini" i.e. celui d'un "voisinage tubulaire de l'infini". C'est là une chose bien familière (due à Lefschetz essentiellement) pour une variété lisse affine de dimension $N \geq 3$. Il est vrai que les multiplicités modulaires ne sont pas affines (sauf pour des petites valeurs de g), mais il suffirait qu'une telle $M_{g,\nu}$ de dimension N (ou plutôt, un revêtement fini convenable) soit réunion de $N - 2$ ouverts affines, donc que $M_{g,\nu}$ ne soit pas "trop proche d'une variété compacte".

N'ayant aucun doute sur ce principe de construction de la tour de Teichmüller, je préfère laisser aux experts de la théorie transcendante, mieux outillés que moi, le soin de prouver le nécessaire (s'il s'en trouve qui soit intéressé), pour expliciter plutôt, avec tout le soin qu'elle mérite, la structure qui en découle pour la tour de Teichmüller par générateurs et relations, dans le cadre discret cette fois et non profini – ce qui revient, essentiellement, à une compréhension complète des quatre multiplicités modulaires $M_{0,4}$, $M_{1,1}$, $M_{0,5}$, $M_{1,2}$, et de leurs groupoïdes fondamentaux par rapport à des "points base" convenablement choisis. Ceux-ci s'offrent tout naturellement, comme les courbes algébriques complexes du type (g, ν) envisagé, qui ont un groupe d'automorphismes (nécessairement fini) plus grand que dans le cas générique (*). En y incluant la sphère holomorphe à trois points marqués (provenant de $M_{0,3}$ i.e. de l'étage 0), on trouve douze "pièces de construction" fondamentales (6 de genre 0, 6 de genre 1) dans un "jeu de Légo-Teichmüller" (grande boîte), où les points marqués sur les surfaces envisagées sont remplacés par des "trous" à bord, de façon à avoir des surfaces à bord, donc des pièces de construction qui peuvent s'assembler par frottement doux comme dans le jeu de Légo ordinaire cher à nos enfants (ou petits-enfants...). Par assemblage on trouve un moyen tout ce qu'il y a de visuel pour construire tout type de surface (ce sont ces assemblages essentiellement qui seront les "points base" pour notre fameuse tour), et aussi de visualiser les "chemins" élémentaires par des opérations tout aussi concrètes telles des "twists", ou des automorphismes des pièces du jeu, et d'écrire les relations fondamentales entre chemins composés. Suivant la taille (et le prix!) de la boîte de construction utilisée, on trouve d'ailleurs de nombreuses descriptions différentes de la tour de Teichmüller par générateurs et relations. La boîte la plus petite est réduite à des pièces toutes identiques, de type $(0,3)$ – ce sont les "pantalons" de Thurston, et le jeu de Légo-Teichmüller que j'essaie de décrire, issu de motivations et de réflexions de géométrie algébrique absolue sur le corps $\mathbf{Q}$, est très proche du jeu de "chirurgie géodésique hyperbolique" de Thurston, dont j'ai appris l'existence l'an dernier par Yves Ladegaillerie. Dans un microséminaire avec Carlos Contou-Carrère et Yves Ladegaillerie, nous avons amorcé une réflexion dont un des objets est de confronter les deux points de vue, qui se complètent mutuellement.

J'ajoute que chacune des douze pièces de construction de la "grande boîte" se trouve munie d'une décomposition cellulaire canonique, stable par toutes les symétries, ayant comme seuls sommets les "points marqués" (ou centres des trous), et comme arêtes certains chemins géodésiques (pour la structure riemanienne canonique sur la sphère ou le tore envisagé) entre certaines paires de sommets (savoir ceux qui se trouvent sur un même

(*) Il faut y ajouter de plus les "points-base" provenant par opérations de recollement de "pièces" du même type en dimension modulaire inférieure. D'autre part, en dimension modulaire 2 (cas de $M_{0,5}$ et $M_{1,2}$), il convient d'exclure les points de certaines familles à un paramètre de courbes admettant un automorphisme exceptionnel d'ordre 2. Ces familles constituent d'ailleurs sur les multiplicités envisagées des courbes rationnelles remarquables, qui me paraissent un ingrédient important de la structure de ces multiplicités.

"lieu réel", pour une structure réelle convenable de la courbe algébrique complexe envisagée). Par suite, toutes les surfaces obtenues dans ce jeu par assemblage sont munies de structures cellulaires canoniques, qui à leur tour (cf. §3 plus bas) permettent de considérer ces surfaces comme associée à des courbes algébriques complexes (et même sur $\overline{\mathbf{Q}}$) canoniquement déterminées. Il y a là un jeu de chassé-croisé typique entre le combinatoire, et l'algébrique complexe (ou mieux, l'algébrique sur $\overline{\mathbf{Q}}$).

La "petite boîte" aux pièces toutes identiques, qui a le charme de l'économie, donnera sans doute une description relativement compliquée pour les relations (compliquée, mais nullement inextricable!). La grande boîte donnera lieu à des relations plus nombreuses (du fait qu'il y a beaucoup plus de points-bases et de chemins remarquables entre eux), mais à structure plus transparente. Je prévois qu'en dimension modulaire 2, tout comme dans le cas plus ou moins familier de la dimension modulaire 1 (avec notamment la description de $\mathrm{Sl}(2, \mathbf{Z})$ par $(\rho, \sigma \,|\, \rho^3 = \sigma^2 \,,\, \rho^4 = \sigma^6 = 1))$, on trouvera un engendrement par les groupes d'automorphismes des trois types de pièces pertinentes, avec des relations simples que je n'ai pas dégagées à l'heure d'écrire ces lignes. Peut-être même trouvera-t-on un principe de ce genre pour tous les $T_{g,\nu}$, ainsi qu'une décomposition cellulaire de $\widehat{M}_{g,\nu}$ généralisant celles qui se présentent spontanément pour $\widehat{M}_{0,4}$ et $\widehat{M}_{1,1}$, et que j'entrevois dès à présent pour la dimension modulaire 2, en utilisant les hypersurfaces correspondant aux diverses structures réelles sur les structures complexes envisagées, pour effectuer le découpage cellulaire voulu.

3. Plutôt que de suivre (comme prévu) un ordre thématique rigoureux, je me suis laissé emporter par ma prédilection pour un thème particulièrement riche et brûlant, auquel je compte me consacrer d'ailleurs prioritairement pendant quelques temps, à partir de la rentrée 84/85. Je reprends donc l'exposé thématique là où je l'ai laissé, tout au début du paragraphe précédent.

Mon intérêt pour les surfaces topologiques commence à poindre en 1974, où je propose à Yves Ladegaillerie le thème de l'étude isotopique des plongements d'un 1-complexe topologique dans une surface compacte. Dans les deux années qui suivent, cette étude le conduit à un remarquable théorème d'isotopie, donnant une description algébrique complète des classes d'isotopie de plongements de tels 1-complexes, ou de surfaces compactes à bord, dans une surface compacte orientée, en termes de certains invariants combinatoires très simples, et des groupes fondamentaux des protagonistes. Ce théorème, qui doit pouvoir s'étendre sans mal aux plongements d'un espace compact quelconque (triangulable pour simplifier) dans une surface compacte orientée, redonne comme corollaires faciles plusieurs résultats classiques profonds de la théorie des surfaces, et notamment le théorème d'isotopie de Baer. Il va finalement être publié, séparément du reste (et dix ans après, vu la dureté des temps...), dans Topology. Dans le travail de Ladegaillerie figure également une description purement algébrique, en termes de groupoïdes fondamentaux, de la catégorie "isotopique" des surfaces compactes X, munies d'un 1-complexe topologique K plongé dans X. Cette description, qui a eu le malheur d'aller à l'encontre du "goût du jour" et de ce fait semble impubliable, a néanmoins servi (et sert encore) comme un guide précieux dans mes réflexions ultérieures, notamment dans le contexte de la géométrie algébrique absolue de caractéristique nulle.

Le cas où (X, K) est une "carte" 2-dimensionnelle, i.e. où les composantes connexes de $X \backslash K$ sont des 2-cellules ouvertes (et où de plus K est muni d'un ensemble fini S de "sommets", tel que les composantes connexes de $K \backslash S$ soient des 1-cellules ouvertes) attire progressivement mon attention dans les années suivantes. La catégorie isotopique de ces cartes admet une description algébrique particulièrement simple, via l'ensemble des "repères" (ou "drapeaux" ou "biarcs") associés à la carte, qui se trouve naturellement muni d'une structure d'ensemble à groupe d'opérateurs, sous le groupe

$$\underline{C}_2 = \left\langle \sigma_0, \sigma_1, \sigma_2 \mid \sigma_0^2 = \sigma_1^2 = \sigma_2^2 = (\sigma_0 \sigma_2)^2 = 1 \right\rangle,$$

que j'appelle le <u>groupe cartographique</u> (non orienté) de dimension 2. Il admet comme sous-groupe d'indice 2 le <u>groupe cartographique orienté</u> engendré par les produits en nombre pair des générateurs, qui peut aussi se décrire comme

$$\underline{C}_2^+ = \left\langle \rho_s, \rho_f, \sigma \mid \rho_s \rho_f = \sigma, \ \sigma^2 = 1 \right\rangle,$$

(avec

$$\rho_s = \sigma_2 \sigma_1 \ , \ \rho_f = \sigma_1 \sigma_0 \ , \ \sigma = \sigma_0 \sigma_2 = \sigma_2 \sigma_0 \ ,$$

opérations de <u>rotation élémentaire</u> d'un repère autour d'un sommet, d'une face et d'une arête respectivement). Il y a un dictionnaire parfait entre la situation topologique des cartes compactes, resp. cartes compactes orientées, d'une part, et les ensembles finis à groupe d'opérateurs $\underline{C}_2$ resp. $\underline{C}_2^+$ de l'autre, dictionnaire dont l'existence était d'ailleurs plus ou moins connue, mais jamais énoncée avec la précision nécessaire, ni développée tant soit peu. Ce travail de fondements est fait avec le soin qu'il mérite dans un excellent travail de DEA, fait en commun par Jean Malgoire et Christine Voisin en 1976.

Cette réflexion prend soudain une dimension nouvelle, avec cette remarque simple que le groupe $\underline{C}_2^+$ peut s'interpréter comme un quotient du groupe fondamental d'une sphère orientée privée de trois points, numérotés

0, 1, 2, les opérations ρ_s, σ, ρ_f s'interprétant comme les lacets autour de ces points, satisfaisant la relation familière

$$l_0 l_1 l_2 = 1,$$

alors que la relation supplémentaire $\sigma^2 = 1$ i.e. $l_1^2 = 1$ signifie qu'on s'intéresse au quotient du groupe fondamental correspondant à un indice de ramification imposé 2 au point 1, qui classifie donc les revêtements de la sphère, ramifiés au plus en les points 0, 1, 2, avec une ramification égale à 1 ou 2 en les points au dessus de 1. Ainsi, les cartes orientées compactes forment une catégorie isotopique équivalente à celle de ces revêtements, soumis de plus à la condition supplémentaire d'être des revêtements finis. Prenant maintenant comme sphère de référence la sphère de Riemann, ou droite projective complexe, rigidifiée par les trois points 0, 1 et ∞ (ce dernier remplaçant donc 2), et se rappelant que tout revêtement ramifié fini d'une courbe algébrique complexe hérite lui-même d'une structure de courbe algébrique complexe, on aboutit à cette constatation, qui huit ans après me paraît encore toujours aussi extraordinaire : toute carte orientée "finie" se réalise canoniquement sur une courbe algébrique complexe! Mieux encore, comme la droite projective complexe est définie sur le corps de base absolue $\mathbf{Q}$, ainsi que les points de ramification admis, les courbes algébriques obtenues sont définies non seulement sur $\mathbf{C}$, mais sur la clôture algébrique $\overline{\mathbf{Q}}$ de $\mathbf{Q}$ dans $\mathbf{C}$. Quant à la carte de départ, elle se retrouve sur la courbe algébrique, comme image inverse du segment réel $[0, 1]$ (où 0 est considéré comme un sommet, et 1 comme milieu d'une "arête pliée" ayant 1 comme centre), lequel constitue dans la sphère de Riemann la "2-carte orientée universelle" (*). Les points de la courbe algébrique X au dessus de 0, de 1 et de ∞ ne sont autres que les sommets, et les "centres" des arêtes et des faces respectivement de la carte (X, K), et les ordres des sommets et des faces ne sont autres que les multiplicités des zéros et des pôles de la fonction rationnelle (définie sur $\overline{\mathbf{Q}}$) sur X, exprimant sa projection structurale vers $\mathbf{P}_{\mathbf{C}}^1$.

Cette découverte, qui techniquement se réduit à si peu de choses, a fait sur moi une impression très forte, et elle représente un tournant décisif dans le

___

(*) Il y a une description analogue des cartes finies non orientées, éventuellement avec bord, en termes de courbes algébriques réelles, plus précisément de revêtement de $\mathbf{P}_{\mathbf{R}}^1$ ramifié seulement en 0, 1, ∞, la surface à bord associée à un tel revêtement étant $X(\mathbf{C})/\tau$, où τ est la conjugaison complexe. La carte non orientée "universelle" est ici le disque, ou hémisphère supérieur de la sphère de Riemann, muni comme précédemment du 1-complexe plongé $K = [0, 1]$.

cours de mes réflexions, un déplacement notamment de mon centre d'intérêt en mathématique, qui soudain s'est trouvé fortement localisé. Je ne crois pas qu'un fait mathématique m'ait jamais autant frappé que celui-là, et ait eu un impact psychologique comparable ([2]). Cela tient sûrement à la nature tellement familière, non technique, des objets considérés, dont tout dessin d'enfant griffonné sur un bout de papier (pour peu que le graphisme soit d'un seul tenant) donne un exemple parfaitement explicite. A un tel dessin se trouvent associés des invariants arithmétiques subtils, qui seront chamboulés complètement dès qu'on y rajoute un trait de plus. S'agissant ici de cartes sphériques, donnant nécessairement naissance à des courbes de genre 0 (qui ne fournissent donc pas des "modules"), on peut dire que la courbe en question est "épinglée" dès qu'on fixe trois de ses points, par exemple trois sommets de la carte, ou plus généralement trois centres de facettes (sommets, arêtes ou faces) – dès lors l'application structurale $f : X \longrightarrow \mathbf{P}^1_{\mathbf{C}}$ peut s'interpréter comme une fonction rationnelle

$$f(z) = P(z)/Q(z) \quad \in \mathbf{C}(z)$$

bien déterminée, quotient de deux polynômes bien déterminés premiers entre eux avec Q unitaire, satisfaisant à des conditions algébriques qui traduisent notamment le fait que f soit non ramifié en dehors des valeurs 0, 1, ∞, et qui impliquent que les coefficients de ces polynômes sont des nombres algébriques; donc leurs zéros sont des nombres algébriques, qui représentent respectivement les sommets et les centres des faces de la carte envisagée.

Revenant au cas général, les cartes finies s'interprétant comme des revêtements sur $\overline{\mathbf{Q}}$ d'une courbe algébrique définie sur le corps premier $\mathbf{Q}$ lui-même, il en résulte que le groupe de Galois Γ de $\overline{\mathbf{Q}}$ sur $\mathbf{Q}$ opère sur la catégorie de ces cartes de façon naturelle. Par exemple, l'opération d'un automorphisme $\gamma \in \Gamma$ sur une carte sphérique donnée par la fonction rationnelle ci-dessus, est obtenue en appliquant γ aux coefficients des polynômes P, Q. Voici donc ce mystérieux groupe Γ intervenir comme agent transformateur sur des formes topologico-combinatoires de la nature la plus élémentaire qui soit, amenant à se poser des questions comme: telles cartes orientées données sont-elles "conjuguées", ou : quelles exactement sont les conjuguées de telle carte orientée donnée ? (il y en a, visiblement, un nombre fini seulement).

J'ai traité quelques cas concrets (pour des revêtements de bas degrés) par des expédients divers, J. Malgoire en a traité quelques autres – je doute qu'il y ait une méthode uniforme permettant d'y répondre à coups d'ordinateurs. Ma réflexion très vite s'est engagée dans une direction plus conceptuelle, pour arriver à appréhender la nature de cette action de Γ. On s'aperçoit d'emblée que grosso modo cette action est exprimée par une certaine action "extérieure" de Γ sur le compactifié profini du groupe cartographique orienté $\underline{C}_2^+$, et cette action à son tour est déduite par passage au quotient de l'action extérieure canonique de Γ sur le groupe fondamental profini $\hat{\pi}_{0,3}$ de $(U_{0,3})_{\overline{\mathbf{Q}}}$, où $U_{0,3}$ désigne la courbe-type de genre 0 sur le corps premier $\mathbf{Q}$, privée de trois points. C'est ainsi que mon attention s'est portée vers ce que j'ai appelé depuis la "géométrie algébrique anabélienne", dont le point de départ est justement une étude (pour le moment limitée à la caractéristique zéro) de l'action de groupes de Galois "absolus" (notamment les groupes $\mathrm{Gal}(\overline{K}/K)$, où K est une extension de type fini du corps premier) sur des groupes fondamentaux géométriques (profinis) de variétés algébriques (définies sur K), et plus particulièrement (rompant avec une tradition bien enracinée) des groupes fondamentaux qui sont très éloignés des groupes abéliens (et que pour cette raison je nomme "anabéliens"). Parmi ces groupes, et très proche du groupe $\hat{\pi}_{0,3}$, il y a le compactifié profini du groupe modulaire $\mathrm{Sl}(2,\mathbf{Z})$, dont le quotient par le centre ± 1 contient le précédent comme sous-groupe de congruence mod 2, et peut s'interpréter d'ailleurs également comme groupe "cartographique" orienté, savoir celui qui classifie les cartes orientées triangulées (i.e. celles dont les faces sont des triangles ou des monogones).

Toute carte finie orientée donne lieu à une courbe algébrique projective et lisse définie sur $\overline{\mathbf{Q}}$, et il se pose alors immédiatement la question : quelles sont les courbes algébriques sur $\overline{\mathbf{Q}}$ obtenues ainsi – les obtiendrait-on toutes, qui sait? En termes plus savants, serait-il vrai que toute courbe algébrique projective et lisse définie sur un corps de nombres interviendrait comme une "courbe modulaire" possible pour paramétriser les courbes elliptiques munies d'une rigidification convenable? Une telle supposition avait l'air à tel point dingue que j'étais presque gêné de la soumettre aux compétences en la matière. Deligne consulté trouvait la supposition dingue en effet, mais sans avoir un contre-exemple dans ses manches. Moins d'un an après, au Congrès International de Helsinki, le mathématicien soviétique Bielyi annonce justement ce résultat, avec une démonstration d'une simplicité déconcertante tenant en deux petites pages d'une lettre de Deligne – jamais sans doute un résultat profond et déroutant ne fut démontré en si peu de lignes!

Sous la forme où l'énonce Bielyi, son résultat dit essentiellement que toute courbe algébrique définie sur un corps de nombres peut s'obtenir comme revêtement de la droite projective ramifié seulement en les points $0, 1, \infty$. Ce résultat semble être passé plus ou moins inaperçu. Pourtant, il m'apparaît d'une portée considérable. Pour moi, son message essentiel a été qu'il y a une identité profonde entre la combinatoire des cartes finies d'une part, et la géométrie des courbes algébriques définies sur des corps de nombres, de l'autre. Ce résultat profond, joint à l'interprétation algébri-co-géométrique des cartes finies, ouvre la porte sur un monde nouveau, in-exploré – et à portée de main de tous, qui passent sans le voir.

C'est près de trois ans plus tard seulement, voyant que décidément les vastes horizons qui s'ouvrent là ne faisaient rien tressaillir en aucun de mes élèves, ni même chez aucun des trois ou quatre collègues de haut vol auxquels j'ai eu l'occasion d'en parler de façon circonstanciée, que je fais un premier voyage de prospection de ce "monde nouveau", de janvier à juin 1981. Ce premier jet se matérialise en un paquet de quelques 1300 pages manuscrites, baptisées "La Longue Marche à travers la théorie de Galois". Il s'agit avant tout d'un effort de compréhension des relations entre groupes de Galois "arithmétiques" et groupes fondamentaux profinis "géométriques". Assez vite, il s'oriente vers un travail de formulation calculatoire de l'opération de $\mathrm{Gal}(\overline{\mathbf{Q}}/\mathbf{Q})$ sur $\widehat{\pi}_{0,3}$, et dans un stade ultérieur, sur le groupe légèrement plus gros $\widehat{\mathrm{Sl}(2,\mathbf{Z})}$, qui donne lieu à un formalisme plus élégant et plus efficace. C'est au cours de ce travail aussi (mais développé dans des notes distinctes) qu'apparaît le thème central de la géométrie algébrique anabélienne, qui est de reconstituer certaines variétés X dites "anabéliennes" sur un corps absolu K à partir de leur groupe fondamental mixte, extension de $\mathrm{Gal}(\overline{K}/K)$ par $\pi_1(X_{\overline{K}})$; c'est alors que se dégage la "conjecture fondamentale de la géométrie algébrique anabélienne", proche des conjectures de Mordell et de Tate que vient de démontrer Faltings ([3]). C'est là aussi que s'amorcent une première réflexion sur les groupes de Teichmüller, et les premières intuitions sur la structure multiple de la "tour de Teichmüller" – les multiplicités modulaires ouvertes $M_{g,\nu}$ apparaissant par ailleurs comme les premiers exemples importants, en dimension > 1, de variétés (ou plutôt, de multiplicités) qui semblent bien mériter l'appellation "anabélienne". Vers la fin de cette période de réflexion, celle-ci m'apparaît comme une réflexion fondamentale sur une théorie alors encore dans les limbes, pour laquelle l'appellation "Théorie de Galois-Teichmüller" me semble plus appropriée que "théorie de Galois" que j'avais d'abord donnée à mes notes.

Ce n'est pas le lieu ici de donner un aperçu plus circonstancié de cet ensemble de questions, intuitions, idées – y compris des résultats palpables, certes. Le plus important me semble celui signalé en passant au par. 2, savoir la fidélité de l'action extérieure de $\mathbb{\Gamma} = \mathrm{Gal}(\overline{\mathbf{Q}}/\mathbf{Q})$ (et de ses sous-groupes ouverts) sur $\widehat{\pi}_{0,3}$, et plus généralement (si je me rappelle bien) sur le groupe fondamental de toute courbe algébrique "anabélienne" (i.e. dont le genre g et le "nombre de trous" ν satisfont l'inégalité $2g + \nu \geq 3$, i.e. telle que $\chi(X) < 0$) définie sur une extension finie de $\mathbf{Q}$. Ce résultat peut être considéré comme essentiellement équivalent au théorème de Bielyi – c'est la première manifestation concrète, par un énoncé mathématique précis, du "message" dont il a été question plus haut.

Je voudrais terminer cet aperçu rapide par quelques mots de commentaire sur la richesse vraiment inimaginable d'un groupe anabélien typique comme le groupe Sl$(2, \mathbb{Z})$ – sans doute le groupe discret infini le plus remarquable qu'on ait rencontré, qui apparaît sous une multiplicité d'avatars (dont certains ont été effleurés dans le présent rapport), et qui du point de vue de la théorie de Galois-Teichmüller peut être considéré comme la "pierre de construction" fondamentale de la "tour de Teichmüller". L'élément de structure de Sl$(2, \mathbb{Z})$ qui me fascine avant tout, est bien sûr l'action extérieure du groupe de Galois $\mathbb{\Gamma}$ sur le compactifié profini. Par le théorème de Bielyi, prenant les compactifiés profinis de sous-groupes d'indice fini de Sl$(2, \mathbb{Z})$, et l'action extérieure induite (quitte à passer également à un sous-groupe ouvert de $\mathbb{\Gamma}$), on trouve essentiellement les groupes fondamentaux de toutes les courbes algébriques (pas nécessairement compactes) définis sur des corps de nombres K, et l'action extérieure de Gal$(\overline{K}/K)$ dessus – du moins est-il vrai que tout tel groupe fondamental apparaît comme quotient d'un des premiers groupes (*). Tenant compte du "yoga anabélien" (qui reste conjectural), disant qu'une courbe algébrique anabélienne sur un corps de nombres K (extension finie de $\mathbb{Q}$) est connue à isomorphisme près quand on connaît son groupe fondamental mixte (ou ce qui revient au même, l'action extérieure de Gal$(\overline{K}/K)$ sur son groupe fondamental profini géométrique), on peut donc dire que toutes les courbes algébriques définies sur des corps de nombres sont "contenues" dans le compactifié profini $\widehat{\mathrm{Sl}(2, \mathbb{Z})}$, et dans la connaissance d'un certain sous-groupe $\mathbb{\Gamma}$ du groupe des automorphismes extérieurs de ce dernier! Passant aux abélianisés des groupes fondamentaux précédents, on voit notamment que toutes les représentations abéliennes l-adiques chères à Tate et consorts, définies par des jacobiennes et jacobiennes généralisées de courbes algébriques définies sur des corps de nombres, sont contenues dans cette seule action de $\mathbb{\Gamma}$ sur le groupe profini anabélien $\widehat{\mathrm{Sl}(2, \mathbb{Z})}$! (^{4})

Il en est qui, face à cela, se contentent de hausser les épaules d'un air désabusé et de parier qu'il n'y a rien à tirer de tout cela, sauf des rêves. Ils oublient, ou ignorent, que notre science, et toute science, serait bien peu de chose, si depuis ses origines elle n'avait été nourrie des rêves et des visions de ceux qui s'y adonnent avec passion.

(*) En fait, il s'agit de quotients de nature particulièrement triviale, par des sous-groupes abéliens produits de "modules de Tate" $\widehat{\mathbb{Z}}(1)$, correspondant à des "groupes-lacets" autour de points à l'infini.

4. Dès le début de ma réflexion sur les cartes bidimensionnelles, je me suis in-téressé plus particulièrement aux cartes dites "régulières", c'est-à-dire celles dont le groupe des automorphismes opère transitivement (et de ce fait, de façon simplement transitive) sur l'ensemble des repères. Dans le cas orienté et en termes de l'interprétation algébrico-géométrique du paragraphe précé-dent, ce sont les cartes qui correspondent à un revêtement galoisien de la droite projective. Très vite aussi, et dès avant même qu'apparaisse le lien avec la géométrie algébrique, il apparaît nécessaire aussi de ne pas exclure les cartes infinies, qui interviennent notamment de façon naturelle comme revêtements universels des cartes finies. Il apparaît (comme conséquence immédiate du "dictionnaire" des cartes, étendu au cas des cartes pas néces-sairement finies) que pour tout couple d'entiers naturels $p, q \geq 1$, il existe à isomorphisme (non unique) près une carte 1-connexe et une seule qui soit de type (p, q) i.e. dont tous les sommets soient d'ordre p et toutes les faces d'ordre q, et cette carte est une carte régulière. Elle se trouve épinglée par le choix d'un repère, et son groupe des automorphismes est alors canoni-quement isomorphe au quotient du groupe cartographique (resp. du groupe cartographique orienté, dans le cas orienté) par les relations supplémentaires

$$\rho_s^p = \rho_f^q = 1.$$

Le cas où ce groupe est fini est le cas "pythagoricien" des cartes régulières sphériques, le cas où il est infini donne les pavages réguliers du plan eu-clidien ou du plan hyperbolique (*). Le lien de la théorie combinatoire avec la théorie "conforme" des pavages réguliers du plan hyperbolique était pressenti, avant qu'apparaisse celui des cartes finies avec les revêtements finis de la droite projective. Une fois ce lien compris, il devient évident qu'il doit s'étendre également aux cartes infinies (régulières ou non): toute carte finie ou non, se réalise canoniquement sur une surface conforme (compacte si et seulement si la carte est finie), en tant que revêtement ramifié de la droite projective complexe, ramifié seulement en les points 0, 1, ∞. La seule difficulté ici était de mettre au point le dictionnaire entre cartes topologiques et ensembles à opérateurs, qui posait quelques problèmes con-ceptuels dans le cas infini, à commencer par la notion même de "carte topologique". Il apparaît nécessaire notamment, tant par raison de co-hérence interne du dictionnaire, que pour ne pas laisser échapper certains cas intéressants de cartes infinies, de ne pas exclure des sommets et des faces d'ordre infini. Ce travail de fondements a été fait également par J. Malgoire

_____

(*) Dans ces énoncés, il y a lieu de ne pas exclure le cas où p, q peuvent prendre la valeur $+\infty$, qu'on rencontre notamment de façon très naturelle comme pavages associés à certains polyèdres réguliers infinis, cf. plus bas.

et C. Voisin, sur la lancée de leur premier travail sur les cartes finies, et leur théorie fournit en effet tout ce qu'on était en droit d'attendre (et même plus...).

C'est en 1977 et 1978, parallèlement à deux cours de C4 sur la géométrie du cube et sur celle de l'icosaèdre, que j'ai commencé à m'intéresser aux polyèdres réguliers, qui m'apparaissent alors comme des "réalisations géométriques" particulièrement concrètes de cartes combinatoires, les sommets, arêtes et faces étant réalisés respectivement comme des points, des droites et des plans dans un espace affine tridimensionnel convenable, avec respect des relations d'incidence. Cette notion de réalisation géométrique d'une carte combinatoire garde un sens sur un corps de base, et même sur un anneau de base arbitraire. Elle garde également un sens pour les polyèdres réguliers de dimension quelconque, en remplaçant le groupe cartographique $\underline{C}_2$ par une variante n-dimensionnelle $\underline{C}_n$ convenable. Le cas $n = 1$, i.e. la théorie des polygones réguliers en caractéristique quelconque, fait l'objet d'un cours de DEA en 1977/78, et fait apparaître déjà quelques phénomènes nouveaux, comme aussi l'utilité de travailler non pas dans un espace ambiant affine (ici le plan affine), mais dans un espace projectif. Ceci est dû notamment au fait que dans certaines caractéristiques (et notamment en caractéristique 2) le centre d'un polyèdre régulier est rejeté à l'infini. D'autre part, le contexte projectif, contrairement au contexte affine, permet de développer avec aisance un formalisme de dualité pour les polyèdres réguliers, correspondant au formalisme de dualité des cartes combinatoires ou topologiques (où le rôle des sommets et des faces, dans le cas $n = 2$ disons, se trouve interchangé). Il se trouve que pour tout polyèdre régulier projectif, on peut définir un hyperplan canonique associé, qui joue le rôle d'un hyperplan à l'infini canonique, et permet de considérer le polyèdre donné comme un polyèdre régulier affine.

L'extension de la théorie des polyèdres réguliers (et plus généralement, de toutes sortes de configurations géométrico-combinatoires, y compris les systèmes de racines...) du corps de base **R** ou **C** vers un anneau de base général, me semble d'une portée comparable, dans cette partie de la géométrie, à l'extension analogue qui a eu lieu depuis le début du siècle en géométrie algébrique, ou depuis une vingtaine d'années en topologie (*), avec l'introduction du langage des schémas et celui des topos. Ma réflexion sporadique sur cette question, pendant quelques années, s'est bornée à dégager quelques principes de base simples, en attachant d'abord mon attention au cas des polyèdres réguliers épinglés, ce qui réduit à un minimum le bagage con-

---

(*) En écrivant cela, je suis conscient que rares sont les topologues, encore aujourd'hui, qui se rendent compte de cet élargissement conceptuel et technique de la topologie, et des ressources qu'elle offre.

ceptuel nécessaire, et élimine pratiquement les questions de rationalité tant soit peu délicates. Pour un tel polyèdre, on trouve une base (ou repère) canonique de l'espace affine ou projectif ambiant, de telle façon que les opérations du groupe cartographique $\underline{C}_n$, engendré par les réflexions fondamentales σ_i ($0 \leq i \leq n$), s'y écrivent par des formules universelles, en termes de n paramètres $\alpha_1, ..., \alpha_n$, qui géométriquement s'interprètent comme les doubles des cosinus des "angles fondamentaux" du polyèdre. Le polyèdre se reconstitue à partir de cette action, et du drapeau affine ou projectif associé à la base choisie, en transformant ce drapeau par tous les éléments du groupe engendré par les réflexions fondamentales. Ainsi le n-polyèdre épinglé "universel" est-il défini canoniquement sur l'anneau de polynômes à n indéterminées

$$\mathbf{Z}[\underline{\alpha}_1, ..., \underline{\alpha}_n],$$

ses spécialisations sur des corps de base arbitraires k (via des valeurs $\alpha_i \in k$ données aux indéterminées $\underline{\alpha}_i$) donnant des polyèdres réguliers correspondant à des types combinatoires divers. Dans ce jeu, il n'est pas question de se borner à des polyèdres réguliers finis, ni même à des polyèdres réguliers dont les facettes soient d'ordre fini, i.e. pour lesquels les paramètres α_i soient des racines d'équations "semicyclotomiques" convenables, exprimant que les "angles fondamentaux" (dans le cas où le corps de base est $\mathbf{R}$) sont commensurables à 2π. Déjà quand $n = 1$, le polygone régulier peut-être le plus intéressant de tous (moralement celui du polygone régulier à un seul côté!) est celui qui correspond à $\alpha = 2$, donnant lieu à une conique circonscrite parabolique, i.e. tangente à la droite à l'infini. Le cas fini est celui où le groupe engendré par les réflexions fondamentales, qui est aussi le groupe des automorphismes du polyèdre régulier envisagé, est fini. Dans le cas du corps de base $\mathbf{R}$ (ou $\mathbf{C}$, ce qui revient au même), et pour $n = 2$, les cas finis sont bien connus depuis l'antiquité – ce qui n'exclut pas que le point de vue schématique y fasse apparaître des charmes nouveaux; on peut dire cependant qu'en spécialisant l'icosaèdre (par exemple) sur des corps de base finis de caractéristique arbitraire, c'est toujours un icosaèdre, avec sa combinatoire propre et le même groupe d'automorphismes simple d'ordre 60 qu'on obtient. La même remarque s'applique aux polyèdres réguliers finis de dimension supérieure, étudiés de façon systématique dans deux beaux livres de Coxeter. La situation est toute autre si on part d'un polyèdre régulier infini, sur un corps tel que $\mathbf{Q}$ disons, et qu'on le "spécialise" sur le corps premier $\mathbf{F}_p$ (opération bien définie pour tout p sauf un nombre fini de nombres premiers). Il est clair que tout polyèdre régulier sur un corps fini est fini – on trouve donc une infinité de polyèdres réguliers finis pour p variable, dont le type combinatoire, ou ce qui revient au même, le groupe des automorphismes, varie de façon "arithmétique" avec p. Cette situation est particulièrement intrigante dans le cas où $n = 2$, où on dispose de la relation explicitée au paragraphe précédent entre 2-cartes combinatoires, et courbes algébriques définies sur des corps de nombres. Dans ce cas, un polyèdre régulier infini défini sur un corps infini quelconque (et de ce fait sur une sous-$\mathbf{Z}$-algèbre à deux générateurs de celui-ci) donne donc naissance à une infinité de courbes algébriques définies sur des corps de nombres, qui sont des revêtements galoisiens ramifiés seulement en 0, 1, ∞ de la droite projective standard. Le cas optimum est bien sûr celui où on part du 2-polyèdre régulier universel, ou plutôt de celui qui s'en déduit par passage au corps des fractions $\mathbf{Q}(\alpha_1, \alpha_2)$ de son anneau de base. Ceci soulève une foule de questions nouvelles, aussi bien des vagues que des précises, dont je n'ai eu le loisir encore d'examiner de plus près aucune – je ne citerai que celle-ci : quelles sont exactement les 2-cartes régulières finies, ou ce qui revient au même, les groupes quotients finis du groupe 2-cartographique qui proviennent de 2-polyèdres réguliers sur des corps finis (*)? Les obtiendrait-on toutes, et si oui : comment ?

Ces réflexions font apparaître en pleine lumière ce fait, qui pour moi était entièrement inattendu, que la théorie des polyèdres réguliers finis, déjà dans le cas de la dimension $n = 2$, est infiniment plus riche, et notamment donne infiniment plus de formes combinatoires différentes, dans le cas où on admet des corps de base de caractéristique non nulle, que dans le cas considéré jusqu'à présent où les corps de base étaient restreints à $\mathbf{R}$, ou à la rigueur $\mathbf{C}$ (dans le cas de ce que Coxeter appelle des "polyèdres réguliers complexes", et que je préfère appeler "pseudo-polyèdres réguliers définis sur $\mathbf{C}$") (**). De plus, il semble que cet élargissement du point de vue doive aussi jeter un jour nouveau sur les cas déjà connus. Ainsi, examinant l'un après l'autre les polyèdres pythagoriciens, j'ai vu se répéter à chaque fois un même petit miracle, que j'ai appelé le <u>paradigme combinatoire</u> du polyèdre envisagé. Vaguement parlant, il peut se décrire en disant que lorsqu'on

---

(*) Ce sont les mêmes d'ailleurs que ceux provenant de polyèdres réguliers sur des corps quelconques, ou algébriquement clos, comme on voit par des arguments de spécialisation standard.

(**) Les pseudo-polyèdres épinglés se décrivent de la même façon que les polyèdres épinglés, avec cette seule différence que les réflexions fondamentales σ_i ($0 \leq i \leq n$) sont remplacées ici par des <u>pseudo-réflexions</u> (que Coxeter suppose de plus d'ordre fini, comme il se borne aux structures combinatoires finies). Cela conduit simplement à introduire pour chacun des σ_i un invariant numérique supplémentaire β_i, de sorte que le n-pseudo-polyèdre universel peut se définir encore sur un anneau de polynômes à coefficients entiers, en les $n + (n + 1)$ variables $\underline{\alpha}_i$ ($1 \leq i \leq n$) et $\underline{\beta}_j$ ($0 \leq j \leq n$).

regarde la spécialisation du polyèdre dans la caractéristique, ou l'une des caractéristiques, la (ou les) plus singulière(s) (ce sont les caractéristiques 2 et 5 pour l'icosaèdre, la caractéristique 2 pour l'octaèdre), on lit, sur le polyèdre régulier géométrique sur le corps fini concerné (F_2 et F_5 pour l'icosaèdre, F_2 pour l'octaèdre) une description particulièrement élégante (et inattendue) de la combinatoire du polyèdre. Il m'a semblé même entrevoir là un principe d'une grande généralité, que j'ai cru retrouver notamment dans une réflexion ultérieure sur la combinatoire du système des 27 droites d'une surface cubique, et ses relations avec le système de racines E_7. Qu'un tel principe existe bel et bien et qu'on réussisse même à le dégager de son manteau de brumes, ou qu'il recule au fur et à mesure où on le poursuit et qu'il finisse par s'évanouir comme une Fata Morgana, j'y trouve pour ma part une force de motivation, une fascination peu communes, comme celle du rêve peut-être. Nul doute que de suivre un tel appel de l'informulé, de l'informe qui cherche forme, d'un entrevu élusif qui semble prendre plaisir à la fois à se dérober et à se manifester – ne peut que mener loin, alors que nul ne pourrait prédire, où ...

Pourtant, pris par d'autres intérêts et tâches, je n'ai pas jusqu'à présent suivi cet appel, ni rencontré personne d'autre qui ait voulu l'entendre, et encore moins le suivre. Mis à part quelques digressions vers d'autres types de structures géométrico-combinatoires, mon travail ici encore s'est borné à un premier travail de dégrossissage et d'intendance, sur lequel il est inutile de m'étendre plus ici (^{5}). Le seul point qui peut-être mérite encore mention, est l'existence et l'unicité de l'hyperquadrique circonscrite à un n-polyèdre régulier donné, dont l'équation peut s'expliciter par des formules simples en termes des paramètres fondamentaux α_i (*). Le cas qui m'intéresse le plus est celui où $n = 2$, et le temps me semble mûr pour réécrire une version nouvelle, en style moderne, du classique livre de Klein sur l'icosaèdre et les autres polyèdres pythagoriciens. Ecrire un tel exposé sur les 2-polyèdres réguliers serait une magnifique occasion pour un jeune chercheur de se familiariser aussi bien avec la géométrie des polyèdres et leurs liens avec les géométries sphérique, euclidienne, hyperbolique, et avec les courbes algébriques, qu'avec le langage et les techniques de base de la géométrie algébrique moderne. S'en trouvera-t-il un un jour pour saisir cette occasion?

(*) Un résultat analogue vaut pour les pseudo-polyèdres. Il semblerait que les "caractéristiques exceptionnelles" dont il a été question plus haut, pour les spécialisations d'un polyèdre donné, sont celles pour lesquelles l'hyperquadrique circonscrite est, soit dégénérée, soit tangente à l'hyperplan à l'infini.

5. Je voudrais maintenant dire quelques mots sur certaines réflexions qui m'ont fait comprendre le besoin de fondements nouveaux pour la topologie "géométrique", dans une direction toute différente de la notion de topos, et indépendante même des besoins de la géométrie algébrique dite "abstraite" (sur des corps et anneaux de base généraux). Le problème de départ, qui a commencé à m'intriguer il doit y avoir une quinzaine d'années déjà, était celui de définir une théorie de "dévissage" des structures stratifiées, pour les reconstituer, par un procédé canonique, à partir de "pièces de construction" canoniquement déduites de la structure donnée. Probablement l'exemple principal qui m'avait alors amené à cette question était celui de la stratification canonique d'une variété algébrique singulière (ou d'un espace analytique complexe ou réel singulier) par la suite décroissante de ses "lieux singuliers" successifs. Mais je devais sans doute pressentir déjà l'ubiquité des structures stratifiées dans pratiquement tous les domaines de la géométrie (que d'autres sûrement ont vu clairement bien avant moi). Depuis, j'ai vu apparaître de telles structures, notamment, dans toute situation de "modules" pour des objets géométriques susceptibles non seulement de variation continue, mais en même temps de phénomènes de "dégénérescence" (ou de "spécialisation") – les strates correspondant alors aux divers "niveaux de singularité" (ou aux types combinatoires associés) pour les objets considérés. Les multiplicités modulaires compactifiées $\widehat{M}_{g,\nu}$ de Mumford-Deligne pour les courbes algébriques stables de type (g, ν) en fournissent un exemple typique et particulièrement inspirant, qui a joué un rôle de motivation important dans la reprise de ma réflexion sur les structures stratifiées, de décembre 1981 à janvier 1982. La géométrie bidimensionnelle fournit de nombreux autres exemples de telles structures stratifiées modulaires, qui toutes d'ailleurs (sauf expédients de rigidification), apparaissent comme des "multiplicités" plutôt que comme des espaces ou variétés au sens ordinaire (les points de ces multiplicités pouvant avoir des groupes d'automorphismes non triviaux). Parmi les objets de géométrie bidimensionnelle donnant lieu à de telles structures modulaires stratifiées de dimension arbitraire, voire de dimension infinie, je citerai les polygones (euclidiens, ou sphériques, ou hyperboliques), les systèmes de droites dans un plan (projectif disons), les systèmes de "pseudodroites" dans un plan projectif topologique, ou les courbes immergées à croisements normaux plus générales, dans une surface (compacte disons) donnée.

L'exemple non trivial le plus simple d'une structure stratifiée s'obtient en considérant une paire (X, Y) d'un espace X et d'un sous-espace fermé Y, en faisant une hypothèse d'équisingularité convenable de X le long de Y, et en supposant de plus (pour fixer les idées) que les deux strates Y et $X \backslash Y$ sont des <u>variétés</u> topologiques. L'idée naïve, dans une telle situation, est de prendre "le" voisinage tubulaire T de Y dans X, dont le bord ∂T devrait être une variété lisse également, fibrée à fibres lisses et compactes sur Y, T lui-même s'identifiant au fibré en cônes sur ∂T associé au fibré précédent. Posant

$$U = X \setminus \mathrm{Int}(T),$$

on trouve une variété à bord dont le bord est canoniquement isomorphe à celui de T. Ceci dit, les "pièces de construction" prévues sont la variété à bord U (compacte si X était compact, et qui remplace en la précisant la strate "ouverte" $X \setminus Y$) et la variété (sans bord) Y, avec comme structure supplémentaire les reliant l'application dite de "recollement"

$$f : \partial U \longrightarrow Y$$

qui est une fibration propre et lisse. La situation de départ (X, Y) se reconstitue à partir de $(U, Y, f : \partial U \to Y)$ par la formule

$$X \cong U \coprod_{\partial U} Y$$

(somme amalgamée sous ∂U, s'envoyant dans U et Y via l'inclusion resp. l'application de recollement).

Cette vision naïve se heurte immédiatement à des difficultés diverses. La première est la nature un peu vague de la notion même de voisinage tubulaire, qui ne prend un sens tant soit peu précis qu'en présence de structures plus rigides que la seule structure topologique, telles la structure "linéaire par morceaux", ou riemanienne (plus généralement, d'espace avec fonction distance); l'ennui ici est que dans aucun des exemples auxquels on pense spontanément, on ne dispose naturellement d'une structure de ce type – tout au mieux d'une classe d'équivalence de telles structures, permettant de rigidifier un tantinet la situation. Si par ailleurs on admet qu'on a pu trouver un expédient pour trouver un voisinage tubulaire ayant les propriétés voulues, qui de plus soit unique modulo un automorphisme (topologique, disons) de la situation, automorphisme qui de plus respecte la structure fibrée fournie par la fonction de recollement, il reste la difficulté de la non-canonicité des choix faits, l'automorphisme en question n'étant visiblement pas unique, quoi qu'on fasse pour le "normaliser". L'idée ici, pour rendre canonique ce qui ne l'est pas, est de travailler systématiquement dans des "catégories isotopiques" associées aux catégories de nature topologique s'introduisant dans ces questions (telle la catégorie des paires admissibles (X, Y) et des homéomorphismes de telles paires, etc.), en gardant les mêmes objets, mais en prenant comme "morphismes" les classes d'isotopie (dans un sens dicté sans ambiguïté par le contexte) d'isomorphismes (voire même, de morphismes plus généraux que des isomorphismes). Cette idée, qui est reprise avec succès dans la thèse de Yves Ladegaillerie notamment (cf. début du par. 3), m'a servi de façon systématique dans toutes mes réflexions ultérieures de topologie combinatoire, quand il s'est agi de formuler avec précision des théorèmes de traduction de situations topologiques, en termes de situations combinatoires. Dans la situation actuelle, mon espoir était d'arriver à formuler (et à prouver!) un théorème d'équivalence entre deux catégories isotopiques convenables, l'une étant la catégorie des "paires admissibles" (X, Y), l'autre celle des "triples admissibles" (U, Y, f) où Y est une variété, U une variété à bord, et $f : \partial U \to Y$ une fibration propre et lisse. De plus, bien sûr, j'espérais qu'un tel énoncé, modulo un travail de nature essentiellement algébrique, s'étendrait de lui-même en un énoncé plus sophistiqué, s'appliquant aux structures stratifiées générales.

Très vite, il apparaissait qu'il ne pouvait être question d'obtenir un énoncé aussi ambitieux dans le contexte des espaces topologiques, à cause des sempiternels phénomènes de "sauvagerie". Déjà quand X lui-même est une variété et Y réduit à un point, on se bute à la difficulté que le cône sur un espace compact Z peut être une variété en son sommet, sans que Z soit homéomorphe à une sphère, ni même soit une variété. Il était clair également que les contextes de structures plus rigides qui existaient à l'époque, tel le contexte "linéaire par morceaux", étaient également inadéquats – une des raisons rédhibitoires communes étant qu'ils ne permettaient pas, pour une paire (U, S) d'un "espace" U et d'un sous-espace fermé S, et une application de recollement $f : S \to T$, de construire la somme amalgamée correspondante. C'est quelques années plus tard que j'étais informé de la théorie de Hironaka des ensembles qu'il appelle, je crois, "semi-analytiques" (réels), qui satisfont à certaines des conditions de stabilité essentielles (sans doute même à toutes) nécessaires au développement d'un contexte utilisable de "topologie modérée". Du coup cela relance une réflexion sur les fondements d'une telle topologie, dont le besoin m'apparaît de plus en plus clairement.

Avec un recul d'une dizaine d'années, je dirais aujourd'hui, à ce sujet, que la "topologie générale" a été développée (dans les années trente et quarante) par des analystes et pour les besoins de l'analyse, non pour les besoins de la topologie proprement dite, c'est-à-dire l'étude des propriétés topologiques de formes géométriques diverses. Ce caractère inadéquat des fondements de la topologie se manifeste dès les débuts, par des "faux problèmes" (au point de vue au moins de l'intuition topologique des formes) comme celle de "l'invariance du domaine", alors même que la solution de ce dernier par Brouwer l'amène à introduire des idées géométriques nouvelles importantes. Aujourd'hui encore, comme aux temps héroïques où on voyait pour la première fois et avec inquiétude des courbes remplir allègrement des carrés et des cubes, quand on se propose de faire de la géométrie topologique dans le contexte technique des espaces topologiques, on se heurte à chaque pas à des difficultés parasites tenant aux phénomènes sauvages. Ainsi, en dehors de cas de (très) basse dimension, il ne peut guère être possible, pour un espace donné X (une variété compacte disons), d'étudier le type d'homotopie (disons) du groupe des automorphismes de X, ou de l'espace des plongements, ou immersions etc. de X dans quelque autre espace Y – alors qu'on sent que ces invariants devraient faire partie de l'arsenal des invariants essentiels associés à X, ou au couple (X, Y), etc., au même titre que l'espace fonctionnel $\underline{\mathrm{Hom}}(X, Y)$ familier en topologie homotopique. Les topologues éludent la difficulté, sans l'affronter, en se rabattant sur des contextes voisins du contexte topologique et moins marqués de sauvagerie que lui, comme les variétés différentiables, les espaces PL (linéaires par morceaux), etc., dont visiblement aucun n'est "bon", i.e. n'est stable par les opérations topologiques les plus évidentes, telles les opérations de contraction-recollement (sans même passer à des opérations du type $X \to \mathrm{Aut}(X)$ qui font quitter le paradis des "espaces" de dimension finie). C'est là une façon de tourner autour du pot! Cette situation, comme tant de fois déjà dans l'histoire de notre science, met simplement en évidence cette inertie quasi-insurmontable de l'esprit, alourdi par des conditionnements d'un poids considérable, pour porter un regard sur une question de fondements, donc sur le contexte même dans lequel on vit, respire, travaille – plutôt que de l'accepter comme un donné immuable. C'est à cause de cette inertie sûrement qu'il a fallu des millénaires pour qu'une idée ou une réalité aussi enfantine que le zéro, un groupe, ou une forme topologique, trouve droit de cité en mathématiques. C'est par elle aussi, sûrement, que le carcan de la topologie générale continue à être traîné patiemment par des générations de topologues, la "sauvagerie" étant portée comme une fatalité inéluctable qui serait enracinée dans la nature même des choses.

Mon approche vers des fondements possibles d'une topologie modérée a été une approche axiomatique. Plutôt que de déclarer (chose qui serait parfaitement raisonnable certes) que les "espaces modérés" cherchés ne sont autres (disons) que les espaces semianalytiques de Hironaka, et de développer dès lors dans ce contexte l'arsenal des constructions et notions familières en topologie, plus celles certes qui jusqu'à présent n'avaient pu être développées et pour cause, j'ai préféré m'attacher à dégager ce qui, parmi les propriétés géométriques de la notion d'ensemble semianalytique dans un espace $\mathbf{R}^n$, permet d'utiliser ceux-ci comme "modèles" locaux d'une notion "d'espace modéré" (en l'occurrence, semianalytique), et ce qui (on l'espère!) rend cette notion d'espace modéré suffisamment souple pour pouvoir bel et bien servir de notion de base pour une "topologie modérée" propre à exprimer avec aisance l'intuition topologique des formes. Ainsi, une fois le travail de fondements qui s'impose accompli, il apparaîtra non une "théorie modérée", mais une vaste infinité, allant de la plus stricte de toutes, celle des "espaces $\overline{\mathbf{Q}}_r$-algébriques par morceaux" (où $\overline{\mathbf{Q}}_r = \overline{\mathbf{Q}} \cap \mathbf{R}$), vers celle qui (à tort ou à raison) m'apparaît comme probablement la plus vaste, savoir celle des "espaces analytiques réels par morceaux" (ou semianalytiques dans la terminologie de Hironaka). Parmi les théorèmes de fondements envisagés dans mon programme, il y a un théorème de comparaison qui, vaguement parlant, dira qu'on trouvera essentiellement les mêmes catégories isotopiques (ou même ∞-isotopiques), quelle que soit la théorie modérée avec laquelle on travaille ([6]). De façon plus précise, il s'agit de mettre le doigt sur un système d'axiomes suffisamment riche, pour impliquer (entre bien autres choses!) que si on a deux théories modérées $\mathcal{T}$, $\mathcal{T}'$ avec $\mathcal{T}$ plus fine que $\mathcal{T}'$ (dans un sens évident), et si X, Y sont deux espaces $\mathcal{T}$-modérés, qui définissent aussi des espaces $\mathcal{T}'$-modérés correspondants, l'application canonique

$$\underline{\mathrm{Isom}}_{\mathcal{T}}(X, Y) \to \underline{\mathrm{Isom}}_{\mathcal{T}'}(X, Y)$$

induit une bijection sur l'ensemble des composantes connexes (ce qui impliquera que la catégorie isotopique des $\mathcal{T}$-espaces est équivalente à celle des $\mathcal{T}'$-espaces), et même, est une équivalence d'homotopie (ce qui signifie qu'on a même une équivalence pour les catégories "∞-isotopiques", plus fines que les catégories isotopiques où on ne retient que le π_0 des espaces d'isomorphismes). Ici les $\underline{\mathrm{Isom}}$ peuvent être définis de façon évidente comme ensembles semisimpliciaux par exemple, pour pouvoir donner un sens précis à l'énoncé précédent. Des énoncés analogues devraient être vrais, en remplaçant les "espaces" $\underline{\mathrm{Isom}}$ par d'autres espaces d'applications, soumises à des conditions géométriques standard, comme celle d'être des plongements, des immersions, lisses, étales, des fibrations etc. Egalement, on s'attend à avoir des énoncés analogues, où X, Y sont remplacés par des systèmes d'espaces modérés, tels ceux qui interviennent dans une théorie de dévissage des structures stratifiées – de telle sorte que dans un sens technique précis, cette théorie de dévissage sera, elle aussi, essentiellement indépendante de la théorie modérée choisie pour l'exprimer.

Le premier test décisif pour un bon système d'axiomes sur une notion de "partie modérée de $\mathbf{R}^n$" me semble la possibilité de prouver de tels théorèmes de comparaison. Je me suis contenté jusqu'à présent de dégager un système d'axiomes plausible provisoire, sans avoir aucune assurance qu'il ne faudra y rajouter d'autres axiomes, que seul un "travail sur pièces" sans doute permettra de faire apparaître. Le plus fort des axiomes que j'ai introduits, et celui sans doute dont la vérification dans les cas d'espèce est (ou sera) la plus délicate, est un axiome de triangulabilité (modérée, il va sans dire) d'une partie modérée de $\mathbf{R}^n$. Je ne me suis pas essayé à prouver en termes de ces seuls axiomes le théorème de comparaison, j'ai eu l'impression néanmoins (à tort ou à raison encore!) que cette démonstration, qu'elle nécessite ou non l'introduction de quelque axiome supplémentaire, ne présentera pas de grosse difficulté technique. Il est bien possible que les difficultés au niveau technique, pour le développement de fondements satisfaisants de la topologie modérée, y inclus une théorie de dévissage des structures modérées stratifiées, soient déjà pour l'essentiel concentrées dans les axiomes, et par suite essentiellement surmontées dès à présent par des théorèmes de triangulabilité à la Lojasiewicz et Hironaka. Ce qui fait défaut, encore une fois, n'est nullement la virtuosité technique des mathématiciens, parfois impressionnante, mais l'audace (ou simplement l'innocence...) pour s'affranchir d'un contexte familier accepté par un consensus sans failles...

Les avantages d'une approche axiomatique vers des fondements de la topologie modérée me semblent assez évidents. Ainsi, pour considérer une variété algébrique complexe, ou l'ensemble des points réels d'une variété algébrique définie sur $\mathbf{R}$, comme un espace modéré, il semble préférable de travailler dans la théorie "$\mathbf{R}$-algébrique par morceaux", voire même la théorie $\overline{\mathbf{Q}}_r$-algébrique par morceaux (où $\overline{\mathbf{Q}}_r = \overline{\mathbf{Q}} \cap \mathbf{R}$) quand il s'agit de variétés définies sur des corps de nombres, etc. L'introduction d'un sous-corps $K \subset \mathbf{R}$ associé à la théorie $\mathcal{T}$ (formé des points de $\mathbf{R}$ qui sont $\mathcal{T}$-modérés, i.e. tels que l'ensemble uniponctuel correspondant le soit) permet d'introduire pour tout point x d'un espace modéré X un corps résiduel $k(x)$, qui est une sous-extension de $\mathbf{R}/K$ algébriquement fermée dans $\mathbf{R}$, et de degré de transcendance fini sur K (majoré par la dimension topologique de X). Quand le degré de transcendance de $\mathbf{R}$ sur K est infini, on trouve une notion de degré de transcendance (ou "dimension") d'un point d'un espace modéré, voisin de la notion familière en géométrie algébrique. De telles notions sont absentes dans la topologie modérée "semianalytique", qui par contre apparaît comme le contexte topologique tout indiqué pour inclure les espaces analytiques réels et complexes.

Parmi les premiers théorèmes auxquels on s'attend dans une topologie modérée comme je l'entrevois, mis à part les théorèmes de comparaison, sont les énoncés qui établissent, dans un sens convenable, l'existence et l'unicité "du" voisinage tubulaire d'un sous-espace modéré fermé dans un espace modéré (compact pour simplifier), les façons concrètes de l'obtenir (par exemple à partir de toute application modérée $X \to \mathbf{R}^+$ admettant Y comme ensemble de ses zéros), la description de son "bord" (alors qu'en général ce n'est nullement une variété à bord!) ∂T, qui admet dans T

un voisinage isomorphe au produit de T par un segment, etc. Moyennant
des hypothèses d'equisingularité convenables, on s'attend à ce que T soit
muni, de façon essentiellement unique, d'une structure de fibré localement
trivial sur Y, admettant ∂T comme sous-fibré. C'est là un des points les
moins clairs dans l'intuition provisoire que j'ai de la situation, alors que la
classe d'homotopie de l'application structurale prévue $T \to Y$ a un sens
évident, indépendamment de toute hypothèse d'équisingularité, comme inverse homotopique de l'application d'inclusion $Y \to T$, qui doit être un
homotopisme. Une façon d'obtenir a posteriori une telle structure serait via
l'hypothétique équivalence de catégories isotopiques envisagée au début, en
tenant compte du fait que le foncteur $(U, Y, f) \longmapsto (X, Y)$ est défini de
façon évidente, indépendamment de toute théorie de voisinages tubulaires.

$\frac{33}{34}$

On dira sans doute, non sans quelque raison, que tout cela n'est peut-
être que rêves, qui s'évanouiront en fumée dès qu'on s'essayera à un travail
circonstancié, voire même dès avant en face de certains faits connus ou
bien évidents qui m'auraient échappé. Certes, seul un travail sur pièces
permettra de décanter le juste du faux et de connaître la substance véritable. La seule chose dans tout cela qui ne fait pour moi l'objet d'aucun
doute, c'est la nécessité d'un tel travail de fondements, en d'autres termes,
la nature artificielle des fondements actuels de la topologie, et des difficultés
que ceux-ci soulèvent à chaque pas. Il est bien possible par contre que la formulation que je donne à une théorie de dévissage des structures stratifiées,
comme un théorème d'équivalence de catégories isotopiques (voire même ∞-
isotopiques) convenables, soit trop optimiste. Je devrais ajouter pourtant
que je n'ai guère de doutes non plus que la théorie de ces dévissages que
j'ai développée il y a deux ans, alors qu'elle reste partiellement heuristique,
exprime bel et bien une réalité tout ce qu'il y a de palpable. Dans une partie
de mon travail, faute de pouvoir disposer d'un contexte "modéré" tout fait,
et pour avoir néanmoins des énoncés précis et démontrables, j'ai été amené à
postuler sur la structure stratifiée de départ des structures supplémentaires
tout ce qu'il y a de plausibles, dans la nature de la donnée de rétractions
locales notamment, qui dès lors permettent bel et bien la construction d'un
système canonique d'espaces, paramétré par l'ensemble ordonné des "drapeaux" Drap(I) de l'ensemble ordonné I indexant les strates, ces espaces
jouant le rôle des espaces (U, Y) de tantôt, reliés entre eux par des applications de plongements et de fibrations propres, qui permettent de reconstituer
de façon tout aussi canonique la structure stratifiée de départ, y compris ces
"structures supplémentaires" ([7]). Le seul ennui, c'est que ces dernières semblent un élément de structure superfétatoire, qui n'est nullement une donnée
dans les situations géométriques courantes, par exemple pour l'espace modulaire compact $\widehat{M}_{g,\nu}$ avec sa "stratification à l'infini" canonique, donnée

$\frac{34}{35}$

par le diviseur à croisements normaux de Mumford-Deligne. Une autre difficulté, moins sérieuse sans doute, c'est que le soi-disant "espace" modulaire est en fait une <u>multiplicité</u> – techniquement, cela s'exprime surtout par la nécessité de remplacer l'ensemble d'indices I pour les strates par une <u>catégorie</u> (essentiellement finie) d'indices, en l'occurrence celle des "graphes MD", qui "paramètrent" les "structures combinatoires" possibles d'une courbe stable de type (g, ν). Ceci dit, je puis affirmer que la théorie de dévissage générale, spécialement développée sous la pression du besoin de <u>cette</u> cause, s'est révélée en effet un guide précieux, conduisant à une compréhension progressive, d'une cohérence sans failles, de certains aspects essentiels de la tour de Teichmüller (c'est à dire, essentiellement de la "structure à l'infini" des groupes de Teichmüller ordinaires). C'est cette approche qui m'a conduit finalement, dans les mois suivants, vers le principe d'une construction purement combinatoire de la tour des groupoïdes de Teichmüller, dans l'esprit esquissé plus haut (cf. par. 2).

Un autre test de cohérence satisfaisant provient du point de vue "topossique". En effet, mon intérêt pour les multiplicités modulaires provenant avant tout de leur sens algébrico-géométrique et arithmétique, c'est aux multiplicités modulaires <u>algébriques</u>, sur le corps de base absolu $\mathbb{Q}$, que je me suis intéressé prioritairement, et à un "dévissage" à l'infini de leurs groupes fondamentaux géométriques (i.e. des groupes de Teichmüller <u>profinis</u>) qui soit compatible avec les opérations naturelles de $\Gamma = \mathrm{Gal}(\overline{\mathbb{Q}}/\mathbb{Q})$. Cela semblait exclure d'emblée la possibilité de me référer à une hypothétique théorie de dévissage de structures stratifiées dans un contexte de "topologie modérée" (ou même de topologie ordinaire, cahin-caha), si ce n'est comme fil conducteur entièrement heuristique. Dès lors se posait la question de

35
36

traduire, dans le contexte des topos (en l'occurrence les topos étales) intervenant dans la situation, la théorie de dévissage à laquelle j'étais parvenu dans un contexte tout différent – avec la tâche supplémentaire, par la suite, de dégager un théorème de comparaison général, sur le modèle des théorèmes bien connus, pour comparer les invariants obtenus (notamment les types d'homotopie de voisinages tubulaires divers) dans le cadre transcendant, et dans le cadre schématique. J'ai pu me convaincre qu'un tel formalisme de dévissage avait bel et bien un sens dans le contexte (dit "abstrait"!) des topos généraux, ou tout au moins des topos noethériens (comme ceux qui s'introduisent ici), via une notion convenable de <u>voisinage tubulaire canonique d'un sous-topos</u> dans un topos ambiant. Une fois cette notion acquise, avec certaines propriétés formelles simples, la description du "dévissage" d'un topos stratifié est considérablement plus simple même dans ce cadre, que dans le cadre topologique (modéré). Il est vrai que là aussi il y a un travail de fondements à faire, notamment pour la notion même de voisinage tubulaire d'un sous-topos – et il est étonnant d'ailleurs que ce travail (pour autant que je sache) n'ait toujours pas été fait, c'est-à-dire que personne (depuis plus de vingt ans qu'il existe un contexte de topologie étale) ne semble en avoir eu besoin; un signe sûrement que la compréhension de la structure topologique des schémas n'a pas tellement progressé depuis le travail d'Artin-Mazur...

Une fois accompli le double travail de dégrossissage (plus ou moins heuristique) autour de la notion de dévissage d'un espace ou d'un topos stratifié, qui a été une étape cruciale dans ma compréhension des multiplicités modulaires, il est d'ailleurs apparu que pour les besoins de ces dernières, on peut sans doute court-circuiter au moins une bonne partie de cette théorie par des arguments géométriques directs. Il n'en reste pas moins que pour moi, le formalisme de dévissage auquel je suis parvenu a fait ses preuves d'utilité et de cohérence, indépendamment de toute question sur les fondements les plus adéquats qui permettent de lui donner tout son sens.

$\frac{36}{37}$ 6. Un des théorèmes de fondements de topologie (modérée) les plus intéressants qu'il faudrait développer, serait un théorème de "dévissage" (encore!) d'une application modérée propre d'espaces modérés,

$$f : X \longrightarrow Y,$$

via une filtration décroissante de Y par des sous-espaces modérés fermés Y^i, tels que au-dessus des "strates ouvertes" $Y^i \setminus Y^{i-1}$ de cette filtration, f induise une fibration localement triviale (du point de vue modéré, il va sans dire). Un tel énoncé devrait encore se généraliser et se préciser de diverses façons, notamment en demandant l'existence d'un dévissage analogue simultané, pour X et une famille finie donnée de sous-espaces (modérés) fermés de X. Egalement la notion même de fibration localement triviale au sens modéré peut se renforcer considérablement, en tenant compte du fait que les strates ouvertes U_i sont mieux que des espaces à structure modérée purement locale, du fait qu'elles sont obtenues comme différence de deux espaces modérés, compacts si Y était compact. Entre la notion d'espace modéré compact (qui se réalise comme un des "modèles" de départ dans un $\mathbf{R}^n$) et celle d'espace "localement modéré" (localement compact) qui s'en déduit de façon assez évidente, il y a une notion un peu plus délicate d'espace "globalement modéré" X, obtenu comme différence $\widehat{X} \setminus Y$ de deux espaces modérés compacts, étant entendu qu'on ne distingue pas entre l'espace défini par une paire $(\widehat{X}, Y)$, et celui défini par une paire $(\widehat{X}', Y')$ qui s'en déduit par une application modérée (nécessairement propre)

$$g : \widehat{X}' \longrightarrow \widehat{X}$$

induisant une bijection $g^{-1}(X) \xrightarrow{\sim} X$, en prenant $Y' = g^{-1}(Y)$. L'exemple naturel le plus intéressant peut-être est celui où on part d'un schéma séparé de type fini sur **C** ou sur **R**, en prenant pour X l'ensemble de ses points complexes ou réels, qui hérite d'une structure modérée globale à l'aide des compactifications schématiques (qui existent d'après Nagata) du schéma de départ. Cette notion d'espace globalement modéré est associée à une notion d'<u>application globalement modérée</u>, qui permet à son tour de renforcer en conséquence la notion de fibration localement triviale, dans l'énoncé d'un théorème de dévissage pour une application $f : X \to Y$ (pas nécessairement propre maintenant) dans le contexte des espaces globalement modérés.

J'ai été informé l'été dernier par Zoghman Mebkhout qu'un théorème de dévissage dans cet esprit avait été obtenu récemment dans le contexte des espaces analytiques réels et/ou complexes, avec des Y^i qui, cette fois, sont des sous-espaces analytiques de Y. Ce résultat rend plausible qu'on dispose dès à présent de moyens techniques suffisamment puissants pour démontrer également un théorème de dévissage dans le contexte modéré, plus général en apparence, mais probablement moins ardu.

C'est le contexte d'une topologie modérée également qui devrait permettre, il me semble, de formuler avec précision un principe général très sûr que j'utilise depuis longtemps dans un grand nombre de situations géométriques, que j'appelle le "<u>principe des choix anodins</u>" – aussi utile que vague d'apparence! Il dit, lorsque pour les besoins d'une construction quelconque d'un objet géométrique en termes d'autres, on est amené à faire un certain nombre de choix arbitraires en cours de route, de façon donc que l'objet obtenu dépend en apparence de ces choix et est donc entaché d'un défaut de canonicité, que ce défaut est sérieux en effet (et pour être levé demande une analyse plus soigneuse de la situation, des notions utilisées, des données introduites etc.) chaque fois que l'un au moins de ces choix s'effectue dans un "espace" qui n'est pas "contractile" i.e. dont le π_0 <u>ou</u> un des invariants supérieurs π_i est non trivial; que ce défaut est par contre apparent seulement, que la construction est "essentiellement canonique" et n'entraînera pas vraiment d'ennuis, chaque fois que les choix faits sont tous "anodins", i.e. s'effectuent dans des espaces <u>contractiles</u>. Quand on essaye dans les cas d'espèce de cerner de plus près ce principe, il semble qu'on tombe à chaque fois sur la notion de "catégories ∞-isotopiques" exprimant une situation donnée, plus fines que les catégories isotopiques (= 0-isotopiques) plus naïves, obtenues en ne retenant que les π_0 des espaces d'isomorphismes qui s'introduisent dans la situation, alors que le point de vue ∞-isotopique retient tout leur type d'homotopie. Par exemple, le point de vue isotopique naïf pour les surfaces compactes à bord orientées de type (g, ν) est "bon" (sans boomerang caché!) exactement dans les cas

que j'appelle "anabéliens" (et que Thurston appelle "hyperboliques") i.e.
distincts de $(0,0)$, $(0,1)$, $(0,2)$, $(1,0)$ – qui sont aussi les cas justement
ou le groupe des automorphismes de la surface a une composante neutre
<u>contractile</u>. Dans les autres cas, sauf le cas $(0,0)$ de la sphère sans trou, il
suffit de travailler avec les catégories 1-isotopiques pour exprimer de façon
satisfaisante par voie algébrique les faits géométrico-topologiques essentiels,
vu que ladite composante connexe est alors un $K(\pi, 1)$. Travailler dans une
catégorie 1-isotopique revient d'ailleurs à travailler dans une bicatégorie,
i.e. avec des $\underline{\mathrm{Hom}}(X, Y)$ qui sont (non plus des ensembles discrets comme
dans le point de vue 0-isotopique, mais) des groupoïdes (dont les π_0 ne sont
autres que les Hom 0-isotopiques). C'est la description en termes purement
algébriques de cette bicatégorie qui est faite dans la dernière partie de la
thèse de Yves Ladegaillerie (cf. par. 3).

Si je me suis étendu ici plus longuement sur le thème des fondements de
la topologie modérée, qui n'est nullement un de ceux auxquels je compte
me consacrer prioritairement dans les années qui viennent, c'est sans doute
justement que je sens qu'il y a là d'autant plus une cause qui a besoin d'être
plaidée, ou plutôt: un travail d'une grande actualité qui a besoin de bras!
Comme naguère pour de nouveaux fondements de la géométrie algébrique,
ce ne sont pas des plaidoyers qui surmontent l'inertie des habitudes acquises,
mais un travail tenace, méticuleux, sans doute de longue haleine, et porteur
au jour le jour de moissons éloquentes.

$\frac{39}{40}$

Je voudrais encore dire quelques mots sur une réflexion plus ancienne
(fin des années 60?), très proche de celle dont il vient d'être question, in-
spirée par les idées de Nash, qui m'avaient beaucoup frappé. Au lieu ici
de définir axiomatiquement une notion de "théorie modérée" via la donnée
de "partie modérée de $\mathbf{R}^n$" satisfaisant à certaines conditions (de stabilité
surtout), c'est à une axiomatisation de la notion de "variété lisse" et du
formalisme différentiable sur de telles variétés que j'en avais, via la donnée, pour chaque entier naturel n, d'un sous-anneau $\mathcal{A}_n$ de l'anneau des
germes de fonctions réelles à l'origine dans $\mathbf{R}^n$. Ce sont les fonctions qui
seront admises pour exprimer les "changements de carte" pour la notion
de $\mathcal{A}$-variété correspondante, et il s'est agi de dégager tout d'abord un système d'axiomes sur le système $\mathcal{A} = (\mathcal{A}_n)_{n \in \mathbf{N}}$ qui assure à cette notion
de variété une souplesse comparable à celle de variété C^∞, ou analytique
réelle (ou de Nash). Suivant le type de constructions familières qu'on tient
à pouvoir effectuer dans le contexte des $\mathcal{A}$-variétés, le système d'axiomes
pertinent est plus ou moins réduit, ou riche. Très peu suffit s'il s'agit seulement de développer le formalisme différentiel, avec la construction de fibrés
de jets, les complexes de De Rham etc. Si on veut un énoncé du type
"quasi-fini implique fini" (pour une application au voisinage d'un point),

40
41

qui est apparu comme un énoncé-clef dans la théorie locale des espaces analytiques, il faut un axiome de stabilité de nature plus délicate, dans le "Vorbereitungssatz" de Weierstrass (*) Dans d'autres questions, un axiome de stabilité par prolongement analytique (dans $\mathbf{C}^n$) apparaît nécessaire. L'axiome le plus draconien que j'ai été amené à introduire, lui aussi un axiome de stabilité, concerne l'intégration des systèmes de Pfaff, assurant que certains groupes de Lie, voire tous, sont des $\mathcal{A}$-variétés. Dans tout ceci, j'ai pris soin de ne pas supposer que les $\mathcal{A}_n$ soient des $\mathbf{R}$-algèbres, donc une fonction constante sur une $\mathcal{A}$-variété n'est "admissible" que si sa valeur appartient à un certain sous-corps K de $\mathbf{R}$ (c'est, si on veut, $\mathcal{A}_0$). Ce sous-corps peut fort bien être $\mathbf{Q}$, ou sa fermeture algébrique $\overline{\mathbf{Q}}_r$ dans $\mathbf{R}$, ou toute autre sous-extension de $\mathbf{R}/\mathbf{Q}$, de préférence même de degré de transcendance fini, ou du moins dénombrable, sur $\mathbf{Q}$. Cela permet par exemple, comme tantôt pour les espaces modérés, de faire correspondre à tout point x d'une variété (de type $\mathcal{A}$) un corps résiduel $k(x)$, qui est une sous-extension de $\mathbf{R}/K$. Un fait qui me semble important ici, c'est que même sous sa forme la plus forte, le système d'axiomes n'implique pas qu'on doive avoir $K = \mathbf{R}$. Plus précisément, du fait que tous les axiomes sont des axiomes de stabilité, il résulte que pour un ensemble S donné de germes de fonctions analytiques réelles à l'origine (dans divers espaces $\mathbf{R}^n$), il existe une plus petite théorie $\mathcal{A}$ pour laquelle ces germes sont admissibles, et que celle-ci est "dénombrable" i.e. les $\mathcal{A}_n$ sont dénombrables, dès que S l'est. A fortiori, K est alors dénombrable, i.e. de degré de transcendance dénombrable sur $\mathbf{Q}$.

L'idée est ici d'introduire, par le biais de cette axiomatique, une notion de fonction (analytique réelle) "élémentaire", ou plutôt, toute une hiérarchie de telles notions. Pour une fonction de 0 variables, i.e. une constante, cette notion donne celle de "constante élémentaire", incluant notamment (dans le cas de l'axiomatique la plus forte) des constantes telles que π, e et une multitude d'autres, en prenant des valeurs de fonctions admissibles (telles l'exponentielle, le logarithme etc.) pour des systèmes de valeurs "admissibles" de l'argument. On sent que la relation entre le système $\mathcal{A} = (\mathcal{A}_n)_{n \in \mathbf{N}}$ et le corps de rationalité K correspondant doit être très étroite, du moins pour des $\mathcal{A}$ qui peuvent être engendrés par un "système de générateurs" S fini – mais il est à craindre que la moindre question intéressante qu'on pourrait se poser sur cette situation soit actuellement hors de portée (^{1}).

41
42

Ces réflexions anciennes ont repris quelque actualité pour moi avec ma

---

(*) Il peut paraître plus simple de dire que les anneaux (locaux) $\mathcal{A}_n$ sont henséliens, ce qui est équivalent. Mais il n'est nullement clair a priori sous cette dernière forme que la condition en question est dans la nature d'une condition de stabilité, circonstance importante comme il apparaîtra dans les réflexions qui suivent.

réflexion ultérieure sur les théories modérées. Il me semble en effet qu'il est possible d'associer de façon naturelle à une "théorie différentiable" $\mathcal{A}$ une théorie modérée T (ayant sans doute même corps de constantes), de telle façon que toute $\mathcal{A}$-variété soit automatiquement munie d'une structure T-modérée, et inversement que pour tout espace T-modéré compact X, on puisse trouver une partie fermée modérée rare Y dans X, telle que $X \setminus Y$ provienne d'une $\mathcal{A}$-variété, et que de plus cette structure de $\mathcal{A}$-variété soit unique tout au moins dans le sens suivant: deux telles structures coïncident dans le complémentaire d'une partie modérée rare $Y' \supset Y$ de X. La théorie de dévissage des structures modérées stratifiées (dont il a été question au par. précédent), dans le cas des strates lisses, devrait d'ailleurs soulever des questions beaucoup plus précises encore de comparaison des structures modérées avec des structures de type différentiable (ou plutôt, $\mathbf{R}$-analytique). Je soupçonne que le type d'axiomatisation proposé ici pour la notion de "théorie différentiable" fournirait un cadre naturel pour formuler de telles questions avec toute la précision et la généralité souhaitables.

7. Depuis le mois de mars de l'an dernier, donc depuis près d'un an, la plus grande partie de mon énergie a été consacrée à un travail de réflexion sur les fondements de l'algèbre (co)homologique non commutative, ou ce qui revient au même, finalement, de l'algèbre homotopique. Ces réflexions se sont concrétisées par un volumineux paquet de notes dactylographiées, destinées à former le premier volume (actuellement en cours d'achèvement) d'un ouvrage en deux volumes à paraître chez Hermann, sous le titre commun "A la Poursuite des Champs". Je prévois actuellement (après des élargissements successifs du propos initial) que le manuscrit de l'ensemble des deux volumes, que j'espère achever en cours d'année pour ne plus avoir à y revenir, fera dans les 1500 pages dactylographiées. Ces deux volumes d'ailleurs sont pour moi les premiers d'une série plus vaste, sous le titre commun "Réflexions Mathématiques", où je compte développer tant soit peu certains des thèmes esquissés dans le présent rapport.

Vu qu'il s'agit d'un travail en cours de rédaction, et même d'achèvement, dont le premier volume sans doute paraîtra cette année et contiendra une introduction circonstanciée, il est sans doute moins intéressant que je m'étende ici sur ce thème de réflexion, et je me contenterai donc d'en parler très brièvement. Ce travail me semble quelque peu marginal par rapport aux thèmes que je viens d'esquisser, et ne représente pas (il me semble) un véritable renouvellement d'optique ou d'approche par rapport à mes intérêts et ma vision mathématiques d'avant 1970. Si je m'y suis résolu soudain, c'est presque en désespoir de cause, alors que près de vingt ans se sont écoulés depuis que se sont posées en termes bien clairs un certain nombre de questions visiblement fondamentales, et mûres pour être menées à leur terme, sans que personne ne les voie, ou prenne la peine de les sonder. Aujourd'hui encore, les structures de base qui interviennent dans le point de vue homotopique en topologie, y compris même en algèbre homologique commutative, ne sont pas comprises, et à ma connaissance, après les travaux de Verdier, de Giraud et d'Illusie, sur ce thème (qui constituent autant de "coups d'envoi" attendant toujours une suite...) il n'y a pas eu d'effort dans ce sens. Je devrais faire exception sans doute pour le travail d'axiomatisation fait par Quillen sur la notion de catégorie de modèles, à la fin des années 60, et repris sous des variantes diverses par divers auteurs. Ce travail à l'époque, et maintenant encore, m'a beaucoup séduit et appris, tout en allant dans une direction assez différente de celle qui me tenait et tient à coeur. Il introduit certes des catégories dérivées dans divers contextes non commutatifs, mais sans entrer dans la question des structures internes essentielles d'une telle catégorie, laissée ouverte également dans le cas commutatif par Verdier, et après lui par Illusie. De même, la question de mettre le doigt sur les "coefficients" naturels pour un formalisme cohomologique non commutatif, au-delà des champs (qu'on devrait appeler 1-champs) étudiés dans le livre de Giraud, restait ouverte – ou plutôt, les intuitions riches et précises qui y répondent, puisées dans des exemples nombreux provenant de la géométrie algébrique notamment, attendent toujours un langage précis et souple pour leur donner forme.

Je reviens sur certains aspects de ces questions de fondements en 1975, à l'occasion (je crois me souvenir) d'une correspondance avec Larry Breen (deux lettres de cette correspondance seront reproduites en appendice au Chap. I du volume 1, "Histoires de Modèles", de la Poursuite des Champs). A ce moment apparaît l'intuition que les ∞-groupoïdes doivent constituer des modèles, particulièrement adéquats, pour les types d'homotopie, les n-groupoïdes correspondant aux types d'homotopie tronqués (avec $\pi_i = 0$ pour $i > n$). Cette même intuition, par des voies très différentes, a été retrouvée par Ronnie Brown à Bangor et certains de ses élèves, mais en utilisant une notion de ∞-groupoïde assez restrictive (qui, parmi les types d'homotopie 1-connexes, ne modélise que les produits d'espaces d'Eilenberg-Mac Lane). C'est stimulé par une correspondance à bâtons rompus avec Ronnie Brown, que j'ai finalement repris une réflexion, commençant par un essai de définition d'une notion de ∞-groupoïde plus large (rebaptisé par la suite "champ en ∞-groupoïdes" ou simplement "champ", sous-entendu: sur le topos ponctuel), et qui de fil en aiguille m'a amené à la Poursuite des Champs. Le volume "Histoire de Modèles" y constitue d'ailleurs une digression entièrement imprévue par rapport au propos initial (les fameux champs étant provisoirement oubliés, et n'étant prévus réapparaître que vers les pages 1000 environ...).

Ce travail n'est pas entièrement isolé par rapport à mes intérêts plus récents. Par exemple, ma réflexion sur les multiplicités modulaires $\widehat{M}_{g,\nu}$ et leur structure stratifiée a relancé une réflexion sur un théorème de Van Kampen de dimension > 1 (un des thèmes de prédilection également du groupe de Bangor), et a peut-être contribué à préparer le terrain pour le travail de plus grande envergure l'année d'après. Celui-ci rejoint également par moments une réflexion datant de la même année 1975 (ou l'année d'après) sur un "complexe de De Rham à puissances divisées", qui a fait l'objet de ma dernière conférence publique, à l'IHES en 1976, et dont le manuscrit, confié je ne me rappelle plus à qui après l'exposé, est d'ailleurs perdu. C'est au moment de cette réflexion que germe aussi l'intuition d'une "schématisation" des types d'homotopie, que sept ans après j'essaye de préciser dans un chapitre (particulièrement hypothétique) de l'Histoire de Modèles.

Le travail de réflexion entrepris dans la Poursuite des Champs est un peu comme une dette dont je m'acquitterais, vis-à-vis d'un passé scientifique où, pendant une quinzaine d'années (entre 1955 et 1970), le développement d'outils cohomologiques a été le Leitmotiv constant, dans mon travail de fondements de la géométrie algébrique. Si la reprise actuelle de ce thème-là a pris des dimensions inattendues, ce n'est pas cependant par pitié pour un passé, mais à cause des nombreux imprévus faisant irruption sans cesse, en bousculant sans ménagement les plans et propos prévus – un peu comme dans un conte des mille et une nuits, où l'attention se trouve maintenue en haleine à travers vingt autres contes avant de connaître le fin mot du premier.

8. J'ai très peu parlé encore des réflexions plus terre-à-terre de géométrie topologique bidimensionelle, associées notamment à mes activités d'enseignant et celles dites de "direction de recherches". A plusieurs reprises, j'ai vu s'ouvrir devant moi de vastes et riches champs mûrs pour la moisson, sans que jamais je réussisse à communiquer cette vision, et l'étincelle qui l'accompagne, à un (ou une) de mes élèves, et à la faire déboucher sur une exploration commune, de plus ou moins longue haleine. A chaque fois jusqu'à aujourd'hui même, après une prospection de quelques jours ou quelques semaines, où je découvrais en éclaireur des richesses insoupçonnées au départ, le voyage tournait court, quand il devenait clair que je serais seul à le poursuivre. Des intérêts plus forts prenaient le pas alors sur un voyage qui, dès lors, apparaissait comme une digression, voire une dispersion, plutôt qu'une aventure poursuivie en commun.

Un de ces thèmes a été celui des polygones plans, centré autour des variétés modulaires qu'on peut leur associer. Une des surprises ici a été l'irruption de la géométrie algébrique dans un contexte qui m'en avait semblé bien éloigné. Ce genre de surprise, lié à l'ubiquité de la géométrie algébrique dans la géométrie tout court, s'est d'ailleurs repéré à plusieurs reprises.

Un autre thème a été celui des courbes (notamment des cercles) immergés dans une surface, avec une attention particulière pour le cas "stable" où les points singuliers sont des points doubles ordinaires (et aussi celui, plus général, où les différentes branches en un point se croisent mutuellement), avec souvent l'hypothèse supplémentaire que l'immersion soit "cellulaire", i.e. donne naissance à une carte. Une variante de situations de ce type est celle des immersions d'une surface à bord non vide, et en tout premier lieu d'un disque (qui m'avait été signalé par A'Campo il y a une dizaine d'années). Au delà de la question de diverses formulations combinatoires de telles situations, qui ne représente plus guère qu'un exercice de syntaxe, je me suis intéressé surtout à une vision dynamique des configurations possibles, avec le passage de l'une à l'autre par déformations continues, qui peuvent se décomposer en composées de deux types d'<u>opérations élémentaires</u> et leurs inverses, à savoir le "<u>balayage</u>" d'une branche de courbe par dessus un point double, et l'<u>effacement</u> ou <u>la création d'un bigône</u>. (La première de ces opérations joue également un rôle-clef dans une théorie "dynamique" des systèmes de pseudo-droites dans un plan projectif réel.) Une des premières questions qui se posent ici est celle de déterminer les différentes <u>classes d'immersions</u> d'un cercle ou d'un disque (disons) modulo ces opérations élémentaires; une autre, celle de voir quelles sont les immersions du bord du disque qui proviennent d'une immersion du disque, et dans quelle mesure les premières déterminent les secondes. Ici encore, il m'a semblé que c'est une étude systématique des variétés modulaires pertinentes (de dimension infinie en l'occurrence, à moins d'arriver à en donner une version purement combinatoire) qui devrait fournir le "focus" le plus efficace, nous forçant en quelque sorte à nous poser les questions les plus pertinentes. Malheureusement, la réflexion sur les questions même les plus évidentes et les plus terre-à-terre est restée à l'état embryonnaire. Comme seul résultat tangible, je peux signaler une théorie de "dévissage" canonique d'une immersion cellulaire stable du cercle dans une surface, en immersions "indécomposables", par "télescopage" de telles immersions. Je n'ai pas réussi malheureusement à voir se transformer mes lumières sur la question en un travail de stage de DEA, ni d'autres lumières (sur une description théorique complète, en termes de groupes fondamentaux de 1-complexes topologiques, des immersions d'une surface à bord qui prolongent une immersion donnée de son bord) en le démarrage d'une thèse de doctorat d'état...

Un troisième thème, poursuivi simultanément depuis trois ans à divers

46
47

niveaux d'enseignement (depuis l'option pour étudiants de première année, jusqu'à trois thèses de troisième cycle actuellement poursuivies sur ce
thème) porte sur la classification topologique-combinatoire des systèmes de
droites ou pseudo-droites. Dans l'ensemble, la participation de mes élèves ici
a été moins décevante qu'ailleurs, et j'ai eu le plaisir parfois d'apprendre par
eux des choses intéressantes auxquelles je n'aurais pas songé. La réflexion
commune, par la force des choses, s'est limitée cependant à un niveau très
élémentaire. Dernièrement, j'ai finalement consacré un mois de réflexion intensive au développement d'une construction purement combinatoire d'une
sorte de "surface modulaire" associée à un système de n pseudo-droites, qui
classifie les différentes "positions relatives" possibles (stables ou non) d'une
$(n + 1)$-ième pseudo-droite par rapport au système donné, ou encore: les
différentes "affinisations" possibles de ce système, par les différents choix
possibles d'une "pseudo-droite à l'infini". J'ai l'impression d'avoir mis le
doigt sur un objet remarquable, faisant apparaître un ordre imprévu dans
des questions de classification qui jusqu'à présent apparaissaient assez chao-
tiques! Mais ce n'est pas le lieu dans le présent rapport de m'étendre plus
à ce sujet.

Depuis 1977, dans toutes les questions (comme dans ces deux derniers
thèmes que je viens d'évoquer) où interviennent des cartes bidimensionelles,
la possibilité de les réaliser canoniquement sur une surface conforme, donc
sur une courbe algébrique complexe dans le cas orienté compact, reste
en filigrane constant dans ma réflexion. Dans pratiquement tous les cas
(en fait, tous les cas sauf celui de certaines cartes sphériques avec "peu
d'automorphismes") une telle réalisation conforme implique en fait une
métrique riemanienne canonique, ou du moins, canonique à une constante
multiplicative près. Ces nouveaux éléments de structure (sans même prendre en compte l'élément arithmétique, dont il a été question au par. 3)
sont de nature à transformer profondément l'aspect initial des questions
abordées, et les méthodes d'approche. Un début de familiarisation avec
les belles idées de Thurston sur la construction de l'espace de Teichmüller,
en termes d'un jeu très simple de chirurgie riemanienne hyperbolique, me
confirme dans ce pressentiment. Malheureusement, le niveau de culture
très modeste de presque tous les élèves qui ont travaillé avec moi pendant
ces dix dernières années ne me permet pas d'aborder avec eux, ne serait-ce
que par allusion, de telles possibilités, alors que l'assimilation d'un langage
combinatoire minimum se heurte déjà, bien souvent, à des obstacles psy-
chiques considérables. C'est pourquoi, à certains égards et de plus en plus
ces dernières années, mes activités d'enseignant ont souvent agi comme un
poids, plutôt que comme un stimulant pour le déploiement d'une réflexion
géométrique tant soit peu avancée, ou seulement délicate.

9. L'occasion me semble propice ici de faire un bref bilan de mon activité enseignante depuis 1970, c'est-à-dire depuis que celle-ci s'effectue dans un cadre universitaire. Ce contact avec une réalité très différente a été pour moi riche en enseignements, d'une portée d'un tout autre ordre d'ailleurs que simplement pédagogique ou scientifique. Ce n'est pas ici le lieu de m'étendre sur ce sujet. J'ai dit aussi au début de ce rapport le rôle qu'a joué ce changement de milieu professionnel dans le renouvellement de mon approche des mathématiques, et celui de mes centres d'intérêt en mathématique. Si par contre je fais le bilan de mon activité enseignante au niveau de la recherche proprement dite, j'aboutis à un constat d'échec clair et net. Depuis plus de dix ans que cette activité se poursuit an par an au sein d'une même institution universitaire, je n'ai pas su, à aucun moment, y susciter un lieu où "il se passe quelque chose" – où quelque chose "passe", parmi un groupe si réduit soit-il de personnes, reliées par une aventure commune. A deux reprises, il est vrai, vers les années 74 à 76, j'ai eu le plaisir et le privilège de susciter chez un élève un travail d'envergure, poursuivi avec élan: chez Yves Ladegaillerie le travail signalé précédemment (par. 3) sur les questions d'isotopie en dimension 2, et chez Carlos Contou-Carrère (dont la passion mathématique n'avait pas attendu la rencontre avec moi pour éclore) un travail non publié sur les jacobiennes locales et globales sur des schémas de bases généraux (dont une partie a été annoncée dans une note aux CR). Ces deux cas mis à part, mon rôle s'est borné, au cours de ces dix ans, à transmettre tant bien que mal des rudiments du métier de mathématicien à quelques vingt élèves au niveau de la recherche, ou tout au moins à ceux parmi eux qui ont persévéré suffisamment avec moi, réputé plus exigeant que d'autres, pour aboutir à un premier travail noir sur blanc acceptable (certaines fois aussi à un travail mieux qu'acceptable et plus qu'un seul travail, fait avec goût et jusqu'au bout). Vu la conjoncture, même parmi les rares qui ont persévéré, plus rares encore seront ceux qui auront l'occasion d'exercer ce métier, et par là, tout en gagnant leur pain, de l'approfondir.

10. Depuis l'an dernier, je sens qu'au cours de mon activité d'enseignant universitaire, j'ai appris tout ce que j'avais à en apprendre et enseigné tout ce que je peux y enseigner, et qu'elle a cessé d'être vraiment utile, à moi-même comme aux autres. M'obstiner sous ces conditions à la poursuivre encore me paraîtrait un gaspillage, tant de ressources humaines que de deniers publics. C'est pourquoi j'ai demandé mon détachement au CNRS (que j'avais quitté en 1959 comme directeur de recherches frais émoulu, pour entrer à l'IHES). Je sais d'ailleurs que la situation de l'emploi est serrée au CNRS comme ailleurs, que l'issue de ma demande est douteuse, et que si un poste m'y était attribué, ce serait au dépens d'un chercheur plus jeune qui resterait

sans poste. Mais il est vrai aussi que cela libérerait mon poste à l'USTL au bénéfice d'un autre. C'est pourquoi je n'ai pas de scrupule à faire cette demande, et s'il le faut à revenir à la charge si elle n'est pas acceptée cette année.

En tout état de cause, cette demande aura été pour moi l'occasion d'écrire cette esquisse de programme, qui autrement sans doute n'aurait jamais vu le jour. J'ai essayé d'être bref sans être sybillin et aussi, après coup, d'en faciliter la lecture et de la rendre plus attrayante, en y adjoignant un sommaire. Si malgré cela elle peut paraître longue pour la circonstance, je m'en excuse. Elle me paraît courte pour son contenu, sachant que dix ans de travail ne seraient pas de trop pour aller jusqu'au bout du moindre des thèmes esquissés (à supposer qu'il y ait un "bout"...), et cent ans seraient peu pour le plus riche d'entre eux!

$\frac{50}{51}$

Derrière la disparité apparente des thèmes évoqués ici, un lecteur attentif percevra comme moi une unité profonde. Celle-ci se manifeste notamment par une source d'inspiration commune, la géométrie des surfaces, présente dans tous ces thèmes, au premier plan le plus souvent. Cette source, par rapport à mon "passé" mathématique, représente un renouvellement, mais nullement une rupture. Plutôt, elle montre le chemin d'une approche nouvelle vers cette réalité encore mystérieuse, celle des "motifs", qui me fascinait plus que toute autre dans les dernières années de ce passé (*). Cette fascination ne s'est nullement évanouie, elle fait partie plutôt de celle du plus brûlant pour moi de tous les thèmes évoqués précédemment. Mais aujourd'hui je ne suis plus, comme naguère, le prisonnier volontaire de tâches interminables, qui si souvent m'avaient interdit de m'élancer dans l'inconnu, mathématique ou non. Le temps des tâches pour moi est révolu. Si l'âge m'a apporté quelque chose, c'est d'être plus léger.

Janvier 1984

(*) Voir à ce sujet mes commentaires dans l'"Esquisse Thématique" de 1972 jointe au présent rapport, dans la rubrique terminale "divagations motiviques" (loc. cit. pages 17-18).

$\frac{51}{52}$ (^{1}) L'expression "hors de portée" ici (et encore plus loin pour une question toute différente), que j'ai laissée passer en allant à l'encontre d'une réticence, me paraît décidément hâtive et sans fondement. J'ai pu constater déjà en d'autres occasions que lorsque des augures (ici moi-même!) déclarent d'un air entendu (ou dubitatif) que tel problème est "hors de portée", c'est là au fond une affirmation entièrement subjective. Elle signifie simplement, à part le fait que le problème est censé ne pas être résolu encore, que celui qui parle est à court d'idées sur la question, ou de façon plus précise sans doute, qu'il est devant elle sans sentiment ni entrain, qu'elle "ne lui fait rien", et qu'il n'a aucune envie de faire quelque chose avec elle – ce qui souvent est une raison suffisante pour vouloir en décourager autrui. Cela n'a pas empêché qu'à l'instar de M. de la Palisse, et au moment même de succomber, les belles et regrettées conjectures de Mordell, de Tate, de Chafarévitch étaient toujours réputées "hors de portée", les pauvres! – D'ailleurs, dans les jours déjà qui ont suivi la rédaction du présent rapport, qui m'a remis en contact avec des questions dont je m'étais quelque peu éloigné au cours de l'année écoulée, je me suis aperçu d'une nouvelle propriété remarquable de l'action extérieure d'un groupe de Galois absolu sur le groupe fondamental d'une courbe algébrique, qui m'avait échappé jusqu'à présent et qui sans doute constitue pour le moins un nouveau pas en avant vers la formulation d'une caractérisation algébrique de Gal($\overline{\mathbb{Q}}/\mathbb{Q}$). Celle-ci, avec la "conjecture fondamentale" (mentionnée au par. 3 ci-dessous) apparaît à présent comme la principale question ouverte pour les fondements d'une "géométrie algébrique anabélienne", laquelle depuis quelques années, représente (et de loin) mon plus fort centre d'intérêt en mathématiques.

(^{2}) Je puis faire exception pourtant d'un autre "fait", du temps où, vers l'âge de douze ans, j'étais interné au camp de concentration de Rieucros (près de Mende). C'est là que j'ai appris, par une détenue, Maria, qui me donnait des leçons particulières bénévoles, la définition du cercle. Celle-ci m'avait $\frac{52}{53}$ impressionné par sa simplicité et son évidence, alors que la propriété de "rotondité parfaite" du cercle m'apparaissait auparavant comme une réalité mystérieuse au-delà des mots. C'est à ce moment, je crois, que j'ai entrevu pour la première fois (sans bien sûr me le formuler en ces termes) la puissance créatrice d'une "bonne" définition mathématique, d'une formulation qui décrit l'essence. Aujourd'hui encore, il semble que la fascination qu'a exercé sur moi cette puissance-là n'a rien perdu de sa force.

(^{3}) Plus généralement, au-delà des variétés dites "anabéliennes" sur des corps de type fini, la géométrie algébrique anabélienne (telle qu'elle s'est dégagée il y a quelques années) amène à une description, en termes de groupes profinis uniquement, de la catégorie des schémas de type fini sur

la base absolue $\mathbf{Q}$ (voire même $\mathbf{Z}$), et par là même, en principe, de la catégorie des schémas quelconques (par des passages à la limite convenables).
Il s'agit donc d'une construction "qui fait semblant" d'ignorer les anneaux
(tels que $\mathbf{Q}$, les algèbres de type fini sur $\mathbf{Q}$ etc.) et les équations algébriques
qui servent traditionnellement à décrire les schémas, en travaillant directe-
ment avec leurs topos étales, exprimables en termes de systèmes de groupes
profinis. Un grain de sel cependant: pour pouvoir espérer reconstituer un
schéma (de type fini sur $\mathbf{Q}$ disons) à partir de son topos étale, qui est un
invariant purement topologique, il convient de se placer, non dans la catégorie des schémas (de type fini sur $\mathbf{Q}$ en l'occurrence), mais dans celle qui
s'en déduit par "localisation", en rendant inversibles les morphismes qui
sont des "homéomorphismes universels", i.e. qui sont finis, radiciels et sur-
jectifs. Le développement d'une telle traduction d'un "monde géométrique"
(savoir celui des schémas, multiplicités schématiques etc.) en termes de
"monde algébrique" (celui des groupes profinis, et systèmes de groupes
profinis décrivant des topos (dits "étales") convenables) peut être considéré
comme un aboutissement ultime de la théorie de Galois, sans doute dans
l'esprit même de Galois. La sempiternelle question "et pourquoi tout ça?"
me paraît avoir ni plus, ni moins de sens dans le cas de la géométrie algébrique anabélienne en train de naître, que pour la théorie de Galois au
temps de Galois (ou même aujourd'hui, quand la question est posée par un
étudiant accablé...) et de même pour le commentaire qui va généralement
avec: "c'est bien général tout ça!".

(^{4}) On conçoit donc aisément qu'un groupe comme $\mathrm{Sl}(2,\mathbf{Z})$, avec sa structure
"arithmétique", soit une véritable machine à construire des représentations
"motiviques" de $\mathrm{Gal}(\overline{\mathbf{Q}}/\mathbf{Q})$ et de ses sous-groupes ouverts, et qu'on obtient
ainsi, au moins en principe, toutes les représentations motiviques qui sont de
poids 1, ou contenues dans un produit tensoriel de telles représentations (ce
qui en fait déjà un bon paquet!). J'avais commencé en 1981 à expérimenter
avec cette machine dans quelques cas d'espèce, obtenant diverses représen-
tations remarquables de $\mathbf{\Gamma}$ dans des groupes $G(\hat{\mathbf{Z}})$, où G est un schéma en
groupes (pas nécessairement réductif) sur $\mathbf{Z}$, en partant d'homomorphismes
convenables

$$\mathrm{Sl}(2,\mathbf{Z}) \longrightarrow G_0(\mathbf{Z}),$$

où G_0 est un schéma en groupes sur $\mathbf{Z}$, et G étant construit à partir de là
comme extension de G_0 par un schéma en groupes convenable. Dans le cas
"tautologique" $G_0 = \mathrm{Sl}(2)_{\mathbf{Z}}$, on trouve pour G une extension remarquable
de $\mathrm{Gl}(2)_{\mathbf{Z}}$ par un tore de dimension 2, avec une représentation motivique
qui "coiffe" celles associées aux corps de classes des extensions $\mathbf{Q}(i)$ et $\mathbf{Q}(j)$
(comme par hasard, les "corps de multiplication complexe" des deux courbes

elliptiques "anharmoniques"). Il y a là un principe de construction qui m'a semblé très général et très efficace, mais je n'ai pas eu (ou pris) le loisir de le dévisser et le suivre jusqu'au bout – c'est là un des nombreux "points chauds" dans le programme de fondements de géométrie algébrique anabélienne (ou de "théorie de Galois", version élargie) que je me propose de développer. A l'heure actuelle, et dans un ordre de priorité sans doute très provisoire, ces points sont:

54
55

a) Construction combinatoire de la Tour de Teichmüller.

b) Description du groupe des automorphismes de la compactification profinie de cette tour, et réflexion sur une caractérisation de $\Gamma = \mathrm{Gal}(\overline{\mathbf{Q}}/\mathbf{Q})$ comme sous-groupe de ce dernier.

c) La "machine à motifs" $\mathrm{Sl}(2, \mathbf{Z})$ et ses variantes.

d) Le dictionnaire anabélien, et la conjecture fondamentale (qui n'est peut-être pas si "hors de portée" que ça!). Parmi les points cruciaux de ce dictionnaire, je prévois le "paradigme profini" pour les corps $\mathbf{Q}$ (cf. b)), $\mathbf{R}$ et $\mathbf{C}$, dont une formulation plausible reste à dégager, ainsi qu'une description des sous-groupes d'inertie de Γ, par où s'amorce le passage de la caractéristique zéro à la caractéristique $p > 0$, et à l'anneau absolu $\mathbf{Z}$.

e) Problème de Fermat.

(^{5}) Je signalerai cependant un travail plus délicat (mis à part le travail signalé en passant sur les complexes cubiques), sur l'interprétation combinatoire des cartes régulières associées aux sous-groupes de congruence de $\mathrm{Sl}(2,\mathbf{Z})$. Ce travail a été développé surtout en vue d'exprimer l'opération "arithmétique" de $\Gamma = \mathrm{Gal}(\overline{\mathbf{Q}}/\mathbf{Q})$ sur ces "cartes de congruences", laquelle se fait, essentiellement, par l'intermédiaire du caractère cyclotomique de Γ. Un point de départ a été la théorie combinatoire du "bi-icosaèdre" développée dans un cours C4 à partir de motivations purement géométriques, et qui (il s'est avéré par la suite) permet d'exprimer commodément l'opération de Γ sur la catégorie des cartes icosaédrales (i.e. des cartes de congruence d'indice 5).

(^{6}) Signalons à ce propos que les classes d'isomorphie d'espaces modérés compacts sont les mêmes que dans la théorie "linéaire par morceaux" (qui n'est pas, je le rappelle, une théorie modérée). C'est là, en un sens, une réhabilitation de la "Hauptvermutung", qui n'est "fausse" que parce que, pour des raisons historiques qu'il serait sans doute intéressant de cerner de plus près, les fondements de topologie utilisés pour la formuler n'excluaient pas les phénomènes de sauvagerie. Il va (je l'espère) sans dire que la nécessité de développer de nouveaux fondements pour la topologie "géométrique" n'exclut nullement que les phénomènes en question, comme toute chose sous le ciel, ont leur raison d'être et leur propre beauté. Des fondements

55
56

plus adéquats ne supprimeront pas ces phénomènes, mais nous permettront de les situer à leur juste place, comme des "cas limites" de phénomènes de "vraie" topologie.

(^{7}) En fait, pour reconstituer ce sytème d'espaces

$$(i_0, \ldots, i_n) \longmapsto X_{i_0 \ldots i_n}$$

contravariant sur Drap(I) (pour l'inclusion des drapeaux), il suffit de connaître les X_i (ou "<u>strates déployées</u>") et les X_{ij} (ou "<u>tubes de raccord</u>") pour $i, j \in I$, $i < j$, et les morphismes $X_{ij} \to X_i$ (qui sont des inclusions "bordantes") et $X_{ij} \to X_j$ (qui sont des fibrations propres, dont les fibres F_{ij} sont appelées "<u>fibres de raccord</u>" pour les strates d'indices i et j). Dans le cas d'une multiplicité modérée cependant, il faut connaître de plus les "<u>espaces de jonction</u>" X_{ijk} ($i < j < k$) et ses morphismes dans X_{ij}, X_{jk}, et surtout X_{ik}, s'insérant dans le diagramme commutatif hexagonal suivant, où les deux carrés de droites sont cartésiens, les flèches $\hookrightarrow$ sont des immersions (pas nécessairement des plongements ici), et les autres flèches sont des fibrations propres:

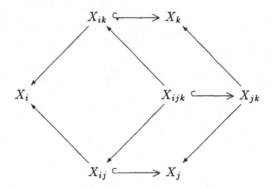

(NB. Ce diagramme définit X_{ijk} en termes de X_{ij} et X_{jk} sur X_j, mais non la flèche $X_{ijk} \to X_{ik}$, car $X_{ik} \to X_k$ n'est <u>pas</u> nécessairement un plongement.)

Dans le cas des espaces modérés stratifiés proprement dits (qui ne sont pas des multiplicités à proprement parler) on peut exprimer de façon commode le "déploiement" de cette structure, i.e. le système des espaces $X_{i_0 \ldots i_n}$, en termes de l'espace modéré X_* somme des X_i, qui est muni d'une <u>structure d'objet ordonné</u> (dans la catégorie des espaces modérés) ayant comme graphe X_{**} de la relation d'ordre la somme des X_{ij} et des X_i (ces derniers constituant la diagonale). Parmi les propriétés essentielles de cette structure ordonnée, relevons seulement ici que $\mathrm{pr}_1 : X_{**} \to X_*$ est une fibration (localement triviale) propre, et $\mathrm{pr}_2 : X_{**} \to X_*$ est un plongement "bordant". On a une interprétation analogue du déployement d'une multiplicité modérée stratifiée, en termes d'une <u>structure de catégorie</u> (remplaçant une simple structure ordonnée) "au sens multiplicités modérées", dont l'application de composition est donnée par les morphismes $X_{ijk} \to X_{ik}$ ci-dessus.

Lieber Herr Faltings*,

Vielen Dank für ihre rasche Antwort und Übersendung der Separata! Ihr Kommentar zur sog. "Theorie der Motive" ist von der üblichen Art, die wohl grossenteils der in der Mathematik stark eingewurzelten Tradition entspringt, nur denjenigen mathematischen Situationen und Zusammenhängen eine (eventuell langatmige) Untersuchung und Aufmerksamkeit zuzuwenden, insofern sie die Hoffnung gewähren, nicht nur zu einem vorläufigen und möglicherweise z.T. mutmasslichen Verständnis eines bisher geheimnisvollen Gebietes zu kommen, wie es in den Naturwissenschaften ja gang und gäbe ist – sondern auch zugleich Aussicht auf die Möglichkeit einer laufenden Absicherung der gewonnenen Einsichten durch stichhaltige Beweise. Diese Einstellung scheint mir nun psychologisch ein ausserordentlich starkes Hindernis zur Entfaltung mathematischer Schauenkraft, und damit auch zum Fortschreiten mathematischer Einsicht im üblichen Sinn, nämlich der Einsicht, die durchdringend oder erschöpfend genug ist, um sich schliesslich "beweisen" zu können. Was meine Erfahrung in mathematischer Arbeit mich immer wieder lehrte, ist dass stets in erster Linie der Beweis aus der Einsicht entspringt, nicht im Gegenteil – und dass die Einsicht in erster Linie aus einem feinfühligen und hartnäckigen Aufspüren der relevanten Wesenheiten und Begriffen entsteht, und deren Wechselbeziehung. Der leitende Faden ist innere Kohärenz des allmählich sich aus dem Dunst lösenden Bildes, und Einklang auch mit der anderweilig Bekannten oder Erahntem – und er leitet um so sicherer, als die "exigence" der Kohärenz eine strengere und feinfühlendere ist.

Um auf Motive zurückzukommen, so existiert meines Wissens keinerlei "Theorie" der Motive, aus dem einfachen Grund, dass niemand sich die Mühe machte, eine solche Theorie auszuarbeiten. Das vorhandene Material an bekannten Tatsachen und an geahnten Zusammenhängen ist von beindruckender Reichhaltigkeit – unverhältnismässig viel mehr, will mir scheinen, als je zu Ausarbeitung einer physikalischen Theorie vorlag! Es existiert z.Z. eine Art "yoga des motifs", das einer Handvoll Eingeweihter geläufig ist, und in manchen Situationen einen sicheren Anhalt gewährt zur Erratung gewisser Zusammenhänge, die sich dann auch bisweilen tatsächlich so oder so beweisen lassen (wie etwa in Ihren letzten Arbeit der Satz über Galois Aktion auf Tateschen Moduln abelscher Mannigfaltigkeiten). Es hat

* Editors' note. This letter is dated June 27, 1983. We are grateful to G. Faltings for giving his permission to reproduce it here.

den Status scheint mir einer Art Geheimwissenschaft – Deligne scheint mir der, dem sie am geläufigsten ist. Seine erste [veröffentlichte] Arbeit, über die Entartung der Leray'schen Spektralfolge für eine glatte eigentliche Abbildung algebraischer Mannigfaltigkeiten über $\mathbb{C}$, entsprang einer einfachen Überlegung über "Gewichte" von Kohomologiegruppen, die zu jener Zeit reine Heuristik ware, heute aber (seit dem Beweis der Weil'schen Vermutungen) sich wohl über beliebigem Grundschema durchführen liesse. Es ist mir auch klar, dass die Deligne'sche Erweiterung der Hodge Theorie weitgehend aus dem ungeschriebenen "Yoga" der Motive schöpfte – nämlich dem Bestreben entsprang, gewisse "Tatsachen" aus jenem Yoga, und ganz besonders die Existenz der Filtration der Kohomologie durch "Gewichte", und zudem die Halbeinfachheit gewisser Aktionen von Fundamentalgruppen, im Rahmen der transzendenten Hodge-Strukturen sicherzustellen.

Nun einige Worte zum "Yoga" der anabelschen Geometrie. Es geht dabei um "absolute" alg. Geometrie, nämlich über Grundkörper (etwa), die endlich erzeugt über dem Primkörper sind. Eine allgemeine Grundidee ist, dass für gewisse, sog. "anabelsche", Schemata X (von endlichem Typ) über K, die Geometrie von X vollständig durch die (profinite) Fundamentalgruppe $\pi_1(X, \xi)$ bestimmt ist (wo ξ ein "geometrischer Punkt" von X ist, etwa mit Wert in einer vorgeschriebenen alg. abg. Hülle $\overline{K}$ von K), zusammen mit der Extra-Struktur, die durch den Homomorphismus

$$(1) \qquad \pi_1(X, \xi) \to \pi_1(K, \xi) = \mathrm{Gal}(\overline{K}/K)$$

gegeben ist, deren Kern die "geometrische Fundamentalgruppe"

$$(2) \qquad \pi_1(\overline{X}, \xi) \qquad (\overline{X} = X \otimes_K \overline{K})$$

ist – also auch die profinite Kompaktifikation der transzendenten Fundamentalgruppe, wenn $\overline{K}$ als Teilkörper des Körpers $\mathbb{C}$ der komplexen Zahlen gegeben ist. Das Bild von (1) ist eine offene Teilgruppe der profinite Galoisgruppe, vom Index 1 dann, genau wenn $\overline{X}$ zusammenhängend ist.

Die erste Frage ist, welche Schemata X als "anabelsch" angesehen werden sollen. Dabei will ich mich jedenfalls auf glattes X beschränken. Völlige Klarheit habe ich nur im Fall $\dim X = 1$. Allenfalls soll die anabelsche Eigenschaft eine rein geometrische sein, nämlich sie hängt nur von $\overline{X}$ über dem alg. abg. Körper $\overline{K}$ (oder dem entsprechenden Schema über beliebiger alg. abg. Erweiterung von $\overline{K}$, etwa $\mathbb{C}$) ab. Zudem soll $\overline{X}$ anabelsch sein wenn und nur wenn die zusammenhängenden Komponente es sind. Schliesslich (im Fall der Dimension 1) ist eine (glatte zusammenhängende) Kurve über $\overline{K}$ anabelsch, wenn ihre EP-charakteristik < 0 ist, oder auch, wenn ihre Fundamentalgruppe nicht abelsch ich; diese letztere

Formulierung trifft jedenfalls in Charakteristik null zu, oder im Fall einer eigentlichen ("kompakten") Kurve – sonst muss die "prim-zu-p" Fundamentalgruppe genommen werden. Andere äquivalente Formulierungen: das Gruppenschema der Automorphismen soll der Dimension null sein, oder auch die Automorphismengruppe soll endlich sein. Ist die Kurve vom Typ (g, ν), wo g das Geschlecht, und ν die Anzahl der "Löcher" oder "Punkte im Unendlichen" ist, so sind die anabelsche Kurven genau die, für die der Typ verschieden ist von

$$(0,0), (0,1), (0,2) \text{ und } (1,0)$$

nämlich

$$2g + \nu > 2 \quad (\text{i.e. } -\chi = 2g - 2 + \nu > 0).$$

Wenn der Grundkörper $\mathbb{C}$ ist, so sind die anabelschen Kurven genau die, deren (transzendente) universelle Überlagerung "hyperbolisch" ist, nämlich isomorph zur Poincaré'schen Halbebene – also genau die, die "hyperbolisch" sind im Sinne von Thurston.

Andernteils sehe ich eine Mannigfaltigkeit jedenfalls dann als "anabelsch" (ich könnte sagen "elementar anabelsch") an, wenn sie sich durch successive glatte Faserungen aus anabelschen Kurven aufbauen lässt. Demnach (einer Bemerkung von M.Artin zufolge) hat jeder Punkt einer glatten Mannigfaltigkeit X/K ein Fundamentalsystem von (affinen) anabelschen Umgebungen.

Schliesslich wurde letztlich meine Aufmerksamkeit immer stärker durch die Modulmannigfaltigkeiten (oder vielmehr die Modulmultiplizitäten) $M_{g,\nu}$ für algebraische Kurven beansprucht, und ich bin so ziemlich überzeugt, dass auch diese als "anabelsch" angesprochen werden dürfen, nämlich dass ihre Beziehung zur Fundamentalgruppe eine ebenso enge ist, wie im Fall von anabelschen Kurven. Ich würde annehmen, dass dasselbe auch für die Modulmultiplizitäten polarisierter abelscher Mannigfaltigkeiten gelten dürfte.

Ein grosser Teil meiner Überlegungen vor zwei Jahren beschränkte sich auf den Fall der Char. Null, den ich nun vorsichtshalber voraussetzen will. Da ich mich seit über einem Jahr nicht mehr mit diesem Fragenkomplexe beschäftigt habe, verlasse ich mich auf meine Gedächtnis, das jedenfalls rascher zugänglich ist als Stapel von Auszeichnungen – ich hoffe, ich webe nicht zuviel Irrtürmer mit hinein! Ein Ausgangspunkt unter anderem war die bekannte Tatsache für Mannigfaltigkeiten X, Y über einem alg. abg. Körper K, wenn Y sich in eine [quasi-]abelsche Mannigfaltigkeit A einbetten lässt, dass eine Abbildung $X \to Y$ bis auf Translation von A bekannt ist, wenn die entsprechende Abbildung für H_1 (etwa ℓ-adisch) bekannt ist.

Daraus folgt in manchen Situationen (etwa, wenn Y "elementar anabelsch" ist) dass bei dominierendem f (nämlich $f(X)$ dicht in Y), f genau bekannt ist, wenn $H_1(f)$ bekannt ist. Doch der Fall von konstanten Abbildungen lässt sich offensichtlich nicht einordnen. Aber gerade der Fall, wo X sich auf einen Punkt reduziert, ist von besonderem Interesse, falls man es auf eine "Bestimmung" der Punkte von Y absieht.

$\frac{3}{4}$

Geht man nun zum Fall eines Körpers K vom endlichen Typ über, und ersetzt man H_1 (nämlich die "abelianisierte" Fundamentalgruppe) durch die volle Fundamentalgruppe, so ergibt sich, falls Y "elementar anabelsch" ist, dass f bekannt ist, wenn $\pi_1(f)$ bekannt ist "bis auf inneren Automorphismus". Wenn ich recht ersinne, genügt es hier, statt der vollen Fundamentalgruppen, mit deren Quotienten zu arbeiten, die sich ergeben, wenn man (2) durch die entsprechenden abelianisierten Gruppen $H_1(\overline{X}, \hat{\mathbf{Z}})$ ersetzt. Der Beweis ergibt sich ziemlich einfach aus dem Mordell-Weilschen Satz, dass die Gruppe $A(K)$ ein endlich erzeugter $\mathbf{Z}$-Modul ist, wo A die "jacobienne généralisée" von Y ist, die der "universellen" Einbettung von Y in einen Torsor unter einer quasi-abelschen Mannigfaltigkeit entspricht. Der springende Punkt ist hier, dass ein Punkt von A über K nämlich ein "Schnitt" von A über K, völlig bestimmt ist durch die entsprechende Spaltung der exakten Sequenz

(3) $$1 \to H_1(\overline{A}) \to \pi_1(A) \to \pi_1(K) \to 1$$

(bis auf inneren Automorphismus), nämlich durch die entsprechende Kohomologieklasse in

$$H^1(K, \pi_1(\overline{A})),$$

wobei $\pi_1(\overline{A})$ auch durch die ℓ-adische Komponente, nämlich den Tate'sche Moduln $T_\ell(\overline{A})$ ersetzt werden kann.

Aus diesem Ergebnis folgt nun leicht das Folgende, das mich vor zwei ein halb Jahren einigermassen verblüffte: es seien K, L zwei Körper vom endlichen Typus (kurz "absolute Körper" genannt), dann ist ein Homomorphismus

$$K \to L$$

völlig bestimmt, wenn man die entsprechende Abbildung

(4) $$\pi_1(K) \to \pi_1(L)$$

der entstprechenden "äusseren Fundamentalgruppen" kennt (nämlich, wenn jene bis auf inneren Automorphismus bekannt ist). Dies klingt stark an die topologische Intuition des Zusammenhanges zwischen $K(\pi, 1)$ Räumen, und deren Fundamentalgruppen – nämlich, Homotopieklassen von Abbildungen der Räume entsprechen eindeutig den entsprechenden Abbildungen zwischen den äusseren Gruppen. Im Rahmen jedoch absoluten alg. Geometrie (nämlich über "absoluten" Körpern) bestimmt die Homotopieklasse einer Abbildung schon die Abbildung. Der Grund dafür scheint mir in der ausserordentlichen Rigidität der vollen Fundamentalgruppe, die wiederum daraus entspringt, dass die [(äussere)] Aktion des "arithmetischen Teils" jener Gruppe, nämlich $\pi_1(K) = \operatorname{Gal}(\overline{K}/K)$, auf dem "geometrischen Teil" $\pi_1(X_{\overline{K}})$, so ausserordentlich stark ist (was sich insbesondere ja auch in den Weil-Deligne'schen Aussagen ausdrückt).

Letztere Aussage ("Der Grund dafür...") kam mir rasch in die Schreibmaschine – ich erinnere jetzt nämlich, dass für die vorangehende Aussage über Homomorphismen von Körpern keineswegs notwendig ist, dass diese "absolut" seien – es genügt, dass beide vom endlichen Typ über einem gemeinsamen Grundkörper k seien, sofern wir uns auf k-Homomorphismen beschränken. Dabei genügt es offensichtlich, sich auf den Fall k alg. abg. zu beschränken. Die erwähnte "Rigidität" hingegen spielt entscheidend ein, wenn wir es darauf absehen, diejenigen Abbildungen (4) zu kennzeichnen, die einem Homomorphismus $K \to L$ entsprechen. In dieser Hinsicht ist folgende Vermutung naheliegend: wenn der Grundkörper k "absolut" ist, dann sind genau die äusseren Homomorphismen (4) "geometrisch", die mit den "Augmentationhomomorphismen" zu $\pi_1(K)$ verträglich sind. [siehe Berichtigung im PS: die Bildgruppe muss darin von endlichen Index sein!] Bei dieser Aussage kann man sich offensichtlich auf den Fall beschränken wenn k der Primkörper ist, also $\mathbb{Q}$ (in der Char. Null). Das "Grundobjekt" der anabelschen alg. Geometrie der Char. Null, dass für den Primkörper $\mathbb{Q}$ steht, ist demnach die Gruppe

$$(5) \qquad \Gamma = \pi_1(\mathbb{Q}) = \operatorname{Gal}(\overline{\mathbb{Q}}/\mathbb{Q}),$$

wo $\overline{\mathbb{Q}}$ die alg. abg. Hülle von $\mathbb{Q}$ in $\mathbb{C}$ sein soll.

Die obige Vermutung kann als die Hauptvermutung der "birationellen" anabelschen alg. Geometrie angesehen werden – sie sagt aus, dass die Kategorie der "absoluten birationellen alg. Mannigfaltigkeiten" in Char. Null sich voll einbetten lässt in die Kategorie der nach Γ augmentierten proendlichen Gruppen. Weiterhin besteht dann noch die Aufgabe, zu einer ("rein algebraischen") Beschreibung der Gruppe Γ zu gelangen, und zudem, zu einem Verständnis, welche Γ-augmentierten proendlichen Gruppen zu einem $\pi_1(K)$ isomorph sind. Ich will vorerst nicht auf diese Fragen eingehen, möchte hingegen eine verwandte und beträchtlich schärfere Vermutung für anabelsche Kurven formulieren, aus der die vorherige folgt. Ich sehe da zwei verschieden erscheinende, aber äquivalente Formulierungen:

1) Es seien X, Y zwei (zusammenhängende, das sei stets vorausgesetzt) anabelsche Kurven über dem absoluten Körper der Char. Null, man betra-

chtet die Abbildung

(6) $\operatorname{Hom}_K(X, Y) \to \operatorname{Hom} \operatorname{ext}_{\pi_1(K)}(\pi_1(X), \pi_1(Y))$,

wo Hom ext die Menge der äusseren Homomorphismen proendlicher Gruppen bezeichnet, und der Index $\pi_1(K)$ Kompatibilität mit den Augmentationen nach $\pi_1(K)$ bedeutet. Nach dem Vorangehenden ist bekannt, dass diese Abbildung injektiv ist. Ich vermute dass sie bijektiv ist. [siehe Berichtigung im P.S.]

2) Diese zweite Form kann als Reformulierung von 1) angesehen werden, im Fall von konstanten Abbildungen von X zu Y. Es sei $\Gamma(X/K)$ die Menge aller K-wertigen Punkte (also "Schnitte") von X über K, man betrachtet die Abbildung

(7) $\Gamma(X/K) \to \operatorname{Hom} \operatorname{ext}_{\pi_1(K)}(\pi_1(K), \pi_1(X))$,

wo die zweite Menge also die Menge aller "Spaltungen" der Gruppenerweiterunge (3) ist (wo A durch X ersetzt wurde – $\pi_1(X) \to \pi_1(K)$ ist in der Tat surjektif, sofern nur X überhaupt einen K-wertigen Punkt besitzt und deshalb X auch "geometrisch zummanenhängend" ist), oder vielmehr die Menge der üblichen Konjugationklassen solcher Spaltungen via Aktion der Gruppe $\pi_1(\overline{X})$. Es ist bekannt, dass (7) injektiv ist, und die Hauptvermutung sagt aus, dass sie bijektiv ist. [siehe unten Berichtigung]

Fassung 1) folgt aus 2), indem man K durch den Funktionenkörper von X ersetzt. Dabei ist es übrigens gleichgültig, ob X anabelsch ist, und wenn ich nicht irre, folgt die Aussage 1) sogar für beliebiges glattes X (ohne $\dim X = 1$ Voraussetzung). Was Y anbelangt, so folgt aus der Vermutung, dass Aussage 1) gültig bleibt, sofern nur Y "elementar anabelsch" ist [siehe Berichtigung im PS], und entsprechend natürlich für Aussage 2). Dies gibt nun im Prinzip, unter Verwendung der Artin'schen Bemerkung, die Möglichkeit einer vollständigen Beschreibung der Kategorie der Schemata vom endlichen Typus über K, "en termes de" $\Gamma(K)$ und Systemen von proendlichen Gruppen. Hier wiederum habe ich etwas rasch in die Schreibmaschine gehauen, da die Hauptvermutung erst noch ausgearbeitet und vervollständigt werden muss, zu einer Aussage darüber, welche nun (bis auf Isomorphismus) die $\Gamma(K)$-augmentierten proendlichen vollständigen Gruppen sind, die von anabelschen Kurven über K stammen. Wenn es nur auf eine Aussage der "pleine fidélité" wie in obigen Formulierungen 1) und 2) ankommt, so dürfte sich folgendes ohne besondere Schwierigkeit aus diesen ableiten lassen, und sogar schon (wenn ich nicht irre) aus der vorangehenden beträchtlich schwächeren birationellen Variante. Es seien nämlich X und Y zwei Schemata, die "wesentlich vom endlichen Typ über $\mathbb{Q}$" sind, z.B. jedes eine vom endlichen Typ über einem (unbestimmt bleibenden) absoluten Körper der Char. Null. X und Y brauchen weder glatt noch zusammenhängend sein, ebensowenig "normal" oder dergleichen – hingegen sollen sie reduziert vorausgesetzt werden. Ich betrachte die entsprechenden etalen Topoi X_{et} und Y_{et}, und die Abbildung

$$(8) \qquad \mathrm{Hom}(X,Y) \to \mathrm{Iskl}\ \mathrm{Hom}_{top}(X_{et}, Y_{et}),$$

wo Hom_{top} die (Menge der) Homomorphismen der Topoi X_{et} in Y_{et} bezeichnet, und Iskl Übergang zu (der Menge der) Isomorphieklassen bedeutet. (Es sei dazu bemerkt, dass die Kategorie $\underline{\mathrm{Hom}}_{top}(X_{et}, Y_{et})$ rigid ist, nämlich dass es stets nur einen Isomorphismus zwischen zwei Homomorphismen $X_{et} \to Y_{et}$ geben kann. Wenn X und Y Multiplizitäten und nicht Schemata sind, muss folgende Aussage durch eine entsprechend feinere, nämlich eine Äquivalenz von Kategorien $\underline{\mathrm{Hom}}(X,Y)$ mit $\underline{\mathrm{Hom}}_{top}(X_{et}, Y_{et})$ ersetzt werden.) Wesentlich ist dabei, dass X_{et} und Y_{et} lediglich als topologische Gebilde, also mit Ausschluss der Strukturgarben, betrachtet werden, während ja das erste Glied von (8) als Iskl $\mathrm{Hom}_{top.ann.}(X_{et}, Y_{et})$ aufgefasst werden kann. Zuerst sei bemerkt, dass aus dem vorhergehendem "Bekannten" ohne Schwierigkeit folgen dürfte, dass (8) injektiv ist. Dazu fällt mir nun allerdings ein, dass ich bei der Beschreibung des zweiten Glieds von (8) ein wichtiges Strukturelement vergass, nämlich X_{et} und Y_{et} sollen als Topoi über der absoluten Basis $\mathbb{Q}_{et}$ aufgefasst werden, die durch die proendliche Gruppe $\Gamma = \pi_1(K)$ (5) völlig beschrieben ist. Also Hom_{top} soll als $\mathrm{Hom}_{top/\mathbb{Q}_{et}}$ gelesen werden. Mit dieser Berichtigung wäre nun die naheliegende Vermutung, dass (8) bijektiv sei. Das kann nun aber aus dem Grund nicht [ganz] zutreffen, weil es ja endliche radizielle Morphismen $Y \to X$ geben kann (sogenannte "universelle Homomorphismen"), die nämlich eine topologische Äquivalenz $Y_{et} \overset{\sim}{\to} X_{et}$ zeitigen, ohne Isomorphismen zu sein, so dass also keine Umkehrung $X \to Y$ existiert, während sie doch für die etalen Topoi existiert. Setzt man X als normal voraus, so vermute ich, dass (8) bijektiv ist. Im allgemeine Fall sollte es zutreffen, dass für jedes ϕ im zweiten Glied sich ein Diagramm bilden lässt

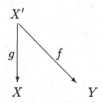

(wo g ein "universeller Homomorphismus" ist), aus dem ϕ im naheliegenden Sinn entspringt. Ich vermute sogar, dass dasselbe gültig bleibt ohne Char.

Null Voraussetzung , nämlich wenn $\mathbb{Q}$ durch $\mathbb{Z}$ ersetzt wird – was damit zusammenhängt, dass die "birationelle" Hauptvermutung auch in beliebiger Charakteristik gültig sein dürfte, sofern wir nur die "absoluten" Körper durch deren "perfekte Hüllen" $K^{p^{-\infty}}$ ersetzen, die ja dasselbe π_1 haben.

Ich fürchte, ich wurde etwas ausschweifig mit jenem Abstecher bezüglich beliebiger Schemata vom endlichen Typus und deren etalen Topoi – interessanter für Sie dürfte eine dritte Formulierung der Hauptvermutung sein, die jene noch leicht verschärft, und die zudem einen besonders "geometrischen" Beiklang hat. Es ist auch die, die ich vor etwa zwei Jahren Deligne mitteilte, und von der er mir sagte, dass aus ihr die Mordell'sche Vermutung folge. Es sei wiederum X eine anabelsche geometrische zusammenhängende Kurve über dem absoluten Körper K der Char. Null, $\tilde{X}$ ihr universell Überdeckung, aufgefasst als ein Schema (aber nicht vom endlichen Typ) über $\overline{K}$, nämlich als universelle Überdeckung der "geometrischen" Kurve $\overline{X}$. Sie steht hier als eine Art algebraisches Analogon zur transzendenten Konstruktion, wo die Überdeckung der Poincaré'schen Halbebene isomorph ist. Ich betrachte aber zudem die Komplettierung $X^\wedge$ von X (die also eine projektive Kurve ist, nicht unbedingt anabelsch, da ja X vom Geschlecht 0 oder 1 sein kann), und deren Normalisierung $\tilde{X}^\wedge$ bezüglich $\tilde{X}$, die eine Art Kompaktifizierung von $\tilde{X}$ darstellt. (Wenn es Ihnen lieber ist, können Sie gerne vorerst X eigentlich, also $X = X^\wedge$ voraussetzen, also $\tilde{X} = \tilde{X}^\wedge$.) Die Gruppe $\pi_1(X)$ kann aufgefasst werden als die Gruppe der X-Automorphismen von $\tilde{X}$, und sie agiert zugleich auf die "Kompaktifizierung" $\tilde{X}^\wedge$. Diese Aktion ist verträglich mit der Aktion auf $\overline{K}$ via $\pi_1(K)$. Ich interessiere mich nun für die entsprechende Aktion

$$\text{Aktion von } \pi_1(X) \text{ auf } \tilde{X}^\wedge(\overline{K}),$$

(die [Menge der] $\overline{K}$-wertigen Punkte, oder was aufs selbe herauskommt, der vom generischen Punkt verschiedenen Punkte von $X^\wedge$), und insbesondere, für eine gegebene Spaltung von (3)

$$\pi_1(K) \to \pi_1(X),$$

für die entsprechende Aktion von der Galois-Gruppe $\pi_1(K)$. Die Vermutung ist nun, dass letztere Aktion (genau) einen Fixpunkt besitzt.

Dass es höchstens einen Fixpunkt geben kann, folgt aus der Injektivität in (7), oder lässt sich jedenfalls auf dem selben Weg mittels dem Mordell-Weilschen Satz beweisen. Unbewiesen ist die Existenz des Fixpunktes, die mehr oder weniger der Surjektivität von (7) äquivalent ist. Dabei geht es mir nun aber auf, dass die Formulierung von der Hauptvermutung via (7), die ich vorhin gab, nur im Fall wo X eigentlich ist eine korrekte ist – und in diesem Fall ist sie mit der dritten (soeben gegebenen) Formulierung tatsächlich äquivalent. Im Falle hingegen, wo X nicht eigentlich ist, also "undendlich ferne Punkte" besitzt, so liefert jeder eine dieser Punkte auf naheliegende Weise ein erkleckliches Paket von Spaltungsklassen (deren Mächtigkeit das des Kontinuums ist), die sich nicht durch im endlichen liegende Punkte erhalten lassen. Diese entsprechen dem Fall eines Fixpunktes in $\widetilde{X}^{\wedge}$, der nicht in $\widetilde{X}$ liegt. Die Einzigkeit des Fixpunktes besagt unter anderen, ausser der Injektivität von (7), dass die "Pakete", die verschiedenen Punkten im Unendlichen entsprechen, leeren Durchschnitt haben, also dass jede Spaltungsklasse, die keinem endlichen Punkt entspricht, einem eindeutig bestimmten Punkt im Unendlichen entspringt.

Anstoss für diese dritte Formulierung der Hauptvermutung waren gewisse transzendenten Überlegungen über Aktion endlicher Gruppen auf algebraischen komplexen Kurven und deren (transzendent definierten) universellen Überlagerungen, die in meinen Überlegungen (in der ersten Hälfte des Jahres 1981, also vor zwei Jahren etwa), über die Aktion von Γ auf gewisse proendliche anabelsche Fundamentalgruppen (insbesondere die von $\mathbb{P}^1 - (0,1,\infty)$), ein entscheidende Rolle spielten. (Diese Rolle war vor allem die eines Leitfadens ins vorerst völlig Unbekannte, da die entsprechenden Aussagen in Char $p > 0$ unbewiesen blieben, und auch heute noch sind.) Um auf die Aktion von Galoisgruppen wie $\pi_1(K)$ zurückzukommen, so erscheint diese in mancher Hinsicht Analog zur Aktion endlicher Gruppen, was sich z.B. auf besonders frappante und präzise Art in der vorherigen Vermutung äussert.

Ich nahm die anabelschen Überlegungen zwischen Dezember 81 und April 82 wieder auf, diesmal aber mit einem verschiedenem Schwergewicht – nämlich dem Bestreben nach einem Verständnis der mannigfaltigen Struktur der [Teichmüllerschen] Fundamentalgruppen $T_{g,\nu}$ (besser, der Fundamentalgroupoids) der Modulmultiplizitäten $\overline{M}_{g,\nu}$, und der Aktion von Γ auf deren proendliche Komplettierung. (An diese Untersuchung möchte ich diesen Herbst wieder anknüpfen, wenn ich mich in diesem Sommer des Aufschreibens von davon etwas sehr entlegenen Überlegungen über Grundlagen kohomologischer bezw. homotopischer Algebra entledigen kann, das mich seit vier Monaten schon beschäftigt.) Ich bitte um Nachsicht um die etwas chaotische Darstellung eines Ideenkreises, der mir zwar sechs Monate lang intensiv im Atem hielt, mit dem ich aber seit zwei Jahren nur sehr flüchtigen Kontakt hatte, wenn überhaupt. Sollte es Sie interessieren, und sich die Gelegenheit ergeben, dass Sie sich in Südfrankreich befinden, wäre es mir ein Vergnügen, mit Ihnen auf diese oder andere Aspekte des "anabelschen Yogas" einzugehen. Es wäre sicherlich auch möglich, Sie für eine Ihnen genehme Zeitspanne zur Universität Montpellier einzuladen, nur fürchte ich, dass die Prozedur bei der heutigen Konjunktur eine etwas langwierige ist, da die Universität selbst für derlei Einladungen z.Z. keine Fonds hat,

eine Einladung also von Paris aus beschlossen bzw. genehmigt werden muss
– was wohl heissen dürfte, dass entsprechende Vorschläge ungefähr ein Jahr
im voraus gemacht werden müssen.

Mit diesen ermunternden Worten will ich nun aber diesen unmässig lang
gewordenen Brief beenden, und Ihnen nur noch erfreuliche Ferien wünschen!
Herzliche Grüsse

Ihr Alexander Grothendieck

PS Beim Nachlesen dieses Briefes fällt mir auf, dass, zugleich wie die zweite
Fassung der Hauptvermutung, auch die Verallgemeinerung auf "elementar
anabelsche" Mannigfaltigkeiten berichtigt werden muss, sorry! Übrigens
sehe ich jetzt, dass die erste Fassung im selben Sinne berichtigt werden
muss – nämlich im Fall wo Y nicht eigentlich ist, ist es notwendig, sich im
ersten Glied von (6) auf nicht-konstante Homomorphismen zu beschränken,
und im zweiten auf Homomorphismen $\pi_1(X) \to \pi_1(Y)$, deren Bild von
endlichem Index (nämlich offen) ist. Im Falle, wo Y durch eine elementar
anabelsche Mannigfaltigkeit ersetzt wird, gilt die Bijektivität von (6), sofern
man sich im ersten Glied auf dominierenden Homomorphismen beschränkt,
und im zweiten dieselbe Beschränkung (Bildgruppe von endlichem Index)
beibehält. Ähnlich muss auch die "birationelle" Fassung berichtigt werden –
nämlich man muss sich auf solche Homomorphismen (4) beschränken, deren
Bildgruppe von endlichem Index ist.

Komme ich nun auf die Abbildung (7) zurück im Fall einer anabelschen
Kurve, so lässt es sich genau angeben, welches die Spaltungsklassen im
zweiten Glied sind, die keinem "endliche" Punkt also keinem Element des
ersten Gliedes entsprechen; und wenn ich recht entsinne, lässt sich eine
solche einfache Kennzeichnung des Bildes von (7) auch auf die allgemeinere
Situation eines "elementar-anabelschen" X ausdehnen. Sofern ich mich nun
entsinne, ist die Kennzeichnung (die natürlich gleichfalls mutmasslich ist
und zwar in beiden Richtungen, "hinreichend" und "genügend") die folgende. Es sei

$$\pi_1(K)^o = \text{Kern von } \pi_1(K) \to \widehat{\mathbb{Z}}^* \text{ (der zyklotomische Charakter)}$$

Gegeben sei eine Spaltung $\pi_1(K) \to \pi_1(X)$, also $\pi_1(K)$ und deshalb auch
$\pi_1(K)^o$ operiert auf $\pi_1(\overline{X})$, der geometrischen Fundamentalgruppe. Die Be-
dingung ist nun, dass die Fixpunktgruppe dieser Aktion nur aus 1 bestehen
soll!

Grothendieck's "Long March through Galois theory"

Leila Schneps

Note. This short text was originally written as a contribution to the "Grothendieck day" which took place in Utrecht on April 12, 1996. It is brief and informal, and was intended to give the audience some very partial idea of what is contained in Grothendieck's long manuscript "La Longue Marche à travers la Théorie de Galois". The close connections between the ideas expressed there and those in parts of the Esquisse and the letter to Faltings in this volume make it relevant to publish it here.

Alexander Grothendieck wrote the *Long March* between January and June 1981. It consists of about 1600 manuscript pages, and nearly as much again in various addenda and developments. About the first 600 pages, consisting of §§1-37, have been read and edited;* the main body of the *Long March* consists of §§1-53. From the table of contents of §§38-53, it seems that they are mainly devoted to the close study of the ideas expressed in the first part, in the cases where $(g, n) = (0, 3)$ and $(1, 1)$. In 1984, Grothendieck wrote and distributed the *Esquisse d'un Programme*, a 57-page text part of which summarizes, sometimes in a more advanced form, the main themes and problems considered in the *Long March*. Both texts are devoted to raising deep questions and examining various approaches to them, and contain few explicit theorems. In reading the *Long March*, it is essential to keep in mind that the text was written entirely by its author for his own use, and not intended for publication; its new TeX format is misleading, and readers are encouraged to consult the facsimile page at the beginning to have an idea of the form of the original manuscript.

The central objects considered in the *Long March* are the moduli spaces of Riemann surfaces of genus g with n marked points, which we denote by $\mathcal{M}_{g,n}$, and the panoply of objects associated to them: curves and their fundamental groups, the fundamental ("Teichmüller") groups and groupoids of the $\mathcal{M}_{g,n}$, morphisms between the various objects, and principally, their relations with the absolute Galois group $\mathrm{Gal}(\overline{\mathbb{Q}}/\mathbb{Q})$, which we denote here by $\Gamma_{\mathbb{Q}}$. The type of problems considered can be gathered into two major themes.

1) *Anabelian geometry.* Let X be an algebraic variety defined over a field

---

* The result of this transcription – the possibility of which was referred to by Grothendieck in the Esquisse as "une compilation de notes pieusement accumulées" – can be obtained by writing to Jean Malgoire at malgoire@math. univ-montp2.fr.

K, and let x be a geometric point of $X_{\overline{K}} = X \otimes \overline{K}$. Then there is a split exact sequence

$$1 \to \hat{\pi}_1(X_{\overline{K}}, x) \to \hat{\pi}_1(X_K, x) \to \text{Gal}(\overline{K}/K) \to 1,$$

where $\hat{\pi}_1(X_{\overline{K}}, x)$ denotes the algebraic fundamental group of X, which is isomorphic to the profinite completion of the topological fundamental group of X, and $\hat{\pi}_1(X_K, x)$ is the *mixed* or *etale* or *arithmetic* fundamental group, isomorphic to the semi-direct product $\hat{\pi}_1(X_{\overline{K}}, x) \rtimes \text{Gal}(\overline{K}/K)$ (see 5.2 of the talk by F. Oort). (Below, we take the liberty of dropping the base point from the notation of the π_1's, but a choice of base point is of course tacitly fixed.) The anabelian question is: *how much information about the isomorphism class of the variety X is contained in the knowledge of the etale fundamental group?* Grothendieck calls varieties which are completely determined by their etale fundamental groups *anabelian varieties*; his "anabelian dream" consists in classifying the anabelian varieties in all dimensions over all fields. Over $\mathbb{Q}$, he expects the moduli spaces $\mathcal{M}_{g,n}$ to be basic anabelian examples. In the particular case of dimension 1 he conjectured that all hyperbolic curves defined over number fields were anabelian varieties, a conjecture which was proved for the punctured projective lines by Nakamura (early 90's), for affine curves by Tamagawa (1995) and finally for all hyperbolic curves defined over number fields by Mochizuki (1996). Mochizuki actually proved this as a consequence of a stronger result, cf. the talk by G. Faltings. Let us remark that there is also a "dimension 0" version of the anabelian conjecture, namely that an infinite finitely generated *field* is determined up to isomorphism by its absolute Galois group. This was proved by Neukirch, Ikeda, Iwasawa and Uchida in the case of global fields, and in full generality by Pop in 1995.

2) *Combinatorial Galois theory.* The other major problem examined in the *Long March* is essentially: can $\mathbb{\Gamma}_{\mathbb{Q}}$ be identified/redefined/characterised by considering only its properties related to its role as automorphism group of the algebraic fundamental groups of the $\mathcal{M}_{g,n}$ respecting the natural "stratified" structure of these groups which comes from the corresponding structure on the spaces themselves (i.e. the divisor at infinity of $\mathcal{M}_{g,n}$ is made up of moduli spaces of smaller dimension).

In the remainder of this text we concentrate on the second theme because the first one will be considered in the talk by G. Faltings. We first give a short guide to the notations in the *Long March*. Armed with this, we explain how Grothendieck first phrased the question and gave an initial conjecture, exciting but probably too strong, then changed point of view, gave a much weaker conjecture, and adduced evidence for it. We give some of the most revealing quotations indicating the direction he believed investigations should take; it is our hope that this short text will make it easier to enter into the ideas of the *Long March*, whose form is far from the usual form of a mathematical text. In the appendix to this note, we give a very brief sketch of some ideas of Drinfel'd which turn out, although he had read neither the *Long March* nor the *Esquisse*, to bear an astonishing resemblance to the approach found there. Instead of citing references along the way, we gather at the end of this note a few surveys and articles (mostly preprints) containing all the salient results mentioned here.

Definitions in a simple form. (1) Let $\pi_{g,n}$ denote the abstract group isomorphic to the fundamental group of a closed topological surface of genus g with n punctures. Fix a system $x_1, \ldots, x_g, y_1, \ldots, y_g, l_1, \ldots, l_n$ of generators (called a *basis*) for $\pi_{g,n}$ satisfying the unique relation

$$[x_1, y_1] \cdots [x_g, y_g] l_1 \cdots l_n = 1. \qquad (*)$$

Let a *discrete loop-group* be a group isomorphic to $\pi_{g,n}$, and a profinite loop-group be a profinite group isomorphic to the profinite completion $\hat{\pi}_{g,n}$. A basis of a profinite loop-group is a set of topological generators satisfying $(*)$.

(2) Let π be a profinite loop-group. Define the group $\mathrm{Aut}_{\mathrm{lac}}(\pi)$ to be the group of (continuous) automorphisms of π which "preserve the inertia groups", which means that if $\phi \in \mathrm{Aut}_{\mathrm{lac}}(\pi)$, then there exists $\lambda \in \mathbb{Z}^*, \sigma \in S_n$ and $\alpha_1, \ldots, \alpha_n \in \pi$ such that

$$\phi(l_i) = \alpha_i l_{\sigma(i)}^{\lambda} \alpha_i^{-1} \qquad (**)$$

for $i = 1, \ldots, n$. Acting by inner automorphisms, the group π corresponds to a subgroup of $\mathrm{Aut}_{\mathrm{lac}}(\pi)$; let $\mathrm{Autext}_{\mathrm{lac}}(\pi)$ denote the quotient. Following Grothendieck, we write $\hat{\mathfrak{S}}(\pi)$ for the group $\mathrm{Aut}_{\mathrm{lac}}(\pi)$ and $\hat{\mathfrak{T}}(\pi)$ for $\mathrm{Autext}_{\mathrm{lac}}(\pi)$.

(3) A *discretification* π_0 of a profinite loop-group π is simply the discrete subgroup generated inside π by the elements of some basis of π (i.e. the image of a basis of $\hat{\pi}_{g,n}$ under some isomorphism $\hat{\pi}_{g,n} \xrightarrow{\sim} \pi$). Let $\mathfrak{S}(\pi_0)$ denote the subgroup of $\hat{\mathfrak{S}}(\pi)$ fixing π_0. The *prediscretification* $\pi_0^{\natural}$ associated to π_0 is the orbit of π_0 under the action of the closure of $\mathfrak{S}(\pi_0)$ inside $\hat{\mathfrak{S}}(\pi)$. This closure is isomorphic to the profinite completion of $\mathfrak{S}(\pi_0)$ and is written $\hat{\mathfrak{S}}(\pi_0^{\natural})$ since it depends only on the prediscretification. Let $\mathfrak{T}(\pi_0)$ and $\hat{\mathfrak{T}}(\pi_0^{\natural})$ denote the corresponding subgroups inside $\hat{\mathfrak{T}}(\pi)$.

(4) There is a natural "cyclotomic character" $\hat{\hat{\mathfrak{T}}}(\pi) \to \hat{\mathbb{Z}}^*$ given by $\phi \mapsto \lambda$. On the subgroups $\mathfrak{T}(\pi_0)$ and $\hat{\mathfrak{T}}(\pi_0^\natural)$, this character takes only the values ± 1. Let $\hat{\mathfrak{T}}^+(\pi_0^\natural)$ be the kernel of this character in $\hat{\mathfrak{T}}(\pi_0^\natural)$.

Ideas and Conjectures. In §26 of the *Long March*, Grothendieck asks the following rather bold questions.

Question 1. Is $\hat{\mathfrak{T}}^+(\pi_0^\natural)$ independent of the prediscretification $\pi_0^\natural$? Equivalently (since conjugation permutes the discretifications) is $\hat{\mathfrak{T}}^+(\pi_0^\natural)$ normal in $\hat{\hat{\mathfrak{T}}}(\pi)$?

Question 2. Choose a discretification π_0 of π and let $\Sigma = \hat{\mathfrak{T}}^+(\pi_0^\natural)$ and $\mathcal{N}_\Sigma = \mathrm{Norm}_{\hat{\hat{\mathfrak{T}}}(\pi)}(\Sigma)$. Is $\mathcal{N}_\Sigma/\Sigma$ isomorphic to $\mathbb{I}_{\mathbb{Q}}$?

Question 3. (motivation for the two previous questions) Given a profinite loop-group of type (g, n), does there exist a discretification π_0 such that we can identify term by term the following exact sequences?

$$1 \to \Sigma \to \mathcal{N}_\Sigma \to \mathcal{N}_\Sigma/\Sigma \to 1$$

$$1 \to \pi_1(\mathcal{M}_{g,n,\overline{\mathbb{Q}}}) \to \pi_1(\mathcal{M}_{g,n,\mathbb{Q}}) \to \mathbb{I}_{\mathbb{Q}} \to 1?$$

If the answer to the first two questions were affirmative, writing $\hat{\mathfrak{T}}^+$ for $\hat{\mathfrak{T}}^+(\pi_0^\natural)$, we would have

$$\hat{\hat{\mathfrak{T}}}(\pi)/\hat{\mathfrak{T}}^+ \simeq \mathbb{I}_{\mathbb{Q}}.$$

This conjecture is very striking. We can rephrase it a little more simply by choosing $\pi_{g,n}$ as a discretification of $\hat{\pi}_{g,n}$ and writing $\hat{\hat{\mathfrak{T}}}_{g,n}$ for $\hat{\hat{\mathfrak{T}}}(\hat{\pi}_{g,n})$, $\hat{\mathfrak{T}}_{g,n}^+$ for $\hat{\mathfrak{T}}^+(\pi_{g,n})$ and $\mathbb{I}_{g,n}$ for $\hat{\hat{\mathfrak{T}}}_{g,n}/\hat{\mathfrak{T}}_{g,n}^+$, where by the hypothetical affirmative answer to the first question, the subgroup $\hat{\mathfrak{T}}_{g,n}^+$ is normal. Then we have

Very strong conjecture. *The $\mathbb{I}_{g,n}$ are isomorphic for all (g, n) such that $2g + n \geq 3$ and they are all isomorphic to $\mathbb{I}_{\mathbb{Q}}$.* *

In §§27 and 28 Grothendieck asks whether it is possible to prove that the $\mathbb{I}_{g,n}$ are all isomorphic. §27 deals with fixed g and decreasing n and §28 deals with finite covers. Although he does succeed in deducing these isomorphisms from certain rather strong conjectures, he begins §29 with the remark "The approach in the previous paragraphs seems, after all, very brutal" and comes to the conclusion that

---

* It follows from Drinfel'd's proof that $GT \neq GT_0$ that this conjecture is wrong for $(0, 3)$ (note that $\hat{\mathfrak{T}}_{0,3}^+ = 1$), but it is not known to be right or wrong for other (g, n).

It is possible that $\hat{\hat{\mathfrak{T}}}(\pi)$ is such a wildly pathological group that it will never be possible to say anything reasonable (and true) about the full group, and we will always be obliged to work with smaller subgroups, not too far from the discrete situation (although with some supplementary "arithmetic" aspects du to $\Gamma_{\mathbb{Q}}$)! Indeed, there is (for $n \neq 0$) in the group $\hat{\mathfrak{T}}(\pi_0)$ a simplicial structure of successive extensions which will be respected by the outer action of the Galois group, and which must be taken into consideration. It is part of the "structure at ∞" of the moduli spaces $\mathcal{M}_{g,n}$, which even for $n = 0$ is doubtless not trivial, and it is possible that this must be taken into account, in order to put a finger on $\Gamma_{\mathbb{Q}}$. (p. 164)

Therefore, at this point, he switches to a different approach: "Without attempting to give a description of $\Gamma_{\mathbb{Q}}$ 'inside $\hat{\hat{\mathfrak{T}}}_{g,n}$', we will proceed more inductively, starting from the presence of $\Gamma_{\mathbb{Q}}$ (for arithmetico-geometric reasons), and trying to unearth properties of this presence, possibly strong enough to end up giving a purely algebraic characterisation." (p. 165)

From here to the end of the available part of the *Long March*, he proceeds to reexamine the whole of the situation as follows: instead of setting $\mathcal{N}_{g,n}$ to be the normalizer of the subgroup $\hat{\mathfrak{T}}^+(\pi_{g,n})$ inside $\hat{\hat{\mathfrak{T}}}_{g,n}$, he considers the exact sequence

$$1 \to \hat{\pi}_{g,n} \to \hat{\pi}_1(\mathcal{U}_{g,n,\mathbb{Q}}) \to \hat{\pi}_1(\mathcal{M}_{g,n,\mathbb{Q}}) \to 1,$$

where $\mathcal{U}_{g,n}$ is the "universal curve" of type (g,n), which is a scheme over $\mathcal{M}_{g,n}$ (actually isomorphic to the moduli space $\mathcal{M}_{g,n+1}$). Since according to this exact sequence, the elements of $\hat{\pi}_1(\mathcal{U}_{g,n,\mathbb{Q}})$ act on the subgroup $\hat{\pi}_{g,n}$ by conjugation, we obtain a homomorphism $\eta : \hat{\pi}_1(\mathcal{U}_{g,n,\mathbb{Q}}) \hookrightarrow \hat{\mathfrak{S}}_{g,n}$; let $M_{g,n}$ denote its image, and let $\mathcal{N}_{g,n}$ be the image of $M_{g,n}$ under the surjection $\hat{\mathfrak{S}}_{g,n} \to \hat{\hat{\mathfrak{T}}}_{g,n}$. The image of the subgroup $\pi_1(\mathcal{U}_{g,n,\overline{\mathbb{Q}}})$ under η is exactly $\hat{\mathfrak{S}}(\pi_{g,n})$, so clearly the quotient $\Gamma_{g,n} := \mathcal{N}_{g,n}/\hat{\mathfrak{T}}^+(\pi_{g,n})$ has a much better chance of looking like $\Gamma_{\mathbb{Q}}$; indeed there is always a surjection $\Gamma_{\mathbb{Q}} \to \Gamma_{g,n}$. Later, in §33bis Grothendieck notes that he thinks he can prove that $\Gamma_{\mathbb{Q}}$ is isomorphic to $\Gamma_{0,3}$ using Belyi's theorem, and in §34 and §35 he does so, essentially describing the beginnings of the theory of dessins d'enfants. The argument is simply that firstly, $\hat{\mathfrak{T}}^+_{0,3} = 1$, so $\Gamma_{0,3} \subset \hat{\hat{\mathfrak{T}}}_{0,3} = \text{Autext}_{\text{lac}}(\hat{\pi}_{0,3})$, and second that it is easily seen to be a consequence of Belyi's theorem that $\Gamma_{\mathbb{Q}}$ acts faithfully on $\pi_{0,3}$. Therefore the surjection $\Gamma_{\mathbb{Q}} \to \Gamma_{0,3}$ cannot have non-trivial kernel. In §28 we find the conjecture:

Conjecture. Is the natural surjection $\Gamma_{\mathbb{Q}} \to \Gamma_{g,n}$ an isomorphism for all (g,n) with $2g + n \geq 3$? In other words, are all the $\mathbb{T}_{g,n}$ isomorphic? Equivalently, is the homomorphism η injective for all such (g,n)?

The sections following §28 are devoted to the behavior of the groups $\mathbb{T}_{g,n}$ under the natural morphisms of erasing a point (§32) and taking finite covers (§33 bis), but always from the point of view of using only a *certain type* of property of these groups. On p. 179, we read

> We propose to try and make precise the type of properties which will characterize $\mathcal{N}_{g,n}$ inside $\mathcal{N}'_{g,n} = \mathrm{Norm}_{\hat{\mathfrak{T}}_{g,n}}(\hat{\mathfrak{T}}^+_{g,n})$, or $\mathbb{T}_{g,n}$ inside $\mathbb{T}'_{g,n} = \mathcal{N}'_{g,n}/\hat{\mathfrak{T}}^+_{g,n}$. We have natural homomorphisms
>
> $$\begin{cases} \mathbb{T}'_{g,n} \to \mathrm{Autext}(\hat{\mathfrak{T}}^+_{g,n}) \\ \mathbb{T}'_{g,n} \to \mathrm{Autext}(\hat{\mathfrak{T}}^{+!}_{g,n} \end{cases}$$
>
> and I presume that $\mathbb{T}_{g,n}$ can be described as the inverse image of a suitable closed subgroup of one or the other of these right-hand sides, i.e. that we can describe it in terms of properties of outer operations on $\hat{\mathfrak{T}}^+_{g,n}$ or $\hat{\mathfrak{T}}_{g,n}$...Of course, we could posit conditions on the outer automorphisms (of $\hat{\mathfrak{T}}^{\prime+!}_{g,n}$, say) which are stable by the successive passages to $\hat{\mathfrak{T}}^{\prime+!}_{g,n-1}$ etc.... but it is not at all clear that this will suffice to describe the $\mathbb{T}_{g,n} \subset \mathbb{T}'_{g,n}$, if only because this condition is empty for the limit case $n = 0$ (if $g \geq 2$; or for the cases $g = 1$, $n = 1$, or $g = 0$, $n = 3$). It is possible that we have to also use properties of the π_1 of the $\mathcal{M}_{g,n,\overline{\mathbb{Q}}}$ linked to compactification. It is only in the case $(g,n) = (0,3)$ that we need not expect any condition of this kind.

Appendix

In his 1991 article [D], Drinfel'd defined the *Grothendieck-Teichmüller group* $\widehat{GT}$ as a group of transformations of quasi-triangular quasi-Hopf algebras, acting as modifications of the *associativity* and *commutativity constraints*. Let us give a revised but equivalent definition of $\widehat{GT}$ here. Let $\hat{F}_2$ denote the profinite free group on two topological generators x and y. Let θ be the automorphism of $\hat{F}_2$ defined by $\theta(x) = y$ and $\theta(y) = x$. Let ω be the automorphism of $\hat{F}_2$ defined by $\omega(x) = y$ and $\omega(y) = (xy)^{-1}$. Let $K(0,5)$ be the pure mapping class group on 5 strands (a quotient of the pure Artin braid group K_5). This group is generated by five elements $x_{i,i+1}$, $i \in \mathbb{Z}/5\mathbb{Z}$; let ρ be the automorphism of the profinite completion $\hat{K}(0,5)$ given by $\rho(x_{i,i+1}) = x_{i+3,i+4}$. There is an injection $\hat{F}_2 \hookrightarrow \hat{K}(0,5)$ given by $x \mapsto x_{12}$ and $y \mapsto x_{23}$; for all $f \in \hat{F}_2$ let $\tilde{f}$ denote its image in $\hat{K}(0,5)$ by this injection.

Definition. Let $\widehat{\underline{GT}}$ be the set

$$\{(\lambda, f) \in \hat{\mathbb{Z}}^* \times \hat{F}_2' \,|\,(I) \quad \theta(f)f = 1;$$
$$(II) \quad \omega^2(fx^m)\omega(fx^m)fx^m = 1 \text{ where } m = (\lambda - 1)/2;$$
$$(III) \quad \rho^4(\tilde{f})\rho^3(\tilde{f})\rho^2(\tilde{f})\rho(\tilde{f})\tilde{f} = 1\}.$$

Under a suitable multiplication law, this set can be made into a monoid; let $\widehat{GT}$ be the group of invertible elements of $\widehat{\underline{GT}}$. In his paper Drinfel'd made the following remarks.

(1) There are injections $\widehat{GT} \hookrightarrow \text{Aut}(\hat{B}_n)$ for all $n \geq 2$, where $\hat{B}_n$ denotes the profinite completion of the Artin braid group on n strands B_n. It is easily seen that this action restricts to the pure Artin braid groups $\hat{K}_n$ and passes to the quotients $K(0,n)$ (the pure mapping class groups, isomorphic to the fundamental groups of the moduli spaces $\mathcal{M}_{0,n}$).

(2) Let $\mathcal{B}_{g,n}$ be the set of base points *near infinity* or *tangential base points* on the moduli space $\mathcal{M}_{g,n}$. These base points are simply connected pieces of the neighborhoods of the points of *maximal degeneration* in the Deligne-Mumford compactification of $\mathcal{M}_{g,n}$. When $g = 0$, it is known that the profinite fundamental (Teichmüller) groupoid $\hat{\pi}_1(\mathcal{M}_{0,n}; \mathcal{B}_{0,n})$ can be equipped with the structure of a subgroupoid of a braided tensor category, i.e. with associativity and commutativity constraints. A similar result on the Teichmüller groupoids in all genera seems to be implied by the results in appendix B of [MS].

(3) Putting (1) and (2) together, Drinfel'd conjectures that $\widehat{GT}$ should be the automorphism group of a suitably defined *Teichmüller tower* consisting of the collection of the $\hat{\pi}_1(\mathcal{M}_{g,n}; \mathcal{B}_{g,n})$ linked by natural homomorphisms coming from degeneration of surfaces of type (g,n) (i.e. considering the part at ∞ of $\mathcal{M}_{g,n}$). We were able to prove this statement on the *genus zero* tower, taking as homomorphisms those given by erasing punctures on the n-punctured Riemann sphere.

(4) Drinfel'd indicated, and Ihara proved that there is a homomorphism $\text{Gal}(\overline{\mathbb{Q}}/\mathbb{Q}) \hookrightarrow \widehat{GT}$; using Belyi's theorem it is seen to be injective.

If (3) is true in all genera, then the group $\widehat{GT}$ possesses all the necessary properties to be an automorphism group of the fundamental groupoids of the moduli spaces respecting the natural morphisms between them, which reflects the "simplicial structure of successive extensions" mentioned in Grothendieck's quotation given above. Even more, by acting on the fundamental groupoids based at the base points in the neighborhood of maximal degeneration, $\widehat{GT}$ also respects the "structure at ∞" of the moduli spaces, mentioned in the same quotation. Therefore, particularly if (3) turns out to be true, a precise formulation of Grothendieck's conjecture is given by

Conjecture. $\widehat{GT} = \mathrm{Gal}(\overline{\mathbb{Q}}/\mathbb{Q})$.

References to preprints

A. Grothendieck, *Esquisse d'un Programme*, this volume

F. Pop, On the anabelian conjecture; results in dimension 0 and 1, this volume.

L. Schneps, On $\widehat{GT}$, a survey, this volume.

P. Lochak, L. Schneps, A cohomological interpretation of the Grothendieck-Teichmüller group, to appear in *Inv. Math.*

References to articles in journals

[D] V.G. Drinfel'd, On quasitriangular quasi-Hopf algebras and a group closely connected with $\mathrm{Gal}(\overline{\mathbb{Q}}/\mathbb{Q})$, *Leningrad Math J.* Vol. 2 (1991), No. 4, 829-860.

[MS] G. Moore, N. Seiberg, Classical and Quantum Conformal Field Theory, *Commun. Math. Phys.*, **123** (1989), 177-254.

The algebraic fundamental group

Frans Oort

An apology. - On 12-IV-1996 a "Grothendieck day" was organized in Utrecht. For non-specialists we tried to explain some of the basic ideas of Grothendieck and their impact on modern mathematics. The text below is the contents of one of the talks. It is written for a general mathematical audience. Simplifications were made in an attempt to make these ideas accessible for a general audience, with the effect that it certainly dilutes the ideas discussed. - The reader should not expect anything more than just an informal exposé, which perhaps would suit better the coffee table than any serious publication.

§1. Introduction.

In 1984 *Alexandre Grothendieck* wrote:
..aujourd'hui je ne suis plus, comme naguère,
le prisonnier volontaire de tâches interminables,
qui si souvent m'avaient interdit de m'élancer dans l'inconnu,
mathématique ou non.
See [2], page 51.

(1.1) [EGA] J. Dieudonné and A. Grothendieck published in 1960 - 1967 eight volumes:
Elements de la géométrie algébrique.
There are 4 chapters, in 8 volumes, together more than 1800 pp. All published as volumes in the series Publ. Math. at the IHES.

In 1960 Grothendieck had planned 12 chapters, at that moment he already had a rather clear picture of what should be contained in the various volumes.

(1.2) [SGA] In 1960 - 1969 Grothendieck, together with many collaborators, had seminars:
Séminaire de géométrie algébrique du Bois Marie.
There are 12 volumes , together more than 6200 pp. Eleven volumes are published as Lecture Notes Math., Springer - Verlag; one volume is published by the North - Holland Publ. Cy.

(1.3) [FGA] In 1957 - 1962 Grothendieck gave 8 talks in the Séminaire Bourbaki:
Fondements de la géométrie algébrique.
There are 8 exposés plus comments, together more than 200 pp. They appeared in one volume: Notes, secr. math. Paris, 1962.

(1.4) How can we understand the citation of 1984 above? Does it really imply that the many pages, the impressive results obtained by Grothendieck, say between 1958 and 1970, belonged to "known territory" for him? Amazing!

(1.5) Considering work by Grothendieck we see his insight, we see the influence of his ideas on present day mathematics. Some of the aspects which we see:

- On several occasions Grothendieck revised existing theory in a superb way.
- Notions previously well known but seemingly unconnected were unified by Grothendieck with the effect that totally new insight became available.
- New, mostly revolutionary, ideas were launched by Grothendieck. We are still reaping the fruits, and trying to see where it all leads to.

Just to give a few examples:

Schemes, methods of sheaves and similar concepts were developed. But Chevalley, Serre and many others stayed in line with classical algebraic geometry over a field; Grothendieck generalized the notion of scheme (locally) to arbitrary rings, and many applications became available.

The fundamental idea of A. Weil to prove the Riemann hypothesis over finite fields via an analogue of the Lefschetz fixed point formula had a decisive influence on the research by Grothendieck. As we see in [1] this was a starting point for Grothendieck to set up algebraic geometry on new foundations.

Serre had the idea of "isotrivial fibre spaces" meeting the desire to describe certain maps in the Zariski topology, see [1], page 104; it took the insight of Grothendieck to see that quite generally "Grothendieck topologies" give the natural framework... Etale cohomology filled in the need for a cohomology theory which could lead to a proof of the Weil conjectures, see [1], page 104.

Fundamental groups and Galois groups existed, Grothendieck realized that arithmetic applications could follow once you see these as two aspects of one theory, to exploit the arithmetic in geometric situations. Below we give one very particular example of this.

Around 1964 Grothendieck had the idea of studying "motives", e.g. see [20].

Does his idea have traceable roots in earlier mathematics? To my feeling this was a completely new insight at that time. Some people however argue that Galois representations were studied already at the time Grothendieck started in algebraic geometry, and that the idea of considering one piece of (co)homology was present at that time; still I feel it is a big step from previous theory to the basic and functorial way Grothendieck approached this concept.

Many more examples can be given. It is an illuminating exercise to study notions introduced and developed by Grothendieck, and to (try to) see where they came from. In some cases one can see the influence of earlier mathematical ideas; in such cases I am impressed by the way Grothendieck gave a new direction to existing concepts. In other cases I do not see where the new ideas had their origin but in the mind of Grothendieck. It seems a beautiful task for us to unravel his ideas together with the historical roots. It is time to start!

(1.6) We try to explain some ideas by Grothendieck (and we will be very far from anything like a complete survey). We have tried to show his work and his ideas between two extremes in time. Some of his ideas around the fundamental group mainly stem from 1960-1964; for Grothendieck this was known territory.... Can we imagine where his ideas, "l'inconnu", in his "longue marche" (1981) might lead us?

(1.7) Remark: In [23] we find a fierce defense by Francesco Severi of "Italian algebraic geometry". It would be nice to have a close historical analysis of the arguments raised by Severi. Certainly one could encounter the following (and much more).

1) By phrasing questions in a certain way, by laying foundations in a certain manner, some aspects are focussed upon, but other do not show up simply because of the way the questions are asked. It would be nice to see what is the influence of the various ways algebraic geometry was approached in this century.

2) As Schappacher and Schoof warn us, see footnote (2) on page 59 of [22], modern schools of algebraic geometry do not always place algebraic geometry performed in the first half of this century in the correct context. Should we talk about "Italian algebraic geometry" (as Severi does)? What do we mean, and how do we place this in the correct historical context?

3) Let us be more concrete about one aspect. In the first page and his foot-note (1) of [23] Severi comments on algebraic geometry over the complex numbers and in the "abstract case". We have seen in the development of algebraic geometry in which way the geometric ideas triggered by questions coming from number theory lead to fundamentally new ideas. It would be nice to describe these historical lines, not in the sense of a discussion about "rigor", but in the spectrum of various questions with different backgrounds about basically the same kind of objects, of the same kind of theory. And we would conclude how much geometry and the abstract case have profited

from the insight which Grothendieck has taught us. Applications in arithmetic geometry (such as Weil conjectures, Ramanujan conjecture, Mordell conjecture, Shafarevich conjecture, Tate conjectures) are unthinkable in the classical style, these really need Grothendieck's foundations of algebraic geometry. It would be nice to have a good historical view on these developments and their background.

§2. Schemes

In [7] we find the word "scheme". Chevalley introduced this name in 1955. Note that in that context only rings contained in a field were considered. It seems natural from a classical point of view. Grothendieck generalized the notion, saw that a theory with less conditions on the objects to be studied gives much more possibilities. Here is the central idea.

Convention. *All rings considered are supposed to be commutative, and are supposed to have an identity element.* NB. the ring $A = \{0\}$ is not excluded. The term "prime ideal" (an ideal such that $x, y \in R$ and $x \cdot y \in I \Rightarrow x \in I$ or $y \in I$) will be used only for proper ideals.

Example. Let V be an affine variety, with coordinate ring A, and let $W \subset V$ be a subvariety. Then $\mathcal{I}(W) = I \subset R$, the set of "polynomial" functions which are zero on W is a prime ideal in A. (Note: in the theory of function algebras this is the usual way to recover points by considering maximal ideals in the ring of functions. By his pre-1958 interests Grothendieck knew that very well.)

Example. Let A and B be commutative rings, $\varphi : A \to B$ a ring homomorphism, and let $J \subset B$ be a prime ideal. Then

$$\varphi^{-1}(J) \subset A$$

is a prime ideal.

From a classical point of view it seems much more natural to consider maximal ideals (corresponding with true geometric points in the classical sense), but note that for a maximal ideal $J \subset B$, the ideal $\varphi^{-1}(J) \subset A$ need not be maximal, e.g. $A = \mathbb{Z} \to \mathbb{Q} = B$, and $J = (0)$. Hence we have to consider all prime ideals as "points", which was counter-intuitive for many people.

(2.1) ± Definition: Let A be a ring (commutative, with $1 \in A$). We define the *affine scheme* $X = \operatorname{Spec}(A)$ as follows: the points of X are the prime ideals of A, the topology on X is the Zariski topology (for $S \subset A$ we define $\mathcal{Z}(S) \subset X$ as the set of ideals containing S, these are the closed sets of

X), and for every point $x \in A$ corresponding with a prime ideal $I \subset A$ we define the local ring $\mathcal{O}_{X,x} := A_I$, the localization of A in I. (See EGA I for more extensive definitions, also see [11], Chapter II; we should talk about sheaves, and so on, to obtain precise definitions).

As is usual in geometry one can define geometric objects (manifolds, abstract varieties, schemes) by atlasses, once charts are given. This can be done: using affine schemes, one defines schemes, which are given locally as affine schemes. Certainly this aspect deserves a longer introduction.

(2.2) Exercise. Place yourself in the situation that you know algebraic geometry à la André Weil very well, and that you know the theory of sheaves (as was studied in the Séminaire Cartan, as in Serre's FAC). Try to find good foundations for algebraic geometry. Would you arrive at FAC, or something like [8], or closer to FGA ?

§3. Galois groups of field extensions, fundamental groups of topological spaces.

(3.1) Galois groups. Let K be a field, and let $K \subset L$ be a finite extension. Let $G = \text{Aut}(L/K)$ be the group of K-automorphisms of L (elements of K are left fixed). One can show that $\#(G) \leq [L : K]$, and in fact:

$$\#(G) = [L : K] \quad \Longleftrightarrow \quad L/K \text{ is normal and separable.}$$

Such a finite extension is called a *finite Galois extension*. For such an extension the Galois correspondence gives a bijective correspondence between intermediate fields $K \subset E \subset L$ and subgroups $G \supset H \supset \{1\}$ by:

$$H \mapsto L^H, \quad E \mapsto \text{Aut}(L/E).$$

An algebraic extension (not necessarily finite) $K \subset L$ is called a Galois extension if it is normal and separable.

(3.2) Fundamental groups. Let X be a topological space, which we suppose to be arcwise connected. Suppose we are given a base point $x \in X$. Loops in X starting and ending at x can be composed; up to homotopy they form a group, denoted by $\pi_1(X, x)$, called the *fundamental group*. In order to avoid confusion below here we shall write $\pi_1^{top}(X, x)$, and say *the topological fundamental group*.

Covering spaces. Suppose moreover that X is locally arcwise connected and locally simply connected. There exists a "universal covering space" $\tilde{X} \to X$ such that $G := \pi_1^{top}(X, x)$ acts on $\tilde{X}/X$, such that for every connected covering $Y \to X$ there exists a subgroup $H \subset G$ such that the covering Y/X fits uniquely in a diagram

$$\tilde{X} \longrightarrow H\backslash\tilde{X} \cong Y \longrightarrow G\backslash\tilde{X} \cong X.$$

This gives a bijection between such coverings and subgroups of $\pi_1^{top}(X, x)$.

Here is a basic example: Let $X = S^1$, the circle, e.g. $X = \{z \in \mathbb{C} \mid |z| = 1\}$. Then

$$\pi_1^{top}(X, x) \cong \mathbb{Z},$$

and the universal covering space can be given by:

$$\tilde{X} \approx \mathbb{R} \quad \longrightarrow \quad X = S^1, \quad \text{by} \quad x \mapsto e^{2\pi\sqrt{-1}\cdot x}$$

Any finite connected covering $Y \to X = S^1$ can be given by a subgroup of finite index of $\pi_1^{top}(X, x)$:

$$Y \cong S^1 \longrightarrow X = S^1 \quad \text{with} \quad s \mapsto s^n = z \in X.$$

(3.3) We summarize: if $K \subset L$ is a finite Galois extension with $G = \text{Gal}(L/K)$, then

$$G \supset H = \text{Aut}_K(E) \supset \{1\} \;\leftrightharpoons\; K \subset L^H = E \subset L,$$

and

$$\tilde{X} \to H\backslash\tilde{X} \cong Y \to G\backslash\tilde{X} \cong X \;\leftrightharpoons\; \{1\} \subset H \subset G = \pi_1^{top}(X, x).$$

(3.4) **Exercise.** Place yourself in the situation that you know these correspondences in Galois theory and in the topological theory of coverings. Try to find a general theory of which these two are special cases.

§4. Profinite completions. Actions defined by an extension.

In this section we gather some easy algebraic tools.

(4.1) Let G be a group. Consider all surjective homomorphisms $G \to I$, where I is a *finite group*, i.e. consider all normal subgroups $H \subset G$ of finite index (and write $G \to H\backslash G \cong I$). The system of such finite quotients forms a *projective system* (sometimes called an *inverse system*). The limit is written as

$$\varprojlim I = \hat{G}.$$

Note that for normal subgroups of finite index $H_I \subset H_J \subset G$ we have a natural homomorphism $\varphi_I^J : J \to I$, and we write:

$$\hat{G} := \{\{x_{H_I}\} \in \prod (H_I \backslash G) \mid I = H_I \backslash G, \text{ and}$$

$$\varphi_I^J(x_J) = x_J \bmod H_I = x_I \ \forall H_I \subset H_J.\}$$

It is called the *profinite* completion of G. There is a canonical homomorphism

$$G \longrightarrow \hat{G}.$$

On $\hat{G}$ we have the profinite topology: every finite quotient $H \backslash G \cong I$ is given the discrete topology (every subset is open), and the projective limit is given the limit topology. This can be characterized by: it is the coarsest topology which makes all projections $\varphi_I : G \to I$ continuous. Or: for any I, for any $a \in I$ take $(\varphi_I)^{-1}(a)$, and use these as a basis for all open sets in the topology on $\hat{G}$. Note that the image of $G \to \hat{G}$ is dense in $\hat{G}$.

(4.2) Examples. 1) Consider $G = \mathbb{Z}$. The finite quotients are $\mathbb{Z} \to \mathbb{Z}/n$ for $n \in \mathbb{Z}_{\geq 0}$. The projective limit is isomorphic to:

$$\hat{\mathbb{Z}} \cong \prod_\ell \mathbb{Z}_\ell,$$

product taken over all prime numbers ℓ; here we write $\mathbb{Z}_\ell$ for the (additive group of) ℓ-adic integers. Note that the natural map $\mathbb{Z} \to \hat{\mathbb{Z}}$ is injective.
2) It can happen that the map $G \to \hat{G}$ is not injective, e.g. $\hat{\mathbb{Q}} = \{0\}$. Below we shall see more interesting examples.
3) Let p be a prime number, let $\mathbb{F}_p$ be the field having p elements, let $k = \overline{\mathbb{F}_p}$ be its algebraic closure (which is an algebraic, separable, normal extension). Every finite extension of $\mathbb{F}_p$ of degree n has a cyclic group $\mathbb{Z}/n$ of automorphisms, and in fact

$$\text{Gal}(k/\mathbb{F}_p) \cong \hat{\mathbb{Z}}.$$

(4.3) Actions defined by an extension. Suppose we have a group π, and a normal subgroup $N \subset \pi$, i.e. an exact sequence

$$1 \to N \longrightarrow \pi \longrightarrow \pi/N =: G \to 1.$$

This extension defines a representation

$$\rho : G \longrightarrow \text{Out}(N);$$

here $\text{Out}(N) := \text{Aut}(N)/\text{Inn}(N)$ is the group of "outer automorphisms", i.e. the factor group of automorphisms of N by the normal subgroup consisting of inner automorphisms. The representation is obtained as follows: for $\sigma \in G$, choose a lifting $\sigma' \in \pi$; it acts by inner conjugation in π on the normal subgroup $N \subset \pi$. In this way we obtain an automorphism

$$\varphi_{\sigma'} \,|_N \colon N \longrightarrow N.$$

Taking another lifting of σ the new action differs from the old one by an inner automorphism of N:

$$\rho(\sigma) := (\varphi_{\sigma'} \,|_N) \bmod \mathrm{Inn}(N) \in \mathrm{Aut}(N)/\mathrm{Inn}(N)$$

is well-defined; this defines a homomorphism ρ.

Such actions play an important role if one considers "arithmetic fundamental groups", i.e. a combination of a geometric fundamental group N and a Galois group G. This is an algebraic way to see how in a family the fundamental group of a base acts on the geometric fundamental group of a fibre.

Remark. A section of $\pi \to \pi/N =: G$ defines a homomorphism $G \to \mathrm{Aut}(N)$.

Suppose H is commutative, and

$$1 \to H \longrightarrow \pi \longrightarrow \pi/H =: G \to 1$$

exact. The we obtain a representation

$$G \longrightarrow \mathrm{Aut}(H).$$

This plays an important role in the theory of algebraic monodromy. For example we can obtain such a situation by choosing $H = N_{Ab}$, the abelianized N, starting with an extension $\pi'/N = G$, and pushing this out, achieving the above exact sequence $\pi/H = G$.

§5. The algebraic fundamental group

Seeing the definition of a Galois group, with the property $K = L^G$ (as above) on the one hand, and the definition of the fundamental group, with the property $X = G\backslash(\tilde{X})$ on the other hand, we can follow Grothendieck in his idea that these are two aspects of one more general concept.

In order to define this concept one needs the notion of "unramified" both in an algebraic setting and in a topological context. An inseparable extension looks like a ramified extension (in the sense that "roots coming together" is the geometric notion of being ramified). This can be combined in one notion.

To this end Grothendieck defined the notion of an *étale morphism* which combines these two aspects;

> *La mer était étale, mais le reflux commençait à se faire sentir.*
>
> V. Hugo, *Les travailleurs de la mer.*

(see [14], page v.)

An étale morphism is a smooth morphism of relative dimension zero. This implies it is flat and unramified (see [11], II.10, and Ex. III.10.3). If a morphism $f : Y \to X$, of finite type, is étale then it is flat, and for every $y \in Y$, with $f(y) =: x \in X$, the extension

$$\mathcal{O}_{X,x}/m_{X,x} \subset \mathcal{O}_{Y,y}/m_{X,x}\mathcal{O}_{Y,y} \quad \text{is a finite separable extension of fields.}$$

Here are some very easy examples which could explain this concept:

1) The morphism $\mathrm{Spec}(\mathbb{C}[S]) \to \mathrm{Spec}(\mathbb{C}[T])$ via $T \mapsto S^2$ is ramified, because

$$\mathbb{C}[S] \otimes_{\mathbb{C}[T]} \mathbb{C}[T]/(T) \cong \mathbb{C}[S]/(S^2),$$

which is not a field. Geometrically: the map $z \mapsto z^2$ on the complex plane is two-to-one outside $z = 0$, but one-to-one at $z = 0$, i.e. ramified at that point.

2) The morphism $\mathrm{Spec}(\mathbb{Z}[\sqrt{-1}]) \to \mathrm{Spec}(\mathbb{Z}) = X$ is ramified at $(2) = x \in X$, and unramified elsewhere. For $p = 5$ we have a factorization $5 = (2+i)\cdot(2-i)$, the prime $p = 3$ is also prime in $\mathbb{Z}[i]$, and above $\mathbb{Z}/(3)$ there are two geometric points in $\mathrm{Spec}(\mathbb{Z}[\sqrt{-1}])$. We see that $\mathbb{Z}[\sqrt{-1}]/(p)$ is a field or a product of fields for prime numbers $p > 2$, and for $p = 2$ this ring has nilpotents:

$$\mathbb{Z}[\sqrt{-1}] \otimes (\mathbb{Z}/2) \cong \mathbb{F}_2[\varepsilon]/(\varepsilon^2).$$

Note that $\mathrm{Spec}(\mathbb{Z}[\frac{1}{2}, \sqrt{-1}]) \to \mathrm{Spec}(\mathbb{Z}[\frac{1}{2}])$ is an étale morphism.

3) The extension $\mathbb{F}_p(t) = K \subset L = \mathbb{F}_p(\sqrt[p]{t})$ is inseparable, and $\mathrm{Spec}(L) \to \mathrm{Spec}(K)$ is not étale.

(5.1) Construction/Properties: Suppose given a connected scheme X, an algebraically closed field Ω, and a (geometric) point $x \in X(\Omega)$. Consider all étale coverings $f : Y \to X$. For a given X we want to classify such coverings. There exists a profinite group $\pi = \pi_1(X, x)$, called the (algebraic) fundamental group of X with base point x, such that there is an equivalence:

$$\text{finite, continuous } \pi - \text{sets} \quad \leftrightarrows \quad \text{étale coverings of } X;$$

in this correspondence to a covering f we associate the set $S := f^{-1}(x)$. We refer to SGA I, or to [15] for details.

Here are two special cases:
If K is a field, with K^s its separable closure, there is an isomorphism:

$$\mathrm{Gal}(K^s/K) \cong \pi_1(X)$$

(one has to specify a base point, etc.).
Let X be an (irreducible) algebraic variety defined over $\mathbb{C}$, then the profinite completion of the topological fundamental group is canonically the algebraic fundamental group:

$$\pi_1^{top}(X(\mathbb{C})) \to \hat{\pi}_1^{top}(X(\mathbb{C})) \cong \pi_1(X).$$

Conclusion. We see that the theory of the algebraic fundamental group combines Galois theory and the theory of (profinitely completed) fundamental groups of algebraic varieties.

(5.2) Properties of the arithmetic fundamental group. Let K be a field (e.g. think of $K = \mathbb{Q}$), and let X be an algebraic variety defined over K. Then we have morphisms

$$X \otimes \overline{K} =: \overline{X} \to X \to \mathrm{Spec}(K).$$

Suppose we have geometric points

$$\overline{a} \in \overline{X}(\Omega), \quad a \in X(\Omega), \quad b : K \to \Omega$$

mapping to each other. Grothendieck proves that there exists an exact sequence:

$$1 \to \pi_1(\overline{X}, \overline{a}) \to \pi_1(X, a) \to \pi_1(K, b) \cong \mathrm{Gal}(K^s/K) \to 1.$$

See SGA I, pp. 251, 268. (Intuitively this is clear: a function field has "two kind of extensions": one can extend the base field, and one can extend the function field by taking coverings! Combination of these two yields all coverings, that is what the theorem says.)

(5.3) Anabelian algebraic geometry. The Galois group of the ground field K operates (via inner conjugation in $\pi_1(X)$) as outer automorphisms on the (profinite completion of the) geometric fundamental group $\pi_1(\overline{X})$. And Grothendieck writes:

"C'est ainsi que mon attention s'est portée vers ce que j'ai appelé depuis la **"géométrie algébrique anabélienne"** *...des groupes fondamentaux qui*

sont très éloignés des groupes abéliens (et que pour cette raison je nomme
"anabéliens")..."
see [2], page 14. Strictly speaking there is no precise definition for the notion
"anabelian", but it indicates a type of questions.

(5.4) Example. Let $X = \mathbb{P}^1_{\mathbb{Q}} - \{0, \infty\}$. What is $= \pi_1(X)$? Note that
$\mathbb{P}^1_{\mathbb{Q}} - \{0, \infty\} = \mathrm{Spec}(\mathbb{Q}[X, \frac{1}{X}])$. Further note that $X(\mathbb{C}) = \mathbb{C}^*$; from this we
easily see that

$$\pi_1^{top}(X(\mathbb{C})) \cong \mathbb{Z}, \quad \text{and} \quad \pi_1(X \otimes \overline{\mathbb{Q}}) \cong \hat{\mathbb{Z}}.$$

We need to know how the Galois group of $\mathbb{Q}$ acts on this. In order to study
this we first consider

An easy example (Kummer theory). Consider the field $M := \mathbb{Q}(t)$, a
purely transcendental extension of $\mathbb{Q}$, let $n \in \mathbb{Z}_{>0}$, and let N be the splitting
field of $X^n - t \in M[X]$. What is $\mathrm{Gal}(N/M)$? Note that we have a tower

$$M := \mathbb{Q}(t) \subset \mathbb{Q}(\zeta_n)(t) \subset N,$$

and

$$\mathrm{Gal}(\mathbb{Q}(\zeta_n)(t)/M) = \mathrm{Gal}(\mathbb{Q}(\zeta_n)/\mathbb{Q}) \cong (\mathbb{Z}/n)^*, \text{ and } \mathrm{Gal}(N/\mathbb{Q}(\zeta_n)(t)) \cong \mathbb{Z}/n.$$

In the exact sequence

$$1 \to \mathrm{Gal}(N/\mathbb{Q}(\zeta_n)(t)) \longrightarrow \mathrm{Gal}(N/M) \longrightarrow \mathrm{Gal}((\mathbb{Q}(\zeta_n)(t)/M) \to 1$$

the cokernel $\mathrm{Gal}(\mathbb{Q}(\zeta_n)(t)/M) \cong (\mathbb{Z}/n)^*$ acts by inner conjugation inside
$\mathrm{Gal}(N/M)$ on the abelian kernel $\mathrm{Gal}(N/\mathbb{Q}(\zeta_n)(t)) \cong \mathbb{Z}/n$. What is this
action? An easy exercise shows it is the natural way $(\mathbb{Z}/n)^*$ acts on $\mathbb{Z}/n$
(check !).

Conclusion. The fundamental group $\pi_1(\mathbb{P}^1_{\mathbb{Q}} - \{0, \infty\})$ fits into an exact
sequence:

$$0 \to \hat{\mathbb{Z}} \longrightarrow \pi_1(\mathbb{P}^1_{\mathbb{Q}} - \{0, \infty\}) \longrightarrow \mathrm{Gal}(\overline{\mathbb{Q}}, \mathbb{Q}) \to 0,$$

and the representation

$$\rho : \mathrm{Gal}(\overline{\mathbb{Q}}/\mathbb{Q}) \longrightarrow \mathrm{Aut}(\hat{\mathbb{Z}})$$

defined by this extension is the **cyclotomic character**.

(5.5) Example. Let $X = \mathbb{P}^1_{\mathbb{Q}} - \{0, 1, \infty\}$. Then the topological fundamental
group of $X(\mathbb{C})$ is a free group on two generators (on three generators with

one relation). The Galois group $\mathrm{Gal}(\overline{\mathbb{Q}}/\mathbb{Q})$ acts on the profinite completion of this group via the exact sequence above. This is a central theme of research, see [10], see [12]. On many occasions Grothendieck emphasized that this is an important object to study.

(5.6) Remark. There was a striking analogy between Galois extensions of a number field and (ramified) covering of an algebraic curve (Galois extension of a function field in one variable), e.g. see [26]. The theory of the algebraic fundamental group by Grothendieck does more than just hinting at an analogy: it unifies both theories as aspects of one theory. The exact sequence above shows how the arithmetic and geometric aspects combine into an action of Galois groups on (the profinite completion of the geometric) fundamental group.

Once you feel that Galois theory and the theory of the fundamental group should be combined, this will come out. It took the insight of Grothendieck to do so, and to predict what kind of applications will follow.

(5.7) Geometric versus algebraic fundamental group: *In general the algebraic fundamental group does not determine the geometric fundamental group.*
1) Serre constructed an example of an algebraic variety X over a number field K plus two embeddings of K into $\mathbb{C}$ such that the geometric fundamental groups of $X_1(\mathbb{C})$ and $X_2(\mathbb{C})$ are not isomorphic, while the algebraic fundamental groups (i.e. their profinite completions) clearly are isomorphic, see [Serre].
2) In general the canonical homomorphism $G \to \hat{G}$ is not injective; in fact this can also happen for a geometric fundamental group, see an example by Catanese and Tovena, [9].

§6. The punctured disc in an algebraic setting.

(6.1) Inertia. Consider $K = \mathbb{C}((t))$ the field of Laurent power series over the field $\mathbb{C}$ of complex numbers (or more generally any local field with an algebraically closed residue field, say, with a residue class field of characteristic zero). Clearly for every $n \in \mathbb{Z}_{>0}$ we have a finite extension

$$K = \mathbb{C}((t)) \subset \mathbb{C}(({}^n\!\sqrt{t})) = L_n$$

which is finite cyclic with $\mathrm{Gal}(L_n/K) \cong \mathbb{Z}/n$. This is "Kummer theory" which applies because the root of unity $\zeta_n \in \mathbb{C} \subset K$ is in the base field.
Moreover:

$$\bigcup_n L_n \text{ is the algebraic closure } \overline{\mathbb{C}((t))}.$$

This is a classical topic; for a proof e.g. see [25], IV.3. We conclude:

$$\pi_1(\mathrm{Spec}(\mathbb{C}((t)))) \cong \hat{\mathbb{Z}}.$$

By the way, we should indicate a base point before we can use the notation "π_1"; here we choose an algebraically closed field Ω, fix $x : K = \mathbb{C}((t)) \to \Omega$, which we write as $x : \mathrm{Spec}(\Omega) \to \mathrm{Spec}(K)$, and in fact we were considering $\pi_1(\mathrm{Spec}(K), x)$.

(6.2) Let us denote the disc and the punctured disc by:

$$D := \{z \in \mathbb{C} \mid \ |z| \leq 1\}, \quad D - \{0\} = D^*.$$

An algebraic extension

$$\mathbb{C}((\sqrt[n]{t})) \supset K = \mathbb{C}((t))$$

gives rise to a map

$$\mathbb{C}^* \xrightarrow{n\ exp} \mathbb{C}^*, \quad s \mapsto s^n,$$

by restriction this gives

$$D^* \xrightarrow{n\ exp} D^*,$$

hence

$$S^1 \xrightarrow{n\ exp} S^1,$$

here $S^1 = \{z \in \mathbb{C} \mid |z| = 1\}$.

Conclusion.

$$\pi_1^{top}(S^1) \cong \mathbb{Z} \ \longrightarrow \ \hat{\mathbb{Z}} \cong \pi_1(\mathrm{Spec}(\mathbb{C}((t)))) = \mathrm{Gal}(\overline{K}/K).$$

This is a particular case of the comparison between topological and algebraic fundamental groups.

Remarks. Instead of D^* we can also consider $\mathbb{C}^*$.
We can consider $\mathrm{Gal}(\overline{\mathbb{Q}((t))}/\mathbb{Q}((t)))$. After what has been said it will be clear what the structure of this group is.

§7. The monodromy theorem.

(7.1) In various settings monodromy has been studied. One of the central results is:
        "the eigenvalues of monodromy are roots of unity,"
i.e. the action is quasi-unipotent. Here we mean that the action of the fundamental group of a parameter space (say around a singular fibre) acts

via matrices on a commutative algebraic object such as a (co)homology group, and the eigenvalues of these are roots of unity. Usually this is called the "algebraic monodromy" when the action is on homology classes of cycles, in contrast with "geometric monodromy" when the action on e.g. loops is considered.

The monodromy theorem in the analytic context is about "Picard-Lefschetz" transformations. It was proved by Landman in his Berkeley PhD-thesis, see [13], page 90, (1.6), Theorem 1, and by Steenbrink in his Amsterdam PhD-thesis, see [24], page 50. Also see [17], page 245, (6.1). Monodromy locally on the fibre around one singularity was studied by Brieskorn, [6], page 113, Satz 4.

We sketch the proof by Grothendieck of the monodromy theorem; details can be found in [21], Appendix, pp. 514-516, also see [16], Section 1, especially Coroll. 1.3.

(7.2) An apology. Quite often when giving a talk, Grothendieck would start by writing X, vertical arrow, S, saying "let X be a scheme over S", and in the rest of the talk results would be given in the greatest possible generality. For the audience not always easy to follow. But then, much later only, you discover the enormous quality of avoiding unnecessary assumptions.

In this talk I try to highlight at least one idea by Grothendieck, but, I will not do this in *his* style. I hope you will see what the basic idea is. Certainly many mathematicians in the audience will see the general form in which this should be put.

(7.3) The monodromy theorem (in simplified form). *Let $K = \mathbb{C}(t)$, and let C_K be an algebraic curve over K, absolutely irreducible, complete, and smooth over K. Let $L := \mathbb{C}((t))$, and $C := C_K \otimes_K L$, and let*

$$\rho_I : I \longrightarrow \text{Aut}(H)$$

be the monodromy representation. Here $I = \text{Gal}(\overline{L}/L)$, and H is a cohomology group attached to C, say $H = H^1_{et}(C, \mathbb{Z}_\ell)$, where we have chosen a prime number ℓ. Let $\gamma \in I(\ell) \cong \mathbb{Z}_\ell$ be a topological generator, and let

$$\rho(\gamma) \in \text{Aut}(H) \cong \text{GL}(2g, \mathbb{Z}_\ell),$$

where g is the genus of C. Then the eigenvalues of the matrix $\rho(\gamma) =: A$ are roots of unity.

Instead of fixing ℓ and considering $H^1_{et}(C, \mathbb{Z}_\ell)$, we could have started with (ordinary) (co)homology.

In the proof there are two ingredients. Grothendieck remarks that all data can be defined over a much smaller field, so that an arithmetic Galois group acts on the geometric situation. Once you see this the Kummer theory discussed in §5 gives extra structure on the monodromy representation, and we shall see that is all we need for this case.

(7.4) Sketch of the proof. Note that C_K is defined over $K = \mathbb{C}(t)$ by a finite number of polynomial equations. There is an extension of finite type $\mathbb{Q} \subset P$ and a curve $D = D_{P((t))}$ such that $D_{P((t))} \otimes \mathbb{C}((t)) \cong C$. We write $G = \mathrm{Gal}(\overline{P((t))}/P((t)))$. We obtain a representation

$$\rho : G \longrightarrow \mathrm{Aut}(H).$$

Note that the restriction of this to $I \cong \mathrm{Gal}((\overline{P((t))}/\overline{P}((t))))$ is ρ_I. We indicate how to obtain this representation. The fundamental group $\pi_1(D)$ gives an exact sequence $\pi_1(D)/\pi_1(D \otimes \overline{P((t))}) = G$. This defines a representation of G on $(\pi_1(D \otimes \overline{P((t))}))_{Ab}$; note that its ℓ–adic part is:

$$(\pi_1(D \otimes \overline{P((t))}))_{Ab}(\ell) \cong H.$$

Thus we derive the ρ indicated above. Further note that the fact that $P/\mathbb{Q}$ is of finite type implies that for any given prime ℓ the field P does not contain all roots of unity of a power of ℓ. Hence $\bigcup_n P(\zeta_{\ell^n}) \supset P$ is an infinite extension. Hence the image of the cyclotomic character in $\mathrm{Aut}(H)$ is infinite. Using the ideas exposed in (5.4), using that $\mathbb{Z} \to \hat{\mathbb{Z}}$ has a dense image we see: there exists an element

$$\sigma \in \mathrm{Gal}(\overline{P((t))}/P((t)))$$

an integer $q \in \mathbb{Z}_{>1}$ and a topological generator $\gamma \in I(\ell)$ such that

$$\sigma' \cdot \gamma \cdot (\sigma')^{-1} \;=\; \gamma^q$$

in $\mathrm{Gal}(\overline{P((t))}/P((t)))$ for any lift σ' of σ.

Applying ρ, and using $\rho(\gamma) = A$ we obtain:

$$S \cdot A \cdot S^{-1} = A^q, \quad \text{with} \quad S := \rho(\sigma').$$

This proves that A and A^q have the same eigenvalues, and raising to the q-th power is a permutation of these eigenvalues. Hence for any such eigenvalue λ it follows that

$$\lambda^{q^{(2g)!}} \;=\; \lambda.$$

$\diamond$

References

[1] A. Grothendieck, The cohomology theory of abstract algebraic varieties. Proceed. ICM Edinburgh 1968; Cambridge Univ. Press, 1960; pp. 103-118.

[2] A. Grothendieck, Esquisse d'un programme. Ms. 57 pp., 1984.

[3] A. Grothendieck, Récoltes et semailles: réflexions et témoignages sur un passé de mathématicien. Manuscript, Montpellier, 1985.

[4] "La longue marche à travers la théorie de Galois." Tome 1 (§§ 1-37). Transcr. manuscr. by A. Grothendieck by J. Malgoire (with P. Lochak & L. Schneps), Univ. de Montpellier II, 1995; 253 pp.

[5] "Grothendieck", article in: Science & Vie **935**, August 1995, pp. 52-57.

[6] E. Brieskorn, Die Mononodromie der isolierten Singularitäten von Hyperflächen. Manuscr. Math. **2** (1970), 103-161.

[7] C. Chevalley, Les schémas I, II. Sém. Cartan - Chevalley, **8** (1955/56), Exp. 5, 6. Notes, Secr. math. Paris 1956.

[8] C. Chevalley, Fondements de la géométrie algébrique. Notes, Secr. Math., Paris, 1958.

[9] Classifications of irregular surfaces, Proceed. Trento 1990 (Ed. E. Ballico, F. Catanese & C. Ciliberto), Lect. Notes Math. 1515, Springer - Verlag 1992, "Trento Example 2": Fundamental groups of algebraic varieties; pp. 136-139.

[10] P. Deligne, Le groupe fondamental de la droite projective moins trois points. In: Galois groups over ℚ (Ed. Y. Ihara, K. Ribet, J-P. Serre). Math. Sc. Res. Inst. Publ. 16. Springer - Verlag, 1989; pp. 79 - 297.

[11] R. Hartshorne, Algebraic geometry. Grad. Texts Math. 52. Springer - Verlag, first printing: 1977.

[12] Y. Ihara, Braids, Galois groups, and some arithmetic functions. Proceed. ICM Kyoto 1990. Math. Soc. Japan, Springer - Verlag, 1991. Vol. I, pp. 99-120.

[13] A. Landman, On the Picard - Lefschetz transformation for algebraic manifolds acquiring general singularities. Transact. Amer. Math. Soc. **181** (1973), 89 - 126.

[14] J. S. Milne, Etale cohomology. Princeton Math. Series 33, Princeton Univ. Press, 1980.

[15] J. P. Murre, Lectures on an introduction to Grothendieck's theory of the fundamental group. Notes, Tata Inst. Fund. Res., Bombay, 1967.

[16] F. Oort, Good and stable reduction of abelian varieties. Manuscr. Math. **11** (1974), 171 - 197.

[17] W. Schmid, Variation of Hodge structures: the singularities of the period map. Invent. Math. **22** (1973), 211-320.

[18] J-P. Serre, Faisceaux algébriques cohérents. Ann. Math. **61** (1955), 197-278, cited as [FAC], [ŒI, 29].

[19] J-P. Serre, Exemples de variétés projectives conjuguées non homéomorphes. C. R. Acad. Sc. Paris **258** (1964), 4194 - 4196 [ŒII, 63].

[20] J-P. Serre, Motifs. In: Journ. arithm. de Luminy, 1989. Astérisque 198-199-200, Soc. Math. France 1991; pp. 333-341.

[21] J-P. Serre & J. Tate, Good reduction of abelian varieties. Ann. Math. **88** (1968), 492-517 [Serre ŒII, 79].

[22] N. Schappacher & R. Schoof, Beppo Levi and the arithmetic of elliptic curves. Math. Intelligencer **18** (1996), 57-68.

[23] F. Severi, La géométrie algébrique Italienne. Sa rigueur, ses méthodes, ses problèmes. Centre Belge Rech. math., Colloq. géom. algébrique. Liège 19-21/XII/1949, pp. 9-55 (1950).

[24] J. H. M. Steenbrink, Limits of Hodge structures. PhD-thesis, Amsterdam, 1974.

[25] R. J. Walker, Algebraic curves. Dover Publ., 1950.

[26] A. Weil, Sur l'analogie entre les corps de nombres algébriques et les corps de fonctions algébriques. Revue Scient. **77** (1939), 104-106 [Collected Papers I, 1939a].

Frans Oort
Mathematisch Instituut
Budapestlaan 6
NL - 3508 TA Utrecht
The Netherlands
email: oort@math.ruu.nl

Etale homotopy type of the moduli spaces of algebraic curves

Takayuki Oda

§1. Main result

We want to show some basic facts on the homotopy type of the moduli stacks of algebraic curves. The notion of stack is now becoming very popular and for some purposes it is indispensable. However, unfortunately, sometimes the claims on stacks are used without proofs. Such manner of "string-math" causes disharmony in the sound universe of mathematics.

We shall write a proof on the problem of the title. As a bonus of writing a proof, we find some new problems on Teichmüller groups. A motivation of this paper is found in a plan of Grothendieck [5]. Our notion of *algebraic stack* is that of Deligne-Mumford [3,§4] throughout this paper.

Let $\mathcal{M}_{g,n}$ be the moduli stack of proper smooth curves of genus g with n distinct (ordered) points. For each scheme S, the category of sections of $\mathcal{M}_{g,n}$ over S is the category of objects:

$\{p : C \to S$, a proper smooth morphism whose fibers are geometrically connected curves of genus g, plus n sections $t_1, \cdots, t_n : S \to C$ of p such that $t_i(s) \neq t_j(s)$ for any $s \in S$, if $i \neq j$ $(1 \leq i, j \leq n)\}$,

the morphisms being the isomorphisms of the above data over S.

It is known that $\mathcal{M}_{g,n}$ is an algebraic stack in the sense of Deligne-Mumford [3]. In fact, if $n = 0$, our $\mathcal{M}_{g,0}$ is M_g^0 of [3], and if $n = 1$, $\mathcal{M}_{g,1}$ is the "universal family" of curves of genus g over $\mathcal{M}_{g,0}$. There is a natural morphism of stacks $\mathcal{M}_{g,n+1} \to \mathcal{M}_{g,n}$ obtained by forgetting the $(n+1)$–th point. It is relatively representable and the fibers are open curves which are isomophic to curves of genus g minus n points.

An algebraic stack $\mathcal{S}$ is a contravariant functor from the category of schemes with etale topology to the category of groupoids. Here a groupoid is a category whose morphisms are all isomorphisms. A stack $\mathcal{S}$ has the descent property with respect to etale morphisms. But the strict descent property used to define etale sheaves is relaxed so that it admits ambiguity up to isomorphisms, incorporated in the very definition of the functor $\mathcal{S}$.

For an algebraic stack $\mathcal{S}$, we can define its etale site $\mathcal{S}_{et}$, consisting of etale morphisms $x : X \to \mathcal{S}$ of stacks where X is a scheme. The existence of at least one *surjective* etale morphism $x : X \to \mathcal{S}$ is a part of the definition of algebraic stack. In $\mathcal{S}_{et}$, the fiber product $X \times_S Y$ exists for any $X, Y \in \mathcal{S}_{et}$ (the existence of such fiber product is another part of the definition of algebraic stacks). Therefore, if we consider simplicial objects with values in $\mathcal{S}_{et}$, the coskeleton functor exists (cf. Artin-Mazur [1, Chap.1]). Thus we can define hypercoverings of $\mathcal{S}$, and their category $HS(\mathcal{S})$. The theorem of Verdier ([10, Th.7.3.2]) which claims that the opposite category $HR(\mathcal{S})^0$ is filtering, is also true for the case of a stack $\mathcal{S}$.

Moreover the same proof shows the following. Replace the schemes X of etale morphisms $x : X \to \mathcal{S}$ by algebraic stacks X in the definition of the site $\mathcal{S}_{et}$. Then we obtain an enlarged site $\mathcal{S}'$ containing $\mathcal{S}$. But the category of hypercoverings in this enlarged site is cofinal with the original one $HR(\mathcal{S})$, because any algebraic stack has a finite etale cover by some scheme.

Note that we can define the etale homotopy type of $\mathcal{S}$ similarly as the case of schemes via the connected component functor

$$\pi_0 : HR(\mathcal{S})^0 \to \text{(simplicial sets)}.$$

We denote this by $(\mathcal{S})_{et}$ which is an object of the category of pro-simplicial sets (cf. [1, Chap.9]).

Let $\Gamma_{g,n}$ be the Teichmüller group of genus g with n punctures. Or, equivalently, as a mapping class group, it is defined as follows (cf. Birman [9, Chap.4]).

Let R be a compact orientable C^∞ surface of genus g and $p_1, \cdots, p_n$ n distinct points on R. We denote by $\text{Diff}^+(R, \{p_1, \cdots, p_n\})$ the group of all orientation preserving diffeomorphisms

$$\varphi : R - \{p_1, \cdots, p_n\} \cong R - \{p_1, \cdots, p_n\}$$

such that $\varphi(p_i) = p_i$ for each point p_i ($1 \le i \le n$). The group $\text{Diff}^+(R, \{p_1, \cdots, p_n\})$ is a topological group with compact open topology. Then we put

$$\Gamma_{g,n} = \pi_0 \text{Diff}^+(R, \{p_1, \cdots, p_n\})$$
$$\cong \text{Diff}^+(R, \{p_1, \cdots, p_n\}) / \text{Diff}_0^+(R, \{p_1, \cdots, p_n\}).$$

Here $\text{Diff}_0^+(R, \{p_1, \cdots, p_n\})$ is the normal subgroup of diffeomorphisms homotopic to the identity. Let $K(\Gamma_{g,n}, 1)$ be the Eilenberg-MacLane space with the fundamental group $\Gamma_{g,n}$, which is weakly equivalent to the classifying space $B_{\Gamma_{g,n}}$ of $\Gamma_{g,n}$.

Let $K(\Gamma_{g,n}, 1)^\wedge$ be the profinite completion of $K(\Gamma_{g,n}, 1)$ in the sense of Artin-Mazur. Then our main result is the following.

Theorem 1. *There is a weak equivalence of pro-simplicial sets:*

$$(\mathcal{M}_{g,n} \otimes_Z \bar{\mathbb{Q}})_{et} \approx K(\Gamma_{g,n}, 1)^{\wedge}.$$

Here $\mathcal{M}_{g,n} \otimes_Z \bar{\mathbb{Q}}$, the extension of scalars of $\mathcal{M}_{g,n}$ to $\bar{\mathbb{Q}}$, is the moduli stack, which represents the restriction of the moduli functor to the sucategory of schemes over $\bar{\mathbb{Q}}$.

§2. Etale homotopy type via simplicial schemes

In the first place, let us recall the result of Cox [2] on the homotopical descent in the context of etale topology. General references on etale homotopy type of simplicial schemes are [2] or Friedlander [4].

Let X be a scheme and $X_\bullet$ be a simplicial scheme which is a hypercovering of X. Then we have the following.

Theorem 2. (Cox [2, Theorem (VI,2)]) *Assume that X is connected. Let $(X)_{et}$ and $(X_\bullet)$ be the etale homotopy types of X and $X_\bullet$ respectively (cf. [4, Chap. 4,5]). Then there is a weak equivalence of pro-simplicial sets:*

$$(X)_{et} \approx (X_\bullet)_{et}.$$

The homotopy machine which is employed to prove the above theorem works well, even if we change the final object X from a scheme to a stack. Hence the following is valid.

Theorem 3. *Let S be an algebraic stack with etale site S_{et}, and let $X_\bullet$ be a simplicial scheme with values in S_{et} which is a hypercovering of S. Then there is a weak equivalence of pro-simplicial sets:*

$$(S)_{et} \approx (X_\bullet)_{et}.$$

Remark. Though it is not necessary in the following argument in the logical sense, we note here that there is another indirect definition of the etale homotopy type of S without using $(S)_{et}$, but setting the etale homotopy type of S to be that of the simplicial scheme $X_\bullet$ which is a hypercovering of S. In fact, in the next section, to define the classical homotopy type of the moduli stacks, we choose this viewpoint. The proof that the latter definition of the etale homotopy type of S does not depend on the choice of the hypercovering $X_\bullet$ is given by a standard argument using bisimplicial schemes.

§3. Classical homotopy type of the moduli stacks

In this section, we work in the category of analytic spaces over $\mathbb{C}$. We consider the moduli stacks in this analytic context.

For each analytic space S, the category of sections $\mathcal{M}_{g,n}^{an}(S)$ over S of the moduli stack $\mathcal{M}_{g,n}^{an}$ is defined in the following way. An object of $\mathcal{M}_{g,n}^{an}(S)$ is a proper smooth morphism $p : C \to S$ of analytic spaces with n sections $t_1, \cdots, t_n : S \to C$ such that for each point $s \in S$, the fiber C_s at s is a compact Riemann surface of genus g, and $t_1(s), \ldots, t_n(s)$ are distinct points of C_s. We define the morphisms of $\mathcal{M}_{g,n}^{an}(S)$ as isomorphisms among these data.

For $\mathcal{M}_{g,n}^{an}$, we can define the site of local isomorphisms $(\mathcal{M}_{g,n}^{an})_{cl}$. Let $x : X \to \mathcal{M}_{g,n}^{an}$ be an object $\mathcal{M}_{g,n}^{an}(X)$. Then x is a local isomorphism if there exists another $s : S \to \mathcal{M}_{g,n}^{an}$ such that $X \times_{\mathcal{M}} S \to S$ is locally isomorphic. Here we denote $\mathcal{M}_{g,n}^{an}$ by $\mathcal{M}$ to simplify the subscript and $X \times_{\mathcal{M}} S$ is the analytic space which represents the functor $\mathrm{Isom}(X \times S; \mathrm{pr}_1^*, \mathrm{pr}_2^*)$. Recall that, in order for $x : X \to \mathcal{M}_{g,n}^{an}$ to be locally isomorphic, it is necessary and sufficient that the corresponding family of curves $C \to X$ plus sections $t_1, \cdots, t_n$ be locally universal in the sense of deformation theory (cf. Mumford [6,§3], but his terminology is *modular families*).

For analytic spaces which are locally isomorphic over $\mathcal{M}_{g,n}^{an}$, we can define fiber products which are also locally isomorphic over $\mathcal{M}_{g,n}^{an}$. Hence we can consider simplicial analytic spaces with values in $(\mathcal{M}_{g,n}^{an})_{cl}$ and can define hypercoverings of $\mathcal{M}_{g,n}^{an}$.

For any simplicial analytic space $X_\bullet$, we can define its geometric realization $|X_\bullet|$ as follows. The space $|X_\bullet|$ is the quotient of the topological space $\coprod_{n=0}^{} X_n \times \Delta[n]$ by the equivalence relation $(x, a(d)) \sim (a(x), d)$ for any $a : \Delta[n] \to \Delta[m]$ in (ST). Here (ST) is the category of simplex types, $X_n \times \Delta[n]$ is given the product topology and

$$\Delta[n] = \{d = (d_0, \cdots, d_n \mid \sum d_i = 1, d_i \geq 0\} \subset \mathbb{R}^{n+1}.$$

Now we can define the classical homotopy type $(\mathcal{M}_{g,n}^{an})_{cl}$ of $\mathcal{M}_{g,n}^{an}$.

Definition. Let $X_\bullet$ be a simplicial analytic space with values in $(\mathcal{M}_{g,n}^{an})_{cl}$ which is a hypercovering of $\mathcal{M}_{g,n}^{an}$. Then we define the classical homotopy type $((\mathcal{M}^{an}))_{cl}$ of $\mathcal{M}_{g,n}^{an}$ as the geometric realization of $|X_\bullet|$.

In order to justify the above definition, one has to show that the weak equivalence class $|X_\bullet|$ does not depend on the choice of hypercovering $X_\bullet$. In fact, let $Y_\bullet$ be another hypercovering of $\mathcal{M}_{g,n}^{an}$ with values in $(\mathcal{M}_{g,n}^{an})_{cl}$. Then we can define a bisimplicial analytic space $Z_{\bullet\bullet}$ by putting

$$Z_{mn} = X_m \times_{\mathcal{M}} Y_n \quad \text{for each} \quad m, n \geq 0.$$

Thus for each m, $Z_{m,\bullet}$ is a hypercovering of X_m. Then the homotopical descent with respect to the classical topology says that the geometric realization $|Z_{m,\bullet}|$ of $Z_{m,\bullet}$ is homotopically equivalent to X_m. Moreover we recall that if $\{m \mapsto |Z_{m,\bullet}|\}$ is the associated simplicial space, the geometric realization of the diagonal analytic space $|\Delta(Z_{\bullet\bullet})|$ is homotopic to $|m \mapsto |Z_{m\bullet}|\}|$ (cf. Quillen [7,p.86, Lemma]). Therefore $|\Delta(Z_{\bullet\bullet})|$ is homotopically equivalent to $|m \mapsto X_m| = |X_\bullet|$. Applying the same argument for $Y_\bullet$, we have the desired weak equivalence $|X_\bullet| = |Y_\bullet|$. Hence $((\mathcal{M}_{g,n}^{an}))_{cl}$ is well-defined.

Fix an embedding $\bar{\mathbb{Q}} \hookrightarrow \mathbb{C}$. Then for any scheme X over $\bar{\mathbb{Q}}$, we denote by X^{an} the complex analytic space associated with the $\mathbb{C}$-valued points $X(\mathbb{C})$.

Let $x : X \to \mathcal{M}_{g,n} \times_{\mathbb{Z}} \bar{\mathbb{Q}}$ be an object of the etale site $(\mathcal{M}_{g,n} \times_{\mathbb{Z}} \bar{\mathbb{Q}})_{et}$, and let $\{p : C \to X$ with sections $t_1, \cdots, t_n : X \to C\}$ be the corresponding object in the category $\mathcal{M}_{g,n} \otimes_{\mathbb{Z}} \bar{\mathbb{Q}}(X)$ of sections over X. Then the associated analytic data $\{p^{an} : C^{an} \to X^{an}$ with sections $t_1^{an}, \cdots, t_n^{an} : X^{an} \to C^{an}\}$ defines an object of $(\mathcal{M}_{g,n}^{an})_{cl}$. Thus we have a functor $(\mathcal{M}_{g,n} \otimes_{\mathbb{Z}} \bar{\mathbb{Q}})_{et} \to (\mathcal{M}_{g,n}^{an})_{cl}$ which defines a continuous morphism of sites

$$(\mathcal{M}_{g,n}^{an})_{cl} \to (\mathcal{M}_{g,n} \otimes_{\mathbb{Z}} \bar{\mathbb{Q}})_{et}.$$

§4. Čech nerve for Teichmüller space

Let $T_{g,n}$ be the Teichmüller space for n-pointed compact Riemann surfaces of genus g. The definition of $T_{g,n}$ via moduli interpretation is given as follows. The only result, which is necessary for us, is that it exists as an analytic space and it is a cell, hence contractible to a point.

The classical theory of Teichmüller space due to the so-called New York school (Ahlfors, Bers etc.), is found in the article of Bers [11]. However their formulation of problems is not convenient for us. Hence some terminology and formulation of the main results are modified according to the articles of A. Weil and Grothendieck quoted in [11].

We first have to define the notion of *marking*. Let $\Pi_{g,n}$ be an abstract group generated by $2g + n$ elements $\alpha_1, \cdots, \alpha_g, \beta_1, \cdots, \beta_g, \gamma_1, \cdots, \gamma_n$ with a defining relation

$$[\alpha_1, \beta_1] \cdots [\alpha_g, \beta_g]\gamma_1 \cdots \gamma_n = 1.$$

Here $[\alpha_i, \beta_i]$ denotes the commutator $\alpha_i \beta_i \alpha_i^{-1} \beta_i^{-1}$ for each i. The group $\Pi_{g,n}$ is isomorphic to the fundamental group of $R - \{p_1, \cdots, p_n\}$, where R is a compact orietable C^∞ surface of genus g and $p_1, \cdots, p_n$ are n distinct points on R. Let N be the minimal normal subgroup of $\Pi_{g,n}$ generated by

$\gamma_1, \cdots, \gamma_n$. Then the quotient group $\Pi_{g,n}/N$ is isomorphic to $\Pi_{g,0}$ which is the surface group of genus g. When $g > 0$, the second integral cohomology group $H^2(\Pi_{g,0}; \mathbb{Z}) \cong \mathbb{Z}$. In this case, we choose a generator e of $H^2(\Pi_{g,0}, \mathbb{Z})$.

Let C^0 be a smooth complex analytic curve (i.e. a Riemann surface) which admits a smooth compactification C such that $C^0 = C - \{p_1, \cdots, p_n\}$ Here $p_1, \cdots, p_n$ are n distinct points on C. Choose a base point p_0 on C^0, and consider the fundamental group $\pi_1(C^0, p_0)$. We define a conjugate class in $\pi_1(C^0, p_0)$ for each p_i.

Let D_i be a small disk with center p_i in C, whose closure does not contain any other p_j $(j \neq i)$. Let us choose a point q_i on the boundary ∂D_i, which is assumed to be a Jordan curve. Let δ_i be the closed loop on C which starts from q_i and ends at q_i, whose orientation is positive with respect to that of C. Let ε_i be a path from p_0 to q_i. Then the composite path $\varepsilon_i \delta_i \varepsilon^{-1}$ defines an element l_i of $\pi_1(C^0, p_0)$. It is easy to check that the conjugacy class $\{l_i\}$ of l_i in $\pi_1(C^0, p_0)$ does not depend on the choice of D_i, q_i and ε_i.

Let us consider an isomorphism of $\pi_1(C^0, p_0)$ with $\Pi_{g,n}$:

$$\varphi : \pi_1(C^0, p_0) = \Pi_{g,n}.$$

satisfying the following conditions:

(i) For each i $(1 \leq i \leq n)$, the conjugacy class $[l_i]$ is mapped to the conjugacy class $[\gamma_i]$ of γ_i in $\Pi_{g,n}$ by φ;

(ii) On passing to the quotient, φ induces an isomorphism:

$$\bar{\varphi} : \pi_1(C, p_0) \cong \Pi_{g,n}/N = \Pi_{g,0}.$$

Moreover, if $g > 0$ the induced isomorphism of cohomology groups

$$H^2(C; \mathbb{Z}) = H^2(\pi_1(C, p_0); \mathbb{Z}) \cong H^2(\Pi_{g,0}; \mathbb{Z})$$

maps the orientation class of C to the given generator e of $H^2(\Pi_{g,0}; \mathbb{Z})$.

Let φ_1, $\varphi_2 : \pi_1(C^0, p_0) \cong \Pi_{g,n}$ be two isomophisms which satisfy the above conditions (i), (ii). Then we call φ_1 and φ_2 equivalent, if there exists an inner automorphism θ of $\Pi_{g,n}$ such that $\varphi_2 = \theta \cdot \varphi_1$. An equivalence class $[\varphi]$ of isomorphisms φ satisfying (i) and (ii) is called a *marking* of C^0. A pair $(C^0, [\varphi])$ is called a marked Riemann surface.

Let $p : C \to S$ be a proper smooth analytic family of curves with sections $t_1, \cdots, t_n : S \to C$ such that for each $s \in S$, the fiber C_s at s is a compact Riemann surface of genus g and $t_i(s) \neq t_j(s)$ if $i \neq j$. Moreover let us consider a locally constant family φ of markings, i.e. an isomorphism up to conjugation of local system of groups

$$\varphi : \Pi_1(C^0/S) = \Pi_{g,n} \times S.$$

Here $C^0 = C - \cup_{i=1}^n t_i(S)$, $\Pi_1(C^0/S)$ is the local system of the fundamental groups of fibers $C_s^0 = C_s - \{t_1(s), \cdots, t_n(s)\}$ $(s \in S)$ of $C^0 - S$,
and $\Pi_{g,n} \times S$ is the constant local system over S with fiber $\Pi_{g,n}$. The induced isomorphism $\bar{\varphi} : \Pi_1(C/S) = \Pi_{g,0} \times S$ should define an isomorphism
$R^2 p_* \mathbb{Z} = H^2(\Pi_{g,0}; \mathbb{Z}) \times S$ which is compatible with orientations at each
fiber C_s.

To each analytic space S, we can associate the category of data $\{p : C \to$
$S; t_1, \cdots, t_n : S \to C; \varphi\}$, with morphisms being isomorphisms. Then it
defines a contravariant functor from the category of analytic spaces to the
category of sets, because any object admits no non-trivial automorphism.
The classical Teichmüller theory tells that this functor is representable by
the Teichmüller space $T_{g,n}$. Via the usual formalism of representable func-
tors, the identity morphism $T_{g,n} \to T_{g,n}$ defines a canonical data

$$\{C_g \to T_{g,n}; \tau_1, \cdots, \tau_n : T_{g,n} \to C_g; \Phi\}$$

which is referred to as the universal family over $T_{g,n}$.

Forget the universal marking Φ, then this canonical data defines an object
of the section category $\mathcal{M}_{g,n}^{an}(T_{g,n})$, i.e. a morphism $\pi : T_{g,n} \to \mathcal{M}_{g,n}^{an}$ of
"analytic stacks".

Now let us consider the fiber product $T_{g,n} \times_{\mathcal{M}} T_{g,n} = \mathrm{Isom}(\pi, \pi)$, where
$\mathcal{M} = \mathcal{M}_{g,n}^{an}$. Since $T_{g,n}$ is a locally universal family of dimension $3g -$
$3 + n$, both projections from $\mathrm{Isom}(\pi, \pi)$ to $T_{g,n}$ are local isomorphisms (cf.
Mumford [6, §3]). Moreover for any $x \in T_{g,n}$ the data over x has no non-
trivial automorphism. Hence $\mathrm{Isom}(\pi, \pi) \to T_{g,n} \times T_{g,n}$ is an immersion.

Let us choose $x, y \in T_{g,n}$ and find the condition when x and y define the
isomorphic object in the category $\mathcal{M}$. For x we have a compact Riemann
surface C_x of genus g with n distinct points $\tau_1(x), \cdots, \tau_n(x)$ on it, and a
marking

$$\Phi_x : \pi_1(C_x - \{\tau_1(x), \cdots, \tau_n(x)\}, \text{base point}) \cong \Pi_{g,n}.$$

For y, we have a similar data.

In order that x and y define the isomorphic object over $\mathcal{M}$, C_x should
be isomorphic to C_y and each point $\tau_i(x)$ should be mapped to $\tau_i(y)$ by
the same isomorphism. For the markings, we have the following freedom of
choice.

Consider automorphisms θ of $\Pi_{g,n}$ satisfying the conditions:

(i) For each i $(1 \le i \le n)$, $\theta(\gamma_i)$ is conjugate to γ_i in $\Pi_{g,n}$;

(ii) On passing to the quotients by N, θ induces an isomorphism $\bar{\theta}$

$$\Pi_{g,0} \cong \Pi_{g,n}/N \overset{\bar{\theta}}{\cong} \Pi_{g,n}/N \cong \Pi_{g,0}.$$

Moreover if $g > 0$, $\bar{\theta}$ induces the identity on the second cohomology group $H^2(\Pi_{g,0}; \mathbb{Z}) \cong \mathbb{Z}$. Note that the inner automorphisms of $\Pi_{g,n}$ satisfy these conditions.

Let $\Sigma_{g,n}$ be the subgroup of $\mathrm{Aut}(\Pi_{g,n})$ consisting of automorphisms satisfying the above conditions (i) and (ii). Then the quotient group of $\Sigma_{g,n}$ by the inner automorphism group $\mathrm{Inn}(\Pi_{g,n})$ is isomorphic to the Teichmüller group $\Gamma_{g,n}$ by the classical theorem of Nielsen. Nowadays it is a direct consequence of obstruction theory, because Riemann surfaces are $K(\pi, 1)$ spaces.

Thus we can finally see that if x and y are isomorphic, for the markings Φ_x and Φ_y there exists a unique class $[\theta] \in \Gamma_{g,n}$ such that $\Phi_y = \theta \cdot \Phi_x$ up to inner automorphisms of $\Pi_{g,n}$. Therefore we have a complete description of $T_{g,n} \times_{\mathcal{M}} T_{g,n} = \mathrm{Isom}(\pi, \pi)$ in $T_{g,n} \times T_{g,n}$:

$$\mathrm{Isom}(\pi, \pi) = \{(x, y) \in T_{g,n} \times T_{g,n} \mid C_x \cong C_y; \tau_i(x) = \tau_i(y) \text{ for each } i$$
$$\text{and } \Phi_y = \theta \cdot \Phi_x \text{ for some } [\theta] \in \Gamma_{g,n}\}.$$

Especially the fibers of the projections $\mathrm{Isom}(\pi, \pi) \to T_{g,n}$ are isomorphic to the discrete set $\Gamma_{g,n}$. Hence we have an isomorphism

$$\mathrm{Isom}(\pi, \pi) = T_{g,n} \times \Gamma_{g,n}$$

by mapping $(x, y) \in \mathrm{Isom}(\pi, \pi)$ to $(x, [\theta])$ where $\theta = \Phi_y \cdot \Phi^{-1}$.

Similarly by induction on m, we have an isomorphism

$$T_{g,n} \times_{\mathcal{M}} \cdots \times_{\mathcal{M}} T_{g,n} = T_{g,n} \times \Gamma_{g,n} \times \Gamma_{g,n} \cdots \Gamma_{g,n}.$$

Now we can prove the main result of this section.

Theorem 4. *Let $\mathrm{cosk}_0^{\mathcal{M}}(T_{g,n})$ be the Čech nerve associated to the locally isomorphic surjective covering $T_{g,n} \to \mathcal{M}_{g,n}^{an}$. Then the geometric realization $|\mathrm{cosk}_0^{\mathcal{M}}(T_{g,n})|$ is homotopically equivalent to the classifying space $|B_{\Gamma_{g,n}}|$ of the Teichmüller group $\Gamma_{g,n}$.*

Proof. The m-th simplex of the Čech nerve $\mathrm{cosk}_0^{\mathcal{M}}(T_{g,n})$ is given by the $(m+1)$-tuple fiber product of the copies of $T_{g,n}$ over $\mathcal{M}_{g,n}^{an}$, which is isomorphic to

$$T_{g,n} \times \Gamma_{g,n} \cdots \times \Gamma_{g,n}.$$

By definition, we can check that $\mathrm{cosk}_0^{\mathcal{M}}(T_{g,n})$ is isomorphic to $T_{g,n} \times B_{\Gamma_{g,n}}$, where $B_{\Gamma_{g,n}}$ is the classifying simplicial set of $\Gamma_{g,n}$. Hence $|\mathrm{cosk}_0^{\mathcal{M}}(T_{g,n})|$ is homeomorphic to $T_{g,n} \times |B_{\Gamma_{g,n}}|$. Since $T_{g,n}$ is contractible to a point, $|\mathrm{cosk}_0^{\mathcal{M}}(T_{g,n})|$ is homeomorphic to $|B_{\Gamma_{g,n}}|$.

Corollary. $((\mathcal{M}_{g,n}^{an}))_{cl}$ *is weakly equivalent to* $|B_{\Gamma_{g,n}}|$.

§5. Comparison theorem and the conclusion of the proof

In this section, first we recall the comparison theorem between etale homotopy type and classical homotopy type of simplicial schemes over $\mathbb{C}$. It is the last step to conclude the proof of Theorem 1.

Theorem 5. (Cox [2,Theorem IV.8], Friedlander [4,Theorem 8.4]) *Let* $X_\bullet$ *be a simplicial scheme of finite type over* $\mathbb{C}$ *and let* $X_\bullet^{an}$ *be the associated simplicial analytic space. Then we have a weak equivalence in the category of pro-simplicial sets*

$$(X_\bullet)_{et} \equiv \mathrm{Sing}(|X_\bullet^{an}|)^\wedge.$$

Here $\mathrm{Sing} : (\text{Top. spaces}) \to (\text{simpl. sets})$ is a functor of singular sets from the category of topological spaces to the category of simplicial sets. And $\wedge$ is the profinite completion functor of Artin-Mazur [1].

Proof of Theorem 1. Let us choose a scheme X of finite type over $\bar{\mathbb{Q}}$ such that there is an etale surjective morphism $x : X \to \mathcal{M}_{g,n} \otimes_{\mathbb{Z}} \bar{\mathbb{Q}}$. In fact such X exists if we consider the moduli stack of curves with Jacobian level structure of level $m \geq 3$.

Let $X_\bullet$ be the Čech nerve $\mathrm{cosk}_0^{\mathcal{M}_{g,n}}(X)$. Then $\mathrm{cosk}_0^{\mathcal{M}_{g,n}}(X)$ is a hypercovering of $\mathcal{M}_{g,n} \otimes_{\mathbb{Z}} \bar{\mathbb{Q}}$ with values in $(\mathcal{M}_{g,n} \otimes_{\mathbb{Z}} \bar{\mathbb{Q}})_{et}$, and the associated analytic version $\mathrm{cosk}_0^{\mathcal{M}}(X_\bullet^{an})$ is a hypercovering of $\mathcal{M}_{g,n}^{an}$ with values in $(\mathcal{M}_{g,n}^{an})_{cl}$. Then the homotopic descent (Theorem 3) of etale homotopy and comparison theorem (Theorem 5) imply that there is a weak equivalence

$$(\mathcal{M}_{g,n} \otimes_{\mathbb{Z}} \bar{\mathbb{Q}})_{et} \cong (\mathrm{cosk}_0^{\mathcal{M}_{g,n}}(X))_{et}$$
$$\cong (\mathrm{cosk}_0^{\mathcal{M}_{g,n}}(X) \otimes_{\mathbb{Q}} \mathbb{C})_{et} \cong \mathrm{Sing}|\mathrm{cosk}_0^{\mathcal{M}}(X^{an})|^\wedge.$$

On the other hand, $|\mathrm{cosk}_0^{\mathcal{M}}(X^{an})|$ is weakly equivalent to $((\mathcal{M}_{g,n}^{an}))_{cl}$, which is in turn weakly equivalent to $|B_{\Gamma_{g,n}}|$ by Theorem 4. Therefore we have an equivalence of pro-simplicial sets

$$(\mathcal{M}_{g,n} \otimes_{\mathbb{Z}} \bar{\mathbb{Q}})_{et} \approx \mathrm{Sing}|B_{\Gamma_{g,n}}|^\wedge \approx (B_{\Gamma_{g,n}})^\wedge,$$

as was to be proved.

Open problem and some remarks

The following problem is kindly suggested by Professors Deligne and Morava.

Problem. By a result of Artin-Mazur ([1, 6.9]), the following two statements are equivalent:

(i) $K(\Gamma_{g,n}, 1)$ is weakly equivalent to $K(\Gamma_{g,n}, 1)^\wedge$.

(ii) The group $\Gamma_{g,n}$ is a *good group* in the sense of Serre (cf. [8, Chap. II, Exercises]).

Do these conditions hold for all (g, n)?

For $g = 0$ or 1, it is easy to see that the group $\Gamma_{g,n}$ is a good group for each n, because these groups are successive extensions of some groups which have finite index free subgroups. For $g = 2$, this follows from a result of Birman-Hilden (cf. Birman [9]), because up to finite center $\Gamma_{2,0}$ is isomorphic to a braid group which is a good group. The groups $\Gamma_{2,n}$ are also good, since they are successive extension of good groups by surface groups or free groups. The author is ignorant of any results for $g > 2$.

In place of the total moduli stack, one can consider the closed substacks $\mathcal{H}_g$ in $\mathcal{M}_{g,0}$ parametrizing only hyperelliptic curves. Then if we replace the Techmüller group by the hyperelliptic modular group which is the centralizer of a chosen hyperelliptic involution in $\Gamma_{g,0}$, a similar result to Theorem 1 is valid with necessary modification. This modular subgroup is a good group.

It seems very interesting problem to consider other various substacks to realize the idea of "Lego" in [5].

References

[1]   M. Artin, B. Mazur, *Etale homotopy*, Lecture Notes in Math. No. **100**, Springer, 1969.

[2]   D.A. Cox, Homotopy theory of simplicial schemes, *Compositio Math.* **39** (1979), 263-296.

[3]   P. Deligne, D. Mumford, The irreducibility of the space of curves of given genus, Publ. Math. I.H.E.S. **36** (1969), 75-110.

[4]   E.M. Friedlander, *Etale homotopy of simplicial schemes*, Annals of Math. Studies, No. 104, Princeton Univ. Press and Univ. of Tokyo Press, 1982.

[5]   A. Grothendieck, *Esquisse d'un programme*, 1984, this volume.

[6]   D. Mumford, Picard groups of moduli problems, in *Arithmetical Algebraic Geometry*, ed. by Schilling, O.F.G., Harper & Row, (1965), 33-81.

[7]   D. Quillen, *Higher algebraic K-theory I*, Lecture Notes in Math. No. **341**, 85-147, Springer, 1973.

[8]   J-P. Serre, *Cohomologie Galoisienne*, Lecture Notes in Math. No. **5**, 1965.

[9]   J.S. Birman, *Braid, links, and mapping class groups*, Annals of Math. Studies. No. **82**, Princeton Univ. Press, 1974.

[10] J.L. Verdier, Exposé V: Cohomologie dans les topos, in *Théorie des Topos et Cohomologie Etale des Schéms, Tome II*, Lecture Notes in Math N. **270**, Springer, 1972.

[11] L. Bers, Uniformization, moduli, and Kleinian groups, *Bull. London Math. Soc.*, **4** (1972), 257-300

School of Mathematical Science, the University of Tokyo
Komaba 3-8-1, Meguro ward, Tokyo 153, JAPAN

The 'Obvious' Part of Belyi's Theorem
and Riemann Surfaces with Many Automorphisms

Jürgen Wolfart

§0. Introduction

Many articles on Grothendieck dessins, Belyi functions, hypermaps on Riemann surfaces etc. have at least in part the character of survey articles or give new access to old material. The reasons for this phenomenon are the rapidly growing interest in the subjects, the non–existence of monographs as common source and reference, and the very different mathematical or physical origins of the people working in this field. Moreover, there are a lot of facts and interrelations in principle well known to experts but often not stated or proved before as explicitly as needed. The present article again belongs to this category of papers and tries to shed new light on some old subjects and to make their connection visible. The first subject is the well–known result of Belyi [Be]:

Theorem 1. *A compact Riemann surface X is isomorphic to the Riemann surface $C(\mathbb{C})$ consisting of the complex points of a nonsingular projective algebraic curve C defined over a number field if and only if there is a nonconstant meromorphic function β on X ramified over at most three points.*

Such functions will be called *Belyi functions*. Of course we may assume that they are ramified over the three points 0, 1 and ∞. The surprisingly simple algorithm found by Belyi to prove the 'only if' part of the theorem is reproduced in many later papers and will not be discussed here, we will care about the 'if' part only. For this proof Belyi and most papers including [CIW] and my old preprint [Wo1] refer to Weil's criteria [We]. As a variant finally coming from the same source one may use the explicit knowledge of the Galois groups of the maximal extensions of $\mathbb{C}(x)$ and $\overline{\mathbb{Q}}(x)$ unramified outside three points (Matzat [Ma]). Both proofs rely on a heavy machinery and — as several people complained to me — it is far from being obvious how to fit the problem precisely to the hypotheses of Weil's criteria (for details see the proof of Lemma 3), and whether or not such a powerful tool is needed. I hope therefore that another version of this proof will be of some use, but the reader will recognize that we do not leave the neighbourhood of Weil's ideas. In this part of the proof — Lemmas 3 and 4 — old–fashioned and elementary but still vital and useful algebraic geometry is needed: Zariski topology, generic points and a specialization argument. Similar ideas often have been used in the literature: Shimura's and Taniyama's proof that abelian varieties with complex multiplication may be defined over a number field ([ST], Prop. 26) is an early example. After completing a first version of this paper I learned that this method has been used together with Riemann's existence theorem by Fried (Lemma 1.2 of [Fr] can serve as a very condensed and general proof of this part of Belyi's theorem) and in special cases recently by Zannier [Za]. In the present paper, certain coverings of X instead of Riemann's theorem are the main topological tool. This part (up to Lemma 2) is taken from [Wo1], [CW] and [CIW]; it is based on the rigidity of triangle groups.

For the part concerning moduli fields, it is a pleasure for me to thank Robert Silhol for an important hint on the literature and Pierre Debes for long discussions clearing the intimate connection of this 'easy half' of Belyi's theorem with the old question about the relation between fields of moduli and fields of definition for algebraic curves (see the Lemmas 5 to 7 and Theorems 3 and 4). For Theorem 4, Pierre Debes and Michel Emsalem worked out a different proof reducing the problem to an analogous question about moduli spaces of coverings [DE]. They also suggested a further possibility for the transcendental descent in the proof of Theorem 1, not discussed in the present paper, namely to replace Weil's method by Grothendieck's generalization ('Morphismes quasi–compacts et fidèlement plats').

As a last subject, the methods explained in this note give a new proof for the essential part of Popp's result on a conjecture of Rauch about Riemann surfaces with many automorphisms (Theorem 5) showing again the close connection of Belyi's Theorem to moduli problems. These Riemann surfaces with many automorphisms turn out to be of particular interest for Galois actions (Theorem 7).

I am grateful to the referee for several useful comments. On the financial side, the work has been supported by a PROCOPE grant.

§1. Belyi surfaces and rigidity

Throughout this paper, we use a terminology recently proposed by David Singerman: A compact Riemann surface X is called a *Belyi surface* if a Belyi function exists on it. In this language, the function–theoretic kernel of the papers [Wo1], [CW] and [CIW] reads as follows.

Lemma 1. *A compact Riemann surface X is a Belyi surface if and only if X is isomorphic to a quotient space $\Gamma \backslash \mathcal{H}$ of the upper half plane $\mathcal{H}$ by some*

*subgroup Γ of finite index in a cocompact Fuchsian triangle group Δ.
For a given Belyi function β on X, the signature $< p, q, r >$ of Δ can be
chosen such that p, q, r are common multiples of the ramification orders of
β over 0, 1 and ∞, respectively. If on the other hand Δ and Γ are given,
β is the canonical map*

$$\Gamma\backslash\mathcal{H} \rightarrow \Delta\backslash\mathcal{H}$$

*where we identify $\Delta\backslash\mathcal{H}$ with the Riemann sphere $\bar{\mathbb{C}}$ and in particular the
fixed points of Δ of order p, q, r with the points 0, 1, ∞, respectively.*

This identification is provided by the Δ–automorphic function J constructed by analytic continuation of the inverse function of a Schwarz triangle function s. This triangle function s can be defined as a biholomorphic mapping of the upper or the lower half plane onto an open hyperbolic triangle with angles π/p, π/q, π/r. The closure of this hyperbolic triangle together with its mirror image obtained by hyperbolic reflection in any of its sides represent a fundamental domain for Δ. The vertices of this triangle (and their Δ–images) are the elliptic fixed points of Δ of order p, q, r respectively, and by an obvious normalization we can always assume J to take the values 0, 1 and ∞ there, respectively. With this normalization it is clear that $\Gamma\backslash\mathcal{H} \rightarrow \Delta\backslash\mathcal{H}$ is a Belyi function, and the other direction of the proof is a consequence of the fact that any branch of $\beta^{-1} \circ J$ has an analytic continuation to $\mathcal{H}$ providing a (ramified) covering map $\mathcal{H} \rightarrow X$ with covering group Γ contained in Δ. The degree of β is the index $(\Delta : \Gamma)$. Therefore, we have to prove

Theorem 2. *Let Γ be a subgroup of finite index in a Fuchsian triangle group Δ. Then the quotient space $\Gamma\backslash\mathcal{H}$ is isomorphic to the Riemann surface consisting of the complex points of an algebraic curve C defined over $\overline{\mathbb{Q}}$.*

Let Y be a smooth projective algebraic curve whose complex points form a Riemann surface isomorphic to $\Gamma\backslash\mathcal{H}$. We can suppose Y to be defined over a finitely generated extension K of $\mathbb{Q}$. The elements σ of the automorphism group $G := \mathrm{Aut}\,\mathbb{C}$ act on the coefficients of the defining equations of Y and on its complex points as well. Call the resulting (again smooth projective algebraic) curve Y^σ. It is defined over $\sigma(K)$ and has a Belyi function β^σ obtained by the action of σ on the coefficients of the Belyi function β on Y. According to Lemma 1, we may identify β with the natural projection $\Gamma\backslash\mathcal{H} \rightarrow \Delta\backslash\mathcal{H}$, and by an extension of K if necessary we may suppose that β is also defined over K. Again by Lemma 1, the existence of β^σ on Y^σ implies that $Y^\sigma(\mathbb{C})$ is complex isomorphic to $\Gamma^\sigma\backslash\mathcal{H}$ for a subgroup Γ^σ of finite index in some triangle group Δ^σ. The Belyi functions β and β^σ have the same degree and the same ramification orders. We may therefore assume that

- Δ^σ has the same signature as Δ , and even $\Delta^\sigma = \Delta$ because triangle groups are uniquely determined by their signatures, up to conjugation in $PSL_2\mathbb{R}$.

The fact that the Belyi functions have the same order implies therefore

$$(\Delta : \Gamma^\sigma) = (\Delta : \Gamma).$$

For reasons of combinatorial group theory, there are only finitely many subgroups of Δ with given index ([MKS], p.102, exercise 19), hence only finitely many non–isomorphic Riemann surfaces $Y^\sigma(\mathbb{C})$. In other words we proved the first part of

Lemma 2. *Let $Y(\mathbb{C})$ be a Belyi surface.*

1. *There is a subgroup U of finite index in the automorphism group $G = \mathrm{Aut}\mathbb{C}$ such that the Riemann surfaces $Y^\sigma(\mathbb{C})$, $\sigma \in U$, are all isomorphic to each other.*

2. *The fixed field $M(Y)$ of U is a number field of degree*

$$[M(Y) : \mathbb{Q}] = (\mathrm{Aut}\mathbb{C} : U).$$

The second part is elementary algebra and noted only for later use. What remains to prove is the following

Lemma 3. *The statement of Lemma 2 implies that there is an algebraic curve C defined over $\overline{\mathbb{Q}}$ such that all $Y^\sigma(\mathbb{C})$, $\sigma \in U$, are isomorphic to $C(\mathbb{C})$.*

Proof. For genus $g = 0$, there is nothing to prove ($C = \mathbb{P}_1$). The case of elliptic curves $g(Y) = 1$ is also rather special. Here, the isomorphism class of $Y(\mathbb{C})$ is uniquely determined by the absolute invariant $j = j(Y) \in K$, and from $j(Y^\sigma) = \sigma(j)$ and Lemma 2 we obtain the U–invariance of j, hence $j \in M(Y) \subset \overline{\mathbb{Q}}$. Since it is well known that we may choose the coefficients of the Weierstrass equation in $\mathbb{Q}(j)$, the case $g = 1$ is settled.

The idea of the proof for higher genus is easily explained in the following special situation. Suppose for simplicity that Y is defined over a purely transcendental extension $K = \mathbb{Q}(\xi)$ of $\mathbb{Q}$, then for some $\sigma \in U$ the number $\eta := \sigma(\xi)$ is algebraically independent of ξ — note that Y^σ is defined over $\mathbb{Q}(\eta)$ — and suppose that the isomorphism

$$f : Y(\mathbb{C}) \to Y^\sigma(\mathbb{C})$$

and its inverse are defined over $\mathbb{Q}(\xi, \eta)$, i.e. that all their coefficients can be written as rational functions in ξ and η with coefficients in $\mathbb{Q}$. (As an example of such a situation, take the isomorphism

$$f \; : \; (x, y, t) \; \mapsto \; (u, v, w) = (x\eta^2\xi, y\eta^3, t\xi^3)$$

of the elliptic curve $ty^2 = x^3 - \xi^6 t^3$ onto the curve $wv^2 = u^3 - \eta^6 w^3$). To indicate the dependence on ξ and η , we write $f_{(\xi,\eta)}$. Now fix ξ and replace η by an arbitrary $z \in \mathbb{C}$. For almost all z , the image of $Y(\mathbb{C})$ under $f_{(\xi,z)}$ is again a nonsingular projective algebraic curve because singularities only occur if certain polynomials in z vanish, and they do not vanish identically as the value $z = \eta$ shows (in the example (1), $z = 0$ is the only exceptional argument). For the same reason, $f_{(\xi,z)}$ defines an isomorphism of $Y(\mathbb{C})$ onto this image curve for almost all z, in particular for some algebraic z

But then the (complex isomorphic) curve $f_{(\xi,z)}(Y(\mathbb{C}))$ consists of the complex points of a curve defined over $\overline{\mathbb{Q}}$. Remark aside: In this special situation, Weil's Theorem 4 [We] applies directly, and our argument can be considered as a very elementary special case of its proof. Unfortunately, its rationality hypothesis on the field of definition of $f_{(\xi,\eta)}$ is not valid in general and its cocycle condition $f_{(\eta,\zeta)} \circ f_{(\xi,\eta)} = f_{(\xi,\zeta)}$ is difficult to verify, therefore the usual reference to [We] for the proof of Theorem 1 is a bit misleading. In general, Y will not be defined over a purely transcendental extension of $\mathbb{Q}$ by one number ξ but of course the field of definition K is a finitely generated extension of $\mathbb{Q}$ or rather of the fixed field $M(Y)$ of U (see Lemma 2). We write $K = M(Y)(\xi_1, \ldots, \xi_n)$ and denote

$$\Xi := (\xi_1, \ldots, \xi_n) \quad \text{and} \quad H := (\eta_1, \ldots, \eta_n) := (\sigma(\xi_1), \ldots, \sigma(\xi_n)) =: \sigma(\Xi) \,.$$

In this notation, Y is defined over $K = M(Y)(\Xi)$ and the curve Y^σ, obtained by applying σ to all coefficients of the defining equations of Y, is defined over $K^\sigma = M(Y)(H)$. We will even write Y^Ξ instead of Y and Y^H instead of Y^σ to make visible that we only have to replace the parameter Ξ by H in passing from Y to Y^σ. To perform the proof we make a very specific choice of σ. By Lemma 2, i.e. by the rigidity of the triangle groups, the group U is uncountable and $K = M(Y)(\Xi)$ and even its algebraic closure $\overline{K}$ is countable. One may suppose that $\xi_1, \ldots, \xi_{n-1}$ are algebraically independent over $\overline{M(Y)}$ and that

$$M(Y)(\Xi) \; = \; M(Y)(\xi_1, \ldots, \xi_{n-1})(\xi_n)$$

is a finite algebraic extension of $M(Y)(\xi_1, \ldots, \xi_{n-1})$. By reasons of cardinality, we can choose $\sigma \in U$ such that

$$\eta_1 = \sigma(\xi_1), \; \ldots, \; \eta_{n-1} = \sigma(\xi_{n-1})$$

are algebraically independent over $\overline{K}$. Then it is an exercise to prove that moreover the algebraic closures $\overline{M(Y)(\Xi)}$ and $\overline{M(Y)(H)}$ are linearly disjoint over $\overline{\mathbb{Q}}$. Recall that $\overline{\mathbb{Q}} = \overline{M(Y)}$; later in Theorem 4 we will treat a slightly more general situation, but always with K countable and U uncountable.

Our assumptions mean that the ξ_ν generate over $\overline{\mathbb{Q}}$ a field of transcendence degree $d = n - 1 > 0$ (if $n = 1$, i.e. if K is a number field, we have nothing to prove), the function field of an affine algebraic variety W of dimension d in an affine space $\mathring{A}^n$ (the algebraic relations between the ξ_ν over $\overline{\mathbb{Q}}$ form the defining equations for W, and Ξ and H are both generic points of W, by the choice of σ the point (Ξ, H) is even a generic point of $W \times W$). The linear disjointness of $\overline{M(Y)(\Xi)}$ and $\overline{M(Y)(H)}$ shows that we can fix Ξ and replace H by a variable Z running over W. As in the special case discussed above, the curve $Y^\sigma(\mathbb{C}) = Y^H(\mathbb{C})$ can be replaced for a non-empty Zariski–open subset V of points $Z \in W(\mathbb{C})$ by some smooth curve $Y^Z(\mathbb{C})$ because singularities only occur if certain polynomials in Z vanish. Again, the point $Z = H \in V$ shows that these polynomials do not vanish identically on W. Since we can find points with algebraic coordinates in this Zariski–open subset V we get curves consisting of the complex points of curves C defined over number fields. We only have to prove that sufficiently many of these curves $Y^Z(\mathbb{C})$, $Z \in V$, are isomorphic to $Y(\mathbb{C})$. This will be done by

Lemma 4. *For all Z in a non-empty Zariski–open subset $\tilde{V}$ of V there are isomorphisms $f_{(\Xi, Z)}$ of $Y(\mathbb{C})$ onto $Y^Z(\mathbb{C})$.*

For curves of genus $g = 0$ we have nothing to prove. For genus $g = 1$ all curves $Y^Z(\mathbb{C})$, $Z \in V$, are isomorphic to $Y(\mathbb{C})$ because the invariant $j^Z := j(Y^Z)$ is a rational function of Z but coincides with $j = j(Y) = j(Y^\Xi)$ for all arguments $Z = \tau(\Xi)$, $\tau \in U$. By Lemma 2, U is so large that j^Z is constant. For genus $g > 1$ we can apply a similar argument using the existence of a moduli space M_g of compact Riemann surfaces of genus g and of a Z–continuous map of the family of curves Y^Z to this moduli space. This map must be a constant map since all Y^σ, $\sigma \in U$, go to the same point of M_g.

A more elementary proof without using M_g may be found by an extension of the previous idea: Recall that for $Z = H$ we have the required isomorphism $f_{(\Xi, H)} := f$, and recall that $g > 1$ implies that f is uniquely determined up to the finite group of automorphisms of $Y(\mathbb{C})$.

1.) This means in particular that f and f^{-1} are defined over $\overline{M(Y)(\Xi, H)}$, otherwise we could find infinitely many different isomorphisms f^τ, $\tau \in U$, fixing Ξ and H.

2.) We denote by ψ_μ, $\mu = 1, \ldots, m$, the coefficients of the components of f and f^{-1}. We know that $\Psi := (\psi_1, \ldots, \psi_m) \in \overline{M(Y)(\Xi, H)}^m$, whence we still can consider (H, Ψ) as a generic point of an algebraic variety $T \subseteq \mathring{A}^{n+m}$ of dimension d, defined over $\overline{M(Y)(\Xi)}$. The defining equations for T contain the defining equations for W and the equations obtained by comparison of coefficients in

$$s_H \circ f_{(\Xi, H)} = t \qquad \text{and} \qquad t \circ f_{(\Xi, H)}^{-1} = s_H,$$

where t and s_H run over a set of defining polynomials for Y and $Y^\sigma = Y^H$ respectively.

3.) Again, fix Ξ and replace H by a variable Z running over W. This in mind, the natural embedding $\overline{M(Y)(\Xi)}(H) \hookrightarrow \overline{M(Y)(\Xi)}(H, \Psi)$ gives a projection of the varieties

$$p : T(\mathbb{C}) \to W(\mathbb{C}) : (Z, Z_m) \mapsto Z$$

as a dimension–preserving morphism whose image clearly contains a non-empty Zariski–open subset $\tilde{V}$ of V: The difference $V - \tilde{V}$ consists only of points where certain polynomials in Z vanish; these are determined by the coefficients of the irreducible equation defining the field extension $\overline{M(Y)(\Xi)}(Z, Z_m)$ over $\overline{M(Y)(\Xi)}(Z)$. Again, $H \in \tilde{V} \neq \emptyset$. We replace the coefficients Ψ of f and f^{-1} by Z_m and, by construction, get isomorphisms $f_{(\Xi, Z)}$ and $f_{(\Xi, Z)}^{-1}$ between $Y(\mathbb{C})$ and $Y^Z(\mathbb{C})$.

Remark 1. Now one may ask if not only C but also the given β as a rational function on $C(\mathbb{C})$ may be defined over a number field. This is not automatically the case; for genus 0 or 1 curves there are too many automorphisms of $C(\mathbb{C})$ changing the field of definition if combined with β. But for genus $g > 1$ the finiteness of the automorphism group implies by similar arguments as above that β is defined over $\overline{\mathbb{Q}}$. These arguments extend to $g = 0$ and 1 if we normalize β e.g. for $g = 0$ by assumptions like $0, 1, \infty \in \beta^{-1}\{0, 1, \infty\}$, thus reducing the number of possibly different β^σ, $\sigma \in U$.

§2. Moduli fields and fields of definition

Some of the results of the previous section can be generalized and expressed in a more sophisticated language. To this aim, different obviously equivalent definitions are available in the literature.

Lemma 5. *Let C be a nonsingular projective algebraic curve defined over a subfield of $\mathbb{C}$. For a field $K \subset \mathbb{C}$, the following properties are equivalent.*

1. K *is the minimal field with the property*

$$\sigma \in \operatorname{Aut}\mathbb{C}, \ \sigma|_K = \operatorname{id} \quad \textit{implies} \quad C(\mathbb{C}) \cong C^\sigma(\mathbb{C}).$$

2. *For all* $\sigma \in \operatorname{Aut}\mathbb{C}, \quad \sigma|_K = \operatorname{id} \iff C(\mathbb{C}) \cong C^\sigma(\mathbb{C})$.

3. K *is the fixed field of* $\quad U := \{\, \sigma \in \operatorname{Aut}\mathbb{C} \mid C(\mathbb{C}) \cong C^\sigma(\mathbb{C})\,\}$.

Then we call K the field of moduli of C and we denote it $M(C)$.

The notation is consistent with that of Lemma 2; since it depends only on the isomorphism class of $C(\mathbb{C})$, we will use it for compact Riemann surfaces as well (all isomorphisms may be considered as defined over $\mathbb{C}$ or as isomorphisms of Riemann surfaces). We call L *a field of definition* for C if $C(\mathbb{C})$ is isomorphic to $Y(\mathbb{C})$ for a curve Y defined over L. Clearly the field of moduli is contained in any field of definition. It is known (see [Ba], [Sh]) that for such a curve C of positive genus g

- the quotients of the coordinates of the corresponding projective point $e \in M_g$ generate the moduli field as an extension of $\mathbb{Q}$, sometimes simply denoted $\mathbb{Q}(e)$.

Here we consider the moduli space M_g of compact Riemann surfaces of genus g according to Baily's construction [Ba] as a Zariski–open subset of a complex projective algebraic subvariety defined over $\mathbb{Q}$ of the (Satake) compactified Siegel quotient space. By taking the normalized period matrix of C or its Jacobian in the Siegel upper half space we get e as the image point in $\mathbb{P}_N(\mathbb{C})$ under the quotient map given by an $(N+1)$–tuple of theta functions of equal weight and multiplier system. In the terminology of moduli fields and fields of definition, we can reformulate the Lemmas 2 and 3 respectively as

Lemma 6. *The moduli field of a Belyi surface is a number field.*

Lemma 7. *Let C be a nonsingular projective algebraic curve defined over a subfield of $\mathbb{C}$. If the moduli field $M(C)$ is a number field, the field of definition can be chosen as a number field, too.*

With Lemma 7, another reformulation of Belyi's Theorem 1 follows immediately:

Theorem 3. *A compact Riemann surface is a Belyi surface if and only if its moduli field is a number field.*

Apparently, the relation between the field of moduli and the fields of definition is crucial. In this direction it is known that

- the field of moduli is a field of definition in the case of curves of genus 0 and 1 (see the beginning of the proof of Lemma 3),

- for hyperelliptic curves C of even genus, $M(C)$ is in general not a field of definition for C (Shimura [Sh]). Another class of such examples for all genera $g > 1$, $g \equiv 1 \bmod 4$, has been found by Earle [Ea].
- For curves C of genus $g > 1$ the field of moduli is a field of definition for the quotient curve $C/\text{Aut}C$ ([Ba], [DE]),
- in particular, the field of moduli is a field of definition if the curve has only the trivial automorphism (this follows directly from [We] as well).

The reader will find further properties in [DE].— An easy generalization of Lemma 7 is the

Theorem 4. *Let C be a nonsingular projective algebraic curve defined over a subfield of $\mathbb{C}$. Then there exists a field of definition for C which is a finite algebraic extension of the moduli field $M(C)$.*

For the proof one has to replace $\overline{\mathbb{Q}} = \overline{M(Y)}$ by the algebraic closure $\overline{M(C)}$ everywhere in the proof of Lemma 3 and 4. Observe that $M(C) (= \mathbb{Q}(e)$, $e \in M_g)$ is finitely generated over $\mathbb{Q}$ and that the fixing subgroup $U \subset \text{Aut}\mathbb{C}$ is still large enough to perform all the steps of the proof.

Remark 2. In the case of Belyi surfaces ($\iff \overline{M(X)} = \overline{\mathbb{Q}}$) one can replace $G = \text{Aut}\mathbb{C}$ by the absolute Galois group $G = \text{Gal}\overline{\mathbb{Q}}/\mathbb{Q}$ without changing $M(X)$ or its G–orbit. In the literature, moduli fields and fields of definition are often considered as fields between a fixed ground field and its algebraic closure, and in this 'relative' situation the fields of moduli may be of course larger than our $M(X)$, see also [DE].

Remark 3. In the context of Belyi functions, the term *moduli field* often means the analogously defined moduli fields $M(X, \beta)$ for *Belyi pairs* (X, β) where β is a specific Belyi function on the Belyi surface X (see e.g. [CG]) or Belyi pairs with a certain marked point (X, β, x_0). In the latter case, the field of moduli is always a field of definition (by an argument of [DE], this follows from [CH]; see also the proof of [Bi], Theorem 2). Using these arguments it is tacitly assumed that β takes the values $0, 1$ or ∞ at these marked points; this means in particular that the base space of the covering defined by β consists of the complex points of a projective line $\mathbb{P}_1$ *defined over $\mathbb{Q}$ and with $\mathbb{Q}$–rational points*. Clearly, the field of moduli $M(X, \beta, x_0)$ may be larger than $M(X)$ or even $M(X, \beta)$. In any case, the action of G on the isomorphism classes of Belyi surfaces or Belyi pairs with marked points is equivariant to the action on the respective moduli fields, not on the fields of definition.

We will see below that this restriction does not apply to an important special class of Riemann surfaces to be introduced now.

§3. Belyi surfaces with many automorphisms

A compact Riemann surface X of genus $g > 1$ is said to have *many automorphisms* if the corresponding point $e = p(X)$ on the moduli space M_g of compact Riemann surfaces of genus g has (in the complex topology) a neighbourhood $U \subset M_g$ with the following property: For any $q \in U$, $q \neq p$, the corresponding Riemann surface $Y(q)$ has an automorphism group strictly smaller than that of X. In other words, the number of automorphisms strictly decreases under proper deformations of X. Riemann surfaces with many automorphisms have several interesting characterizations [Po], e.g. as isolated singularities of the moduli space M_g if $g > 3$; we can add the following one:

Lemma 8. *The compact Riemann surface X of genus $g > 1$ has many automorphisms if and only if the normalizer Δ of its universal covering group $\Gamma \subset PSL_2\mathbb{R}$ is a cocompact Fuchsian triangle group. In particular, X is a Belyi surface with the canonical projection*

$$\Gamma\backslash\mathcal{H} \to \Delta\backslash\mathcal{H}$$

as a Belyi function β. For the moduli fields of X and of the Belyi pair (X, β) one has

$$M(X) = M(X, \beta).$$

For the proof observe that any automorphism of X lifts to an automorphism of the universal covering space $\mathcal{H}$. The lifted automorphisms form a group $\Delta \subset PSL_2\mathbb{R}$ containing the universal covering group Γ of X as a normal subgroup. On the other hand, any $\sigma \in PSL_2\mathbb{R}$ normalizing Γ induces an automorphism of $X \cong \Gamma\backslash\mathcal{H}$. Therefore, the automorphism group A of X is isomorphic to Δ/Γ where Δ is the normalizer of Γ in $PSL_2\mathbb{R}$. By the assumption on g and the compactness of X both Γ and Δ are cocompact. Now a small deformation of X into another Riemann surface Y (or more precisely their corresponding points in M_g) lifts to a deformation of Γ into the universal covering group Ω of Y using an appropriate (Chabauty-) topology on the space of Fuchsian groups of constant signature and its continuous projection to M_g; for the precise definition of this deformation space, its projection onto the moduli space and the existence of local sections for this projection see [Ha]. The order of the automorphism group of X is preserved under this deformation if and only if Δ can be deformed to a group Σ containing Ω as a normal subgroup with

$$\Delta/\Gamma \cong \Sigma/\Omega.$$

Proper deformations of X correspond to deformations of Γ (hence of Δ as well) not induced by $PSL_2\mathbb{R}$–conjugations which exist for Fuchsian groups of all signatures with the only exception of the triangle groups. So the characterization of the triangle groups by their rigidity property implies the characterization of Riemann surfaces with many automorphisms by triangle groups. Since the canonical projection β is uniquely determined by X and its automorphism group (which is Galois–invariant e.g. by [JSt]) the equality for the moduli fields follows. Using Theorem 2 we get moreover

Theorem 5. *The moduli field of a compact Riemann surface X of genus $g > 1$ with many automorphisms is a number field. In particular, X is isomorphic to the Riemann surface $C(\mathbb{C})$ consisting of the complex points of a nonsingular projective algebraic curve C defined over a number field.*

According to the above mentioned result of Baily, this theorem has also an interpretation in terms of the coordinates of the corresponding point $e = p(X) \in M_g$. In particular, as in the work of Rauch [Rau] and Popp [Po] one can read the (reduced) automorphism group of X as a subgroup of the Siegel modular group giving the algebraicity of the ratios of the coordinates of e as a consequence of theta relations.

Coming back to the first subject of this note, Lemma 8 gives the 'only if' part of

Theorem 6. *A compact Riemann surface X of genus $g > 1$ has many automorphisms if and only if there exists a Belyi function β defining a normal covering*

$$\beta : X \to \mathbb{P}_1(\mathbb{C}) .$$

To prove the 'if' part remember that if β defines a normal covering, it has the same ramification order p in every point of $\beta^{-1}\{0\}$, and as well constant ramification orders q, r above $1, \infty$ respectively. In Lemma 1 we can therefore take Δ with signature $< p, q, r >$ and obtain a representation of X as quotient space $\Gamma\backslash\mathcal{H}$ with a torsion–free normal subgroup Γ of Δ. By consequence, Γ is the universal covering group of X and Δ/Γ is the covering group of β, hence an automorphism group of X. The only missing point is that we do not know that Δ is the normalizer of Γ. But the normalizer at least contains Δ and is itself a Fuchsian group; it is well known that these properties imply that the normalizer is again a triangle group whence X has many automorphisms.

It should be noted that there are examples of curves with many automorphisms whose covering groups are normal subgroups of different triangle groups: the Fermat curves of exponent $n > 3$ have as universal covering groups the commutator subgroup $\Gamma = [\Delta, \Delta]$ of the triangle group Δ with signature $< n, n, n >$, but Γ is normal also in the triangle groups with signatures

$$< 3, 3, n > , \quad < 2, n, 2n > \quad \text{and} \quad < 2, 3, 2n > \, .$$

The last one is the normalizer whose quotient by Γ gives the full automorphism group.

Remark 4. In general, it is reasonable to consider Belyi pairs and their moduli fields instead of Belyi surfaces because Belyi functions are by no means unique. The situation is very different for Riemann surfaces with many automorphisms where the quotient map by the automorphism group of X defines precisely one 'canonical' Belyi function (Lemma 8). According to Pierre Debes and Michel Emsalem [DE], Theorem 6 together with results of Coombes and Harbater [CH] implies the following simplification. As in the opposite case with trivial automorphism group

- the moduli field is a field of definition if X has many automorphisms.

Readers with experience in this field may reformulate the statement of Theorem 6 in terms of regular dessins, maps or hypermaps, see [BI] and [JS]. It should be mentioned that M. Streit in his recent Ph.D. thesis ([St1], [St2]) developed a very clever method for the determination of equations for such curves representing Riemann surfaces with many automorphisms. According to Remark 2, we can assume G to be $\mathrm{Gal}\overline{\mathbb{Q}}/\mathbb{Q}$ if we consider its action on the set of these curves defined over number fields or equivalently on $M_g(\overline{\mathbb{Q}})$, the algebraic points of the moduli space (or on the corresponding regular maps, hypermaps, dessins, drawings, ...). For this action, the Riemann surfaces with many automorphisms are of particular interest because the action of G on arbitrary Belyi pairs (Y, β) can be traced back to the action on the set of Riemann surfaces with many automorphisms: normalizing the cover map β and applying Theorem 6 and Lemma 8 we get the first part of

Theorem 7. *Let* (Y, β) *be any Belyi pair. If its normal cover* X *has genus* $g(X) > 1$*, then* X *has many automorphisms and* β *is a factor of the canonical Belyi function*

$$X \rightarrow \mathbb{P}_1(\mathbb{C}) \, .$$

For every $\sigma \in G$*, the Riemann surface* X^σ *is the corresponding cover for* (Y^σ, β^σ)*.*

Remark 5. In a few cases (genus $g(X) = 0$ or 1 and some special ramification orders of β) the normal cover X has genus $g(X) = 0$ or 1. The genus 0 case can be treated by means of the finite spherical triangle groups of H.A. Schwarz' list, and in the euclidean genus 1 case one may extend the notion 'many automorphisms' dividing out the obvious automorphisms coming from the translations on $\mathbb{C}$. However, a canonical Belyi function does not exist in this case: the euclidean covering group Γ of X is contained in infinitely many groups all isomorphic to one Δ (of signature $< 2, 3, 6 >$ or $< 2, 4, 4 >$). In a recent contribution of Singerman and Syddall [SS] these elliptic curves are treated in detail.

For another proof of Theorem 7 and another application of this principle to replace arbitrary Belyi surfaces by those with many automorphisms, see [Wo2].— The genus g and the automorphism group A are invariant under the action of G (see e.g. [JSt]) and as well the ramification orders of the Belyi function β in Theorem 6. Since the length of the G-orbit coincides with the degree of the moduli field of these curves (Remark 3) one obtains the following upper bound.

Theorem 8. *The degree of the moduli field for the curve C in Theorem 5 is bounded by the maximal number $n(\Delta, A)$ of homomorphisms h of the triangle group Δ normalizing the universal covering group Γ (see Lemma 8) onto the automorphism group A such that the kernels of h are torsion-free Fuchsian groups not conjugated in $PSL_2\mathbb{R}$. By Remark 4, $n(\Delta, A)$ is also an upper bound for the degree of the minimal field of definition for C.*

(With some additional translation work it should be possible to derive this result also from bounds found by Matzat [Ma], Kap.II, õõ1 and 2.) If Δ is a maximal triangle group, the condition 'not conjugated in $PSL_2\mathbb{R}$' simply means that the kernels are pairwise different or equivalently that different homomorphisms h cannot be obtained from each other by combination with automorphisms of A. Note that the genus of all these curves $\ker h \backslash \mathcal{H}$ is uniquely determined by the order of A and the signature $< p, q, r >$ of Δ as

$$g = 1 + \frac{|A|}{2}\left(1 - \frac{1}{p} - \frac{1}{q} - \frac{1}{r}\right).$$

The bounds $n = n(\Delta, A)$ are explicitly known for many Δ and A and turn out to be relatively small in comparison with the genus.

Examples. As far as I know, the first systematic results in this direction are the representations of the fractional linear groups $A = PSL_2(q)$ as *Hurwitz groups*, i.e. as homomorphic images of Δ with signature $< 2, 3, 7 >$ found by Macbeath [Ma]. He determined the upper bound n of Theorem 8 for all possible Δ-images $PSL_2(q)$:

$$\begin{cases} n = 1 & \text{for } q = 7 \text{ (Klein's quartic, } g = 3) \\ n = 1 & \text{for } q = p^3, \ p \equiv \pm 2, \pm 3 \bmod 7, \ p \text{ prime}, \\ n = 3 & \text{for } q = p \equiv \pm 1 \bmod 7, \ p \text{ prime}. \end{cases}$$

This means that for $q = 7$, $8\,(g = 7)$, $27\,(g = 118)$ the (unique) curves C can be defined over $\mathbb{Q}$, and that both for $q = 13\,(g = 14)$ and $q = 29\,(g = 146)$ each of the three curves can be defined over an at most cubic field. For a long list of references to more recent work on $n(\Delta, A)$ see [J]. Jones' (and Silver's) work on Ree groups and Suzuki groups [J], [JSi] gives even a weak but interesting lower bound for the degree of the field of moduli of the curves with these automorphism groups: In these cases, the respective maps occur in *chiral pairs*, i.e. antiholomorphic involutions give non–isomorphic curves. Therefore, the curves have non–real moduli fields and a fortiori, they cannot be defined over real number fields.

References

[Ba]   W.L.Baily, On the theory of theta functions, the moduli of abelian varieties and the moduli of curves, *Ann. of Math.* **75** (1962), 342–381.

[BI]   M.Bauer, C.Itzykson, Triangulations, pp.179–236 in *The Grothendieck Theory of Dessins d'Enfants*, LMS Lecture Note Series 200, ed. L. Schneps, Cambridge University Press, 1994.

[Be]   G.Belyï, On Galois extensions of a maximal cyclotomic field, *Math. USSR Izv.* **14**, No.2 (1980), 247–256.

[Bi]   B.Birch, Noncongruence Subgroups, Covers and Drawings, pp. 25–46 in *The Grothendieck Theory of Dessins d'Enfants*, LMS Lecture Note Series 200, ed. L. Schneps, Cambridge University Press, 1994.

[CG]   J.-M.Couveignes, L.Granboulan, Dessins from a geometric point of view, pp. 79–113 in *The Grothendieck Theory of Dessins d'Enfants*, LMS Lecture Note Series 200, ed. L. Schneps, Cambridge University Press, 1994.

[CH]   K. Coombes, D. Harbater, Hurwitz families and arithmetic Galois groups, *Duke Math. J.* **52** (1985), 821–839.

[CIW] P.Beazley Cohen, C. Itzykson, J.Wolfart, Fuchsian Triangle Groups and Grothendieck Dessins. Variations on a Theme of Belyi, *Commun. Math. Phys.* **163** (1994), 605–627.

[CW]   P.Beazley Cohen, J.Wolfart, Dessins de Grothendieck et variétés de Shimura, *C.R. Acad. Sci. Paris* **315**, Sér. I (1992), 1025–1028.

[DE]   P.Debes, M.Emsalem, On Fields of Moduli of Curves, preprint.

[Ea]   C.J.Earle, On the Moduli of Closed Riemann Surfaces with Symmetries, pp. 119–130 in L.V.Ahlfors et al.: Advances in the Theory of Riemann Surfaces, *Ann. of Math. Studies* **66**, Princeton 1971.

[Fr]   M.D.Fried, Fields of Definition of Function Fields and Hurwitz Families — Groups as Galois Groups, *Comm. Alg.* **5** (1977), 17–82.

[Ha]   W.J.Harvey, Spaces of discrete groups, pp.295–348 in W.J.Harvey, Discrete groups and automorphic functions, Proc. Univ. Cambridge 1975, *Academic Press* 1977.

[J]   G.A.Jones, Ree Groups and Riemann Surfaces, *J. of Algebra* **165** (1994), 41–62.

[JSi]   G.A.Jones, S.A.Silver, Suzuki Groups and Surfaces, *J. London Math. Soc. (2)* **48** (1993), 117–125.

[JS]   G.A.Jones, D.Singerman, Maps, Hypermaps and Triangle Groups, pp. 115–145 in *The Grothendieck Theory of Dessins d'Enfants*, LMS Lecture Note Series 200, ed. L. Schneps, Cambridge University Press, 1994.

[JSt]   G.A.Jones, M.Streit, Galois Groups, Monodromy Groups and Cartographic Groups, to appear in *Geometric Galois Actions II.*

[Mac]   A.M.Macbeath, Generators of the linear fractional groups, pp. 14–32 in W.J.Leveque, E.G Straus: Number Theory, Proc. Sympos. in Pure Math. **12**, *Amer. Math. Soc.* 1969.

[MKS]   W. Magnus, A. Karrass, D. Solitar, Combinatorial Group Theory, *Dover* 1976.

[Ma]   B.H.Matzat, Konstruktive Galoistheorie, *LNM* **1284**, *Springer Verlag* 1987.

[Po]   H.Popp, On a conjecture of H. Rauch on theta constants and Riemann surfaces with many automorphisms, *J. Reine Angew. Math.* **253** (1972), 66–77.

[Rau]   H.E.Rauch, Theta constants on a Riemann surface with many automorphisms. Symposia Mathematica **III**, 305–322, *Academic Press* 1970.

[Sh]   G.Shimura, On the Field of Rationality for an Abelian Variety, *Nagoya Math. J.* **45** (1972), 167–178.

[ST]   G.Shimura, Y.Taniyama, Complex multiplication of abelian varieties and its applications to number theory. *Publ. Math. Soc. Japan* **6** (1961).

[SS]   D.Singerman, R.I.Syddall, Belyï Uniformization of Elliptic Curves, preprint.

[St1]   M.Streit, Darstellungstheorie für Hypermaps und kanonisches Modell algebraischer Kurven, Dissertation, Frankfurt 1995.

112   Jürgen Wolfart

[St2] M.Streit, Homology, Belyï Functions and Canonical Curves, *Manuscr. Math.* **90** (1996), 489–509.

[We]   A.Weil, The field of definition of a variety, *Amer. J. Math.* **78** (1956), 509–524.

[Wo1] J.Wolfart, Mirror-invariant triangulations of Riemann surfaces, triangle groups and Grothendieck dessins: Variations on a theme of Belyi, preprint Frankfurt 1992.

[Wo2] J.Wolfart, Triangle groups and Jacobians of CM type, in preparation.

[Za]   U.Zannier, On Davenport's bound for the degree of $f^3 - g^2$ and Riemann's Existence Theorem, *Acta Arithmetica* **72** (1995), 107–137.

Mathematisches Seminar der J.W.Goethe Universität
Robert–Mayer–Str. 6–10
D-60054 Frankfurt a.M.
e-mail: wolfart@math.uni-frankfurt.de

Glimpses of Grothendieck's Anabelian Geometry

Florian Pop

§0. Introduction

The idea of Grothendieck's anabelian geometry is that under certain "anabelian" hypotheses geometry and arithmetic of schemes are encoded in their étale fundamental group. More precisely, let S be some connected base scheme, and $\mathcal{A}_S$ a category of connected S-schemes and isomorphisms of S-schemes. Let $\mathcal{G}_S$ be the category of the pro-finite groups with augmentation $G \to \pi_1(S)$ to the (étale) fundamental group $\pi_1(S)$ of S and outer $\pi_1(S)$-isomorphisms. (Here, a morphism between two objects $\varphi : G \to \pi_1(S)$ and $\varpi : H \to \pi_1(S)$ in $\mathcal{G}_S$ is the following concept: First, a $\pi_1(S)$-isomorphism of φ and ϖ is a pro-finite group isomorphism $\Phi : G \to H$ such that $\varpi \Phi$ and φ are conjugated by some element of $\pi_1(S)$. It is clear that $\mathrm{Inn}(G)$ and $\mathrm{Inn}(H)$ act by composition from the left, respectively right, on the space of all $\pi_1(S)$-homomorphisms from G to H, and one has $\mathrm{Inn}(G) \circ \Phi = \widetilde{\Phi} = \Phi \circ \mathrm{Inn}(H)$. By definition, a $\mathcal{G}_S$-morphism is a class $\widetilde{\Phi}$ as defined/introduced above.) There exists a covariant functor

$$\mathcal{A}_S \to \mathcal{G}_S \quad \text{by} \quad (X \to S) \mapsto \big(\pi_1(X) \to \pi_1(S)\big).$$

We say that $\mathcal{A}_S$ is S-*anabelian* if for every pair of objects X and Y of $\mathcal{A}_S$ the canonical mapping

$$\mathrm{Hom}_{\mathcal{A}_S}(X, Y) \cong \mathrm{Hom}_{\mathcal{G}_S}\big(\pi_1(X), \pi_1(Y)\big)$$

is a bijection. We will say that an S-scheme X is S-anabelian if the category consisting of X itself endowed with all the S-automorphisms of X is an anabelian category. In the case $S = \mathbb{Z}$ we will simply speak about anabelian objects, hence not making any reference to the base.

Such ideas are not completely new, and a first approximation of this kind of development could be the following:

In the middle Twenties, Artin–Schreier showed that the hypothesis of being *finite* is a very restrictive one for the absolute Galois group. More precisely they showed:

Theorem. *(Artin–Schreier) Let K be a field with G_K finite and $\neq 1$. Then $G_K \cong G_{\mathbb{R}}$. Moreover, K has a total ordering, hence it is real closed.*

It seems that a similar question for the *p-adics* goes back to Krull. The answer in the *p*-adic case is also known, but the proofs are much more difficult:

Theorem. *(Neukirch, Pop, Efrat, Koenigsmann) Let $\mathbb{K}|\mathbb{Q}_p$ be a finite extension, and K be a field which has $G_K \cong G_{\mathbb{K}}$. Then K is p-adically closed.*

Let us explain a quite unexpected consequence of the above theorem for the Galois theory of global fields as initiated by Neukirch in the late Sixties. Let K and L be number fields, and $\Phi : G_K \to G_L$ an isomorphism. By the *p*-adic Artin–Schreier Theorem above (proved by Neukirch for subfields of $\overline{\mathbb{Q}}$), it follows that Φ maps decomposition groups of places of K onto decomposition groups of places of L. Thus, the isomorphism Φ gives rise functorially to a bijection $\varphi : \mathbb{P}(K) \to \mathbb{P}(L)$ between the spaces of places of K and L, a bijection which preserves the local ramification and inertia numerical invariants. The same is true also for global function fields. Let's call this fact the "local theory" for global fields.

In particular, in the number field case, Φ defines via φ an *arithmetical equivalence* of L and K. Thus, by classical results in number theory, which nevertheless involves *analytic methods*, Neukirch got the following:

Under the supplementary hypothesis that K is normal over $\mathbb{Q}$ it follows that $K \cong L$. In particular, every normal subgroup of $G_{\mathbb{Q}}$ is a characteristic subgroup.

The following questions were then at hand:

1) $G_{\mathbb{Q}}$ has only inner automorphisms (hence in particular, every normal subgroup is *a priori* characteristic).

2) If K, L are number fields with $G_K \cong G_L$ then $K \cong L$.

The first question was answered in positive by Ikeda, and finally, Iwasawa (unpublished) and Uchida gave a spectacular answer to both questions above by proving an equivalent form of the fact that the category of all number fields is, in our terminology above, *anabelian*.

Shortly after, Uchida proved the same even for the category of all global fields.

This is a highly non-trivial and deep fact!...

§1. Conjectures/Some Results

We now come, following Grothendieck, to discussing the question of *which interesting categories* of arithmetic schemes are anabelian. By the remarks above we already know, although it is unclear whether Grothendieck was aware of this fact, that the category of all global fields is anabelian.

First, a typical source of anabelian objects should be: Let X be a scheme of finite type over $\mathbb{Z}$ or more special, over a finitely generated field. Then **conjecturally** *every regular point of X has a basis of Zariski open subsets which are anabelian objects.*

There are clear conjectures of what the ≤ 1 dimensional anabelian objects should be only, but there is no clear idea about what the higher dimensional anabelian objects could/should be.

Conjecture 0.
(0-dimensional anabelian geometry: Birational anabelian geometry)
The category of all finitely generated infinite fields and field isomorphisms is anabelian, i.e., if K and L are such fields then the map

$$\operatorname{Isom}(L^{\mathrm{i}}, K^{\mathrm{i}}) \longrightarrow \operatorname{Out}(G_K, G_L)$$

is a bijection, where the superscript ^{i} *means inseparable closure.*

Conjecture 1.
(1-dimensional anabelian geometry: Anabelian curves)
The category of all hyperbolic curves over finitely generated fields and scheme isomorphisms is anabelian, i.e., if X and Y are such curves, then there exists a functorial bijection

$$\operatorname{Isom}(X^{\mathrm{i}}, Y^{\mathrm{i}}) \cong \operatorname{Out}\big(\pi_1(X), \pi_1(Y)\big)$$

where ^{i} *means normalization in the inseparable closure of the field of rational functions.*

Looking at the structure of the geometric fundamental group of a hyperbolic curve in the situation from Conjecture 1 above, it follows that it is actually encoded in the fundamental group of the curve. Hence one could combine the two conjectures above by saying that *the category of all finitely generated infinite fields and hyperbolic curves over finitely generated fields is an anabelian category.*

Unfortunately, it is not clear what **Conjecture** d should be. A first observation is that for $d \geq 2$ the dimension of the variety in discussion is in general not encoded in the fundamental group. Indeed, by "general" hyper-plane sections arguments, one can find fundamental groups of surfaces as "representatives" for every fundamental group isomorphy type. Thus, the situation above where one has a "dimension ≤ 1 anabelian conjecture" is very special, and one cannot expect similar conjectures in higher dimensions. In particular, for higher dimensional anabelian conjectures, d must be part of the hypothesis.

One expects that the *moduli spaces* $M_{g,n}$ of n-pointed smooth genus g curves, as well as the (category of all) *Artin neighborhoods* with hyperbolic curves as fibers are anabelian (in particular, the dimension of an Artin neighborhood should be encoded in the fundamental group).

We next come to give a short presentation of the known results as well as, to the best of our knowledge, the contributors.

Theorem 0. *The Conjecture 0 above is true.*

As we saw above, the proof of Theorem 0 in the global field case evolved in the middle of the Seventies from the work of Neukirch, Ikeda, Iwasawa, Uchida. The main tools were Neukirch's "local theory" for global fields, and the theory of arithmetical equivalence of number fields, respectively the global class field theory in the global function field case.

The next important step in this direction was the solution to Conjecture 0 above in the case of arithmetical function fields, i.e., function fields of one variable over number fields. This was done by Pop [P1] in the middle of the Eighties, and M. Spiess [Sp] gave a new proof in the middle of the Nineties.

Finally, the general case was settled in the middle of the Nineties by Pop [P3].

We want also to draw the reader's attention to a quite ambitious approach by Bogomolov [B]. Namely he works in the following context: Let κ be an algebraically closed field, K a function field of at least 2 variables over κ, and ℓ a prime number $\neq \operatorname{char}(\kappa)$. Let $G_K^{(\ell)}$ be the pro-ℓ quotient of G_K and $G_K^{(\ell)}(3)$ its third term in the lower central series. Then K should be encoded in the quotient $G_K^{(\ell)}/G_K^{(\ell)}(3)$. See the concluding remarks for some information on this.

We next come to the known results concerning the anabelian curves.

Theorem 1'. *(Tamagawa) Conjecture 1 is true in the affine case.*

This is a result of A. Tamagawa [T]. He actually proved the conjecture for *affine, hyperbolic curves over finite fields*, and from this he deduced Conjecture 1 in the *characteristic zero case*.

Special cases of the above Theorem, at least in the *weak form* were known before: First, Nakamura [N1] showed (beginning of the Nineties) that the isomorphy type of a punctured projective line over a finitely generated field of characteristic zero (at least two points erased) is encoded in its fundamental group, i.e., we can read off the erased points from the fundamental group. Further, Pop (unpublished) remarked that every smooth point of a curve over a finitely generated field has a basis of neighborhoods whose isomorphy type can be read off from their fundamental group. (This can be deduced from the birational conjecture for function fields of one variable over finitely generated fields and Faltings's proof of the Shafarevich conjecture.)

Theorem 1″. *(Mochizuki) Conjecture 1 is true in the characteristic zero case.*

This is a result of Mochizuki [M2]. Actually he proves a much stronger result from which one can deduce the Theorem above, as well as the birational version for function fields of one variable over p-adic fields; see also the discussion in the last subsection.

§2. Outline of the Proofs

Proof of Conjecture 0

First, since the absolute Galois group of finitely generated infinite fields K and L has a trivial center, one checks without pain that Conjecture 0 is equivalent to the following:

For every isomorphism $\Phi : G_K \to G_L$ there exists a unique field isomorphism $\phi' : \overline{L} \to \overline{K}$ which defines the given Φ by the rule $\Phi(g) = \phi'^{-1} g \phi'$.

The general idea of the proof is to first find an isomorphism $\phi : L \to K$ and then, hoping that everything is functorial, to show the uniqueness property of ϕ'.

The first step in the proof is to develop a *local theory* which should generalize the "local theory" developed for global fields as we have already mentioned above.

As in the global field case, the "local theory" for general finitely generated fields has the aim of recovering information of local nature from the Galois information. The idea is the following: Let K be a finitely generated infinite field, and $X \to \mathbb{Z}$ be a model of K, i.e, some normal, separated, reduced scheme of finite type over the integers having function field isomorphic to K. We define the Kronecker dimension $d = \mathrm{Kr.dim}(K)$ of K to be the dimension of X. (N.B., if $\mathrm{cd}_\ell(K)$ denotes the ℓ-cohomological dimension of K then $\mathrm{Kr.dim}(K) = \mathrm{cd}_\ell(K) - 1$ for every ℓ different from the characteristic of K. Hence, if K, L are finitely generated infinite fields with $G_K \cong G_L$ then $\mathrm{Kr.dim}(K) = \mathrm{Kr.dim}(L)$.) For every $1 \leq k \leq d$ let X^k be the set of points of X of codimension k. Hence, X^1 is the space of prime Weil divisors of X, and X^d is the space of closed points of X. The aim of a local theory is to "describe"the spaces X^k in Galois terms, or to recognize them in G_K. Unfortunately, *this is not possible!* for the simple reason that these spaces are not birational invariants of X, whereas all the information in G_K is of birational nature.

To overcome this difficulty we consider the family of all proper models X_i of K. It builds a projective system. We denote by $\mathfrak{X}_K$ the projective limit of this projective system, and by $\mathfrak{X}_K^k$ the spaces of points of $\mathfrak{X}_K$ of codimension k. (By definition, a point $\mathfrak{x} = (x_i)$ of $\mathfrak{X}_K$ has codimension k is for all sufficiently large i the point x_i has codimension k in X_i. N.B., the codimension of the points x_i is a decreasing function of i). Clearly, by their construction, the spaces $\mathfrak{X}_K^k$ are birational invariants of X.

The birational interpretation of $\mathfrak{X}_K^k$ is the following: $\mathfrak{X}_K^k$ is the space of all valuations v^k on K having $\mathrm{Kr.dim}\big(\kappa(v^k)\big) = \mathrm{Kr.dim}(K) - k$, where $\kappa(v^k)$ is the residue field of v^k. In particular, $\mathfrak{X}_K^1$ is the space of all Zariski prime divisors of K.

The main step in the local theory is that the space of Zariski prime divisors $\mathfrak{X}_K^1$ defined above is encoded in G_K, hence the isomorphism Φ defines a functorial bijection

$$\varphi^1 : \mathfrak{X}_K^1 \to \mathfrak{Y}_L^1$$

between the spaces of Zariski prime divisors of K and L. The first proof of this (by Pop) used *local-global principles* of Hasse type for cohomology groups of finitely generated fields, see Hasse–Brauer–Noether, Tate, Roquette, Lichtenbaum, Saito, Kato, Jannsen. Meanwhile there exist other, I would say more elementary approaches by Bogomolov, respectively Koenigsmann, which are in progress.

The next step in the proof is to gain "more geometry" from the Galois group. In order to do that let us call a set D of Zariski prime divisors *geometric* if there exists a quasi-projective, normal model of K such that D is the set of prime Weil divisors of X. The essential fact is now that the local bijection φ^1 maps geometric sets of prime divisors onto geometric sets of prime divisors. In the course of the proof one needs the existence of *regular, complete models* for finitely generated fields over a base field. This is known to exist in characteristic zero, by Hironaka, but not known to exist in positive characteristic. One can nevertheless get through by just using the weaker form of "desingularizing" by alterations, as done by A. de Jong.

Finally, by induction on the $d = \mathrm{Kr.dim}$ (N.B. the residue fields of Zariski prime divisors have Kr.dim equal to $d-1$), and using in a quite technical way the inductive hypothesis, one shows that Φ gives rise via Kummer Theory to a multiplicative isomorphism ϕ of $L^\times$ onto $K^\times$, which finally turns out to be also additive.

Proof of Conjecture 1 (the known cases)

Since the fundamental group of a hyperbolic curve has trivial center, we can reformulate Conjecture 1 as in the birational case as follows:

For every isomorphism $\Phi : \pi_1(X) \to \pi_1(Y)$ there exists a unique isomorphism of schemes $\phi' : X^{i'} \to Y^{i'}$ which defines the given Φ by the rule $\Phi(g) = \phi'^{-1} g \phi'$, where $^{i'}$ denotes the universal cover of the normalization in the maximal pure inseparable extension of the function field.

Let κ be an arbitrary field, and X a hyperbolic curve over κ, hence a smooth, geometrically integral curve with non-commutative geometric fundamental group. Let $\overline{X} = X \times \overline{\kappa}$ denote the base change to the algebraic closure of κ. Finally let Z be the smooth completion of X (which is defined over some pure inseparable extension of the base field), and denote by g the genus of Z and by r the cardinality of $\overline{Z}\backslash\overline{X}$.

Choosing a geometric point of X (which we are not going to specify), one gets an exact sequence

$$1 \to \pi_1(\overline{X}) \to \pi_1(X) \to G_\kappa \to 1.$$

Starting with a κ-rational point of Z one can construct a section of $\pi_1(X) \to G_\kappa$. Moreover, if the κ-rational point to start with lies in X, then the conjugacy class of the constructed section depends functorially on the point. Nevertheless, **it is one of the main open questions of the theory** to describe the sections s of $p : \pi_1(X) \to G_\kappa$ which are defined by κ-rational points of the smooth completion Z of X. This represents the *regular* correspondent of the birational "local theory" described above.

At this place one should mention a result of Nakamura [Na1]. Namely using the weight filtration on the π_1^{ab} with respect to Frobenius elements, as introduced by Deligne, he showed the following:

In the above situation let κ be finitely generated over $\mathbb{Q}$. Let C be a non-trivial procyclic subgroup of $\pi_1(\overline{X})$ such that there exists a section s of p

with $s(G_\kappa)$ normalizing C and acting on it via the cyclotomic character of G_κ. Then C is contained in an inertia group at a κ-rational point of the smooth completion Z of X.

In other words, if s and C are as above, then $s(G_\kappa) \cdot C$ is contained in a decomposition group at a κ-rational point in $Z \backslash X$.

Clearly, conjugated sections define the same κ-rational point of Z, but there are "many" non-conjugated sections defining the same κ-rational point of $Z \backslash X$. In order to classify the latter conjugacy classes one can use the more sophisticated notion of *tangential base point* as defined/introduced by Deligne [D3], Ihara [I2].

a) *Tamagawa's Result*

The main novelty Tamagawa brings into the game is to develop a good *regular local theory* (in the sense explained above) for smooth, hyperbolic curves over *finite fields*. Namely, with the notations from above he first remarks that given a section s of p (in the case of a finite base field κ) the fact that s is defined by a κ-rational point of Z can be characterized as follows:

1) First, for every open subgroup H containing $s(G_\kappa)$, let Z_H be the cover of Z which is defined by H (N.B., Z_H is defined over κ, and it is geometrically integral). Then one has: s is defined by a κ rational point if and only if Z_H has κ-rational points for all H as above.

2) Second, the fact that Z_H has κ-rational points can be read off from the Galois cohomology of H, more precisely, using the Lefschetz trace formula for the action of the Frobenius of κ on the abelian quotient $\overline{H}^{\mathrm{ab}}$ of the geometric part $\overline{H}$ of H, etc.

Nevertheless, in order to do that one has first to discover the various numerical invariants of X and Z in the fundamental group, in particular, to single out the Frobenius of κ in G_κ. In this process Tamagawa answers, among other things, a question by Harbater [Ha], 4.2.

The next step in the proof is to use, à la Uchida, the class field for X and get in this way the multiplicative group of the function field $K = \kappa(X)$ of X together with the divisors of all the functions. The major difficulty left is then to gain the additive structure, a step which is much more difficult than in the global function field case (as done by Uchida [U1]).

One finally gets an isomorphism $\phi^i : X^i \to Y^i$ which defines the isomorphism $\Phi : \pi_1(X) \to \pi_1(Y)$ in the way indicated at the beginning of this Section.

Let us now consider the general case of hyperbolic curves $X \to \kappa$ and $Y \to \lambda$ over some finitely generated infinite base fields κ and λ. One first shows that every isomorphism $\Phi : \pi_1(X) \to \pi_1(Y)$ is compatible with the projections to $\pi_1(X) \to G_\kappa$ and $\pi_1(Y) \to G_\lambda$, hence it induces and isomorphisms $\tilde{\Phi} : G_\kappa \to G_\lambda$. By Theorem 0 we can suppose that $\kappa = \lambda$. We consider a model S over $\mathbb{Z}$ for κ, and curves $\mathcal{X}$ and $\mathcal{Y}$ over S having X, respectively Y as generic fibers. Replacing S by a purely inseparable cover of it, and then by some affine open sub-scheme if necessary, we can finally assume that both $\mathcal{X}$ and $\mathcal{Y}$ are smooth S-curves. By the above result of Tamagawa it follows that $\mathcal{X}$ and $\mathcal{Y}$ have isomorphic special fibers at all the closed points of S. Using globalization techniques it follows that X and Y are functorially isomorphic, and the corresponding isomorphism induces the isomorphism Φ in the desired way.

b) *Mochizuki's Result*

In the general context above let the base field k be a finite extension of $\mathbb{Q}_p$. Let $\varphi : \pi_1(\overline{X}) \to \overline{\Delta}_X$ denote the maximal p-quotient of $\pi_1(\overline{X})$. Then $\ker(\varphi)$ is a characteristic subgroup of $\pi_1(\overline{X})$, hence it is a normal subgroup in $\pi_1(X)$. Setting $\Delta_X = \pi_1(\overline{X})/\ker(\varphi)$ we get a canonical exact sequence

$$1 \to \overline{\Delta}_X \to \Delta_X \to G_k \to 1.$$

The main result of Mochizuki is now the following:

Theorem. *Let $k|\mathbb{Q}_p$ be a finite extension. Then for every smooth, geometrically integral, hyperbolic curves X and Y over k one has a functorial bijection*

$$\mathrm{Isom}_k(X,Y) \to \mathrm{Out}_{G_k}(\Delta_X, \Delta_Y).$$

The theorem above represents a sharpening of Grothendieck's conjecture in characteristic zero in two directions: First, one works over a p-adic instead of a finitely generated field, and second, one replaces the full geometric fundamental group by its maximal pro-p quotient. From the result above, one deduces Theorem 1 above in the *characteristic zero case,* and second, a *relative pro-p version* of the birational conjecture for function fields of one variable over p-adic fields.

The idea of the proof in Mochizuki's approach, say in the case of complete curves, is as follows:

First, the local theory consists in giving a functorial description of a special class of L-rational points of X which correspond to the "non-degenerate" sections of the canonical projection $\Delta_{X_L} \to G_L$. Here L is a complete discrete valued field extension of k having residue field a function field of one variable over the residue field of k, and X_L is the base change of X to L. One of the main points in this process is the identification of the Pic-part in the second étale cohomology group of X with coefficients in $\mathbb{Z}_p(1)$. This is a quite technical and intricate point in the proof.

The "globalization" is a nice application of the p-adic Hodge theory and duality. Namely, since the curves in discussion are hyperbolic, their ℓ-adic cohomology (N.B., we are in characteristic zero) is encoded in the fundamental group of the curve in discussion, once the cyclotomic character is known. Thus, via the comparison theorems of Faltings et al, it follows that the sheaf of holomorphic differentials $\mathrm{H}^0(X, \Omega^1)$ is encoded in the canonical projection $\Delta_1(X) \to G_k$. In particular, if P is any point of X, the evaluation map at P defines a point $\omega(P)$ in $\mathbb{P}_k^{g-1}$ which lies in the image of X via the "canonical embedding" $X \to \mathbb{P}_K^{g-1}$ (if X is hyperbolic, one has to be a little bit more careful!), where g is the genus of X. Choosing a point P which is defined by a "non-degenerate" section as above, it follows that P is a generic point of X. One recovers X as the schematic closure of $\omega(P)$.

Finally observe that all the facts explained above are compatible with G_k-isomorphisms of hyperbolic (complete) curves.

§3. Some final comments

Although nowhere clearly stated, one has the belief that the "anabelian" geometry is not a "nilpotent" one, hence being anabelian is related to the interaction of primes in the fundamental group. Nevertheless, in the light of some newer results one has to rethink this point of view to some extent.

First, one can show that in the birational case, an "arithmetic" version of an assertion à la Bogomolov's one can be proved: Let K be a finitely generated field with Kr.dim$(K) \geq 2$, and ℓ a prime number different from char(K). Suppose that $\mu_\ell \subset K$. Then the isomorphy type of K is functorially encoded in the pro-ℓ quotient of G_K.[*] However, we do not know whether this is the case also for *global fields*. Nevertheless, as a first step in the proof of the case Kr.dim ≥ 2, one shows that denoting by U_0 the global units of a global field K_0, the group $K_0^\times / U_0$ is functorially encoded in the pro-ℓ quotient of G_K.

The next point we want to make is Mochizuki's result that the p-adic Galois action on the pro-p quotient of the geometric fundamental group already determines functorially the hyperbolic curve in discussion. One

---

[*] It seems that an even stronger assertion can be proved.

cannot compare directly this situation with the pro-ℓ one from above, because here the action of the whole absolute Galois group of the base field is needed, hence we do not have a "pure" pro-p situation. But probably under some certain supplementary hypothesis, already the maximal p-quotient $\pi_1^{(p)}(X)$ of the fundamental group of $X \to k$ could functorially encode X. By the way, for the birational version, i.e., for function fields of one variable K over p-adic fields containing μ_p, one can show that the pro-p quotient of the absolute Galois group of K functorially determines the function field under discussion.

Even more essential, some *Lie versions* of the anabelian conjectures for curves, as suggested by Deligne, appear quite plausible; see Nakamura [N2], [N3]. From there also some hints for the higher dimensional situation could evolve.

Finally, we would like to recall again the *Section conjecture*. It asserts that for complete, smooth, geometrically integral, hyperbolic curves over finitely generated fields $X \to \kappa$ the κ-*rational points of X are in a functorial bijection with the conjugacy classes of the sections* of the canonical projection $\pi_1(X) \to G_\kappa$. Even a weaker form of this, namely to *give group theoretic conditions on the sections* above which define rational points of the curve, would be an important progress here. As explained above, this problem was successfully attacked and solved by Tamagawa in the case of a finite base field κ, but it is a complete mystery in the remaining cases.

A weaker version of the Section conjecture, which is still open as well, is its *birational version*. It is the following: In the situation above let K denote the function field of the complete curve $X \to \kappa$. One asks whether the conjugacy classes of the sections of the canonical projection $G_K \to G_\kappa$ correspond to the κ-rational points of X.

Naturally, it is very desirable to understand the higher dimensional versions of the Section Conjecture, but we have a long way ahead to that...

References

[B]   F.A. Bogomolov, F. A., On two conjectures in birational algebraic geometry, in *Algebraic Geometry and Analytic Geometry*, ICM-90 Satellite Conference Proceedings, ed A. Fujiki et al., Springer-Verlag, Tokyo, 1991.

[D1]   P. Deligne, Théorie de Hodge I, in *Actes ICM Nice*, t.I, 425–430, Gauthier-Villars Paris, 1970.

[D2]   ——, La conjecture de Weil I, *Publ. Math. IHES* **43** (1974), 273–307.

[D3]   ——, Le groupe fondamental de la droite projective moins trois points, Galois groups over $\mathbb{Q}$, *Math. Sci. Res. Inst. Publ.* **16**, 79–297, Springer 1989.

[Dr]   V.G. Drinfeld, On quasi-triangular quasi-Hopf algebras and a group closely connected with $\mathrm{Gal}(\overline{\mathbb{Q}}/\mathbb{Q})$, *Leningrad Math. J.* **2**(4) (1991), 829–860.

[F1]   G. Faltings, Endlichkeitssätze für abelsche Varietäten über Zahlkörpern, *Inv. Math.* **73** (1983), 349–366.

[F2]   ——, p-adic Hodge Theory, *J. of Amer. Math. Soc.* **1** (1988), 255–299.

[G1]   A. Grothendieck, Letter to Faltings (June 1983), this volume.

[G2]   ——, Esquisse d'un programme (1984), this volume.

[Ha]   D. Harbater, Fundamental groups of curves in characteristic p, Proceedings of the ICM, Zürich 1994, Birkhäuser Verlag, Basel 1995.

[H]   H. Hironaka, Resolution of singularities of an algebraic variety over a field of characteristic zero, *Ann. of Math.* **79** (1964), 109–203; 205–326.

[I1]   Y. Ihara, On Galois representations arising from towers of covers of $\mathbb{P}^1 \backslash \{0, 1, \infty\}$, *Inv. Math.* **86** (1986), 427–459.

[I2]   ——, Braids, Galois groups and some arithmetic functions, Proc. ICM Kyoto 1990, 99-120.

[Ik]   M. Ikeda, Completeness of the absolute Galois group of the rational number field, *J. reine angew. Math.* **291** (1977), 1–22.

[K]   H. Koch, Die Galoissche Theorie der p-Erweiterungen, *Math. Monogr.* **10**, Berlin 1970.

[Ko]   J. Koenigsmann, From p-rigid elements to valuations (with a Galois characterization of p-adic fields), *J. reine angew. Math.*, **465** (1995), 165–182.

[Km] K. Komatsu, A remark on Neukirch's conjecture, *Proc. Japan Acad.* **50** (1974), 253–255.

[M1] Sh. Mochizuki, The profinite Grothendieck conjecture for closed hyperbolic curves over number fields, Preprint 1995.

[M2] ———, The local pro-p Grothendieck conjecture for hyperbolic curves, Preprint 1995.

[Na1] H. Nakamura, Galois rigidity of the étale fundamental groups of punctured projective lines, *J. reine angew. Math.* **411** (1990) 205–216.

[Na2] ———, Galois rigidity of pure sphere braid groups and profinite group calculus, *J. Math. Sci. Univ. Tokyo* **1** (1994), 71–136.

[Na3] ———, Galois rigidity of profinite fundamental groups, Sugaku Exposition, to appear.

[N1] J. Neukirch, Über eine algebraische Kennzeichnung der Henselkörper, *J. reine angew. Math.* **231** (1968), 75–81.

[N2] ———, Kennzeichnung der p-adischen und endlichen algebraischen Zahlkörper, *Inv. Math.* **6** (1969), 269–314.

[N3] ———, Über die absoluten Galoisgruppen algebraischer Zahlkörper, Astérisque **41–42** (1977), 67–79.

[O] T. Oda, A note on ramification of the Galois representation of the fundamental group of an algebraic curve I, *J. Number Theory* (1990), 225–228.

[Pa] A.N. Parshin, Finiteness Theorems and Hyperbolic Manifolds, in The Grothendieck Festschrift III, ed. P. Cartier et al., PM Series Vol 88, Birkhäuser, Boston Basel Berlin, 1990.

[P1] F. Pop, On the Galois theory of function fields of one variable over number fields, *J. reine angew. Math.* **406** (1990), 200–218.

[P2] ———, On Grothendieck's conjecture of birational anabelian geometry, *Ann. of Math.* **138** (1994), 145–182.

[P3] ———, On Grothendieck's conjecture of birational anabelian geometry II, Preprint Series Arithmetik II, No 16, Heidelberg 1995.

[Sp] M. Spiess, An arithmetic proof of Pop's Theorem concerning Galois groups of function fields over number fields, *J. reine angew. Math.* (to appear).

[T] A. Tamagawa, The Grothendieck conjecture for affine curves, Thesis, Kyoto University 1995; to appear in *Comp. Math.*

[Ta] J. Tate, Endomorphisms of Abelian Varieties over Finite Fields, *Inv. Math.* **2** (1966), 134–144.

[U1]   K. Uchida, Isomorphisms of Galois groups of algebraic function fields, *Ann. of Math.* **106** (1977), 589–598.

[U2]   ———, Isomorphisms of Galois groups of solvably closed Galois extensions, *Tohoku Math. J.* **31** (1979), 359–362.

[U3]   , ———, Homomorphisms of Galois groups of solvably closed Galois extensions, *J. Math. Soc. Japan* **33**, No.4, 1981.

[V]   V.A. Voevodski, Galois representations connected with hyperbolic curves, *Math. USSR Izv.* **39** (1992), 1281–1291.

[W]   R. Ware, Valuation Rings and Rigid Elements in Fields, *Can. J. Math.* **33** (1981), 1338–1355.

Some illustrative examples for

anabelian geometry in high dimensions

Yasutaka Ihara and Hiroaki Nakamura

§1. Introduction

In [G1-2], Grothendieck conjectured that smooth (possibly non-complete) irreducible *hyperbolic* curves X over a finitely generated field (say, over a fixed algebraic number field k) are determined uniquely by their algebraic fundamental groups $\pi_1(X)$ (which are naturally extensions of $G_k = \mathrm{Gal}(\bar{k}/k)$ by $\pi_1(\overline{X})$, where $\bar{k}$ is an algebraic closure of k and $\overline{X} = X \times \bar{k}$). This means, for example, that the cross ratio of four k-rational points $a_1, \ldots, a_4$ on $\mathbb{P}^1$ should be determined by $\pi_1(\mathbb{P}^1 - \{a_1, \ldots, a_4\})$. This conjecture of Grothendieck was first proved in the case of genus 0 ([N1]), then by Tamagawa in the case of arbitrary non-complete curves ([T1]), and finally, by Mochizuki [M1][M2] in all cases (even over local fields). On the other hand, Pop has proved that finitely generated fields K over a prime field (say, over $\mathbb{Q}$) are determined uniquely by their absolute Galois groups $G_K = \mathrm{Gal}(\bar{K}/K)(= \pi_1(\mathrm{Spec}\,K))$ [P]. So, if we call, after Grothendieck, a class $\mathfrak{X}$ of algebraic varieties *anabelian*, when the functor

$$\mathfrak{X} \ni X \longmapsto \pi_1(X) \in \{\text{Profinite groups}\}$$

(where, by definition, the only morphisms are isomorphisms (and those modulo inner automorphisms)) is fully faithful, then all hyperbolic curves over k, as well as spectra of local rings at the generic points of any irreducible algebraic varieties over k, form anabelian classes. By using the term "anabelian", Grothendieck seems to suggest that one could expect such phenomena to occur even in higher dimensional situations, and that whether this holds or not would be tightly related to whether the geometric part $\pi_1(\overline{X})$ of $\pi_1(X)$ is "far from" being an abelian group.

Thus, the main problems are (i) to find wider classes of anabelian varieties ("only" the higher dimensional case is left open), and (ii) to see how the geometry of X is reflected in the group theory of $\pi_1(X)$. This note is to give two types of examples in the higher dimensional case, one being an "alarming example", and the other, "supporting".

Among the first things to note is that Pop's theorem suggests a possibility that every irreducible variety over k has a non-empty open subvariety which is "anabelian" (including Artin type neighborhoods of each nonsingular closed point [AGV] Exp.XI, Prop.3.3, cf. also [N4])[1]. Another is that by Lefschetz' theorem, general hyperplane cuts of quasi-projective varieties of dimension > 2 leave fundamental groups invariant, warning us that one should choose a good model from each equivalence class of varieties having the same "anabelian" fundamental groups. (Should they always be $K(\pi, 1)$?) About hyperbolicity, since varieties of general type and hyperbolic manifolds (in the sense of [Ii][Ko]) are natural higher dimensional generalizations of hyperbolic curves, one asks what additional conditions would be necessary in order that these varieties be anabelian. As a test, we examine whether the "automorphism group" of $\pi_1(X)$ (one can put several different senses in this) is canonically isomorphic to $\operatorname{Aut}_k X$. We shall show that some Shimura varieties of the most classical split type, namely the Hilbert modular varieties (e.g. surfaces) and the Siegel modular varieties, *cannot* be anabelian[2]. Finally, we shall also show, as a "favorable anabelian example", that if X is a braid configuration of a hyperbolic curve, then 'Aut'$\pi_1(X)$ corresponds bijectively with $\operatorname{Aut}_k X$, by combining a previous result of the second named author (partly with Takao) and the (above mentioned) result of Tamagawa and Mochizuki. Of course, these are only two (types of) examples, but it might be worthwhile to keep them in mind (somewhere in the corner) in studying "higher anabelian varieties".

§2. Group of self-equivalence classes

In this article, we only consider smooth algebraic varieties X over a number field k which is embedded in the field $\mathbb{C}$ of complex numbers. Given such an X/k, identify $\pi_1(X)$ with the total Galois group of the tower of all finite etale coverings of X. Then, since the tower consists of the constant field extension and the geometric covering extension, we have a canonical exact sequence of Galois groups:

$$(2.1) \qquad 1 \to \pi_1(\overline{X}) \to \pi_1(X) \overset{p_{X/k}}{\to} G_k \to 1.$$

Here, the kernel group $\pi_1(\overline{X})$ classifies the geometric covering extensions over $\overline{X} = X \times \bar{k}$ and is isomorphic to the profinite completion of the usual

---

[1] In Grothendieck [G2] p.3, one reads "Andernteils sehe ich eine Mannigfaltigkeit jedenfalls dann als "anabelsch" (ich könnte sagen "elementar anabelsch") an, wenn sie sich durch successive (glatte) Faserungen aus anabelschen Kurven aufbauen lässt. Demnach (einer Bemerkung von M.Artin zufolge) hat jeder Punkt einer glatten Mannigfaltigkeit X/K ein Fundamentalsystem von (affinen) anabelschen Umgebungen."

[2] Unexpectedly from a (vague) statement by Grothendieck "ich würde annehmen, dass dasselbe (i.e., anabelianity) auch für die Modulmultiplizitäten polarisierter abelscher Mannigfaltigkeiten gelten dürfte" ([G2] p.3).

discrete fundamental group of the associated manifold $X(\mathbb{C})$. Every auto-morphism $f \in \mathrm{Aut}_k X$ induces an element of $\mathrm{Aut}_{\overline{k}}(\overline{X})$, which in turn induces (by extension and conjugation) an automorphism of $\pi_1(X)$ determined up to inner automorphisms by elements of $\pi_1(\overline{X})$. This automorphism of $\pi_1(X)$ reduces to the identity on G_k, as f is defined over k. In other words, we have a natural homomorphism

$$\mathrm{Aut}_k X \longrightarrow E_k(X) \stackrel{def}{:=} \mathrm{Aut}_{G_k}(\pi_1(X))/\mathrm{Int}\pi_1(\overline{X}),$$

where $\mathrm{Aut}_{G_k}(\pi_1(X)) := \{\alpha \in \mathrm{Aut}(\pi_1(X)) \mid p_{X/k} = p_{X/k} \circ \alpha\}$ and $\mathrm{Int}\pi_1(\overline{X})$ is its subgroup of the inner automorphisms by the elements of $\pi_1(\overline{X})$. The first naive criterion for anabelianity of X would be to check whether the group $E_k(X)$ recovers (or, in the strongest sense, is isomorphic to) $\mathrm{Aut}_k X$.

Another candidate for approximating $\mathrm{Aut}_k X$ in terms of π_1 is what is called the *Galois centralizer*: The exact sequence (2.1) induces the exterior Galois representation

$$\varphi_{X/k} : G_k \longrightarrow \mathrm{Out}\pi_1(\overline{X})$$

which associates to each $\sigma \in G_k$ the outer class of the conjugate actions on $\pi_1(\overline{X})$ by the preimages of σ in $\pi_1(X)$. The *Galois centralizer* $\mathrm{Out}_{G_k}\pi_1(\overline{X})$ is, by definition, the centralizer of the image of $\varphi_{X/k}$ in $\mathrm{Out}\pi_1(\overline{X})$. Every element of $E_k(X)$ gives, by restriction, an element of $\mathrm{Out}_{G_k}\pi_1(\overline{X})$; thus, having obtained a diagram of group homomorphisms

$$\mathrm{Aut}_k X \to E_k(X) \stackrel{\mathrm{res}}{\to} \mathrm{Out}_{G_k}\pi_1(\overline{X}),$$

we may also ask whether $\mathrm{Out}_{G_k}\pi_1(\overline{X})$ recovers $\mathrm{Aut}_k X$.

The two groups $E_k(X)$ and $\mathrm{Out}_{G_k}\pi_1(\overline{X})$ are isomorphic if the center Z of $\pi_1(\overline{X})$ is trivial. In fact, we have the following exact sequence (a profinite version of Wells' exact sequence, see [N3] 1.5.5)

$$(2.2) \qquad 1 \to H^1_{cont}(G_k, Z) \to E_k(X) \to \mathrm{Out}_{G_k}\pi_1(\overline{X}) \to H^2_{cont}(G_k, Z),$$

where the arrows are homomorphisms except for the last one which is merely a mapping of sets preserving origins, and G_k acts on Z by conjugation. As a third candidate for approximating $\mathrm{Aut}_k X$, we define $\bar{E}_k(X)$ to be the image of the homomorphism $E_k(X) \to \mathrm{Out}_{G_k}\pi_1(\overline{X})$.

A test for anabelianity. *If X/k deserves to be called anabelian, then at least one of $\mathrm{Out}_{G_k}\pi_1(\overline{X})$, $E_k(X)$, $\bar{E}_k(X)$ should coincide with $\mathrm{Aut}_k X$.*

§3. Locally symmetric spaces

We shall give two "alarming examples" of varieties which are "hyperbolic" in the usual sense but not "anabelian" in the sense of the above test. First, we discuss the case of Siegel modular varieties, and then the Hilbert modular case.

Example (S). Let $A_{g,n}$ be the Siegel modular variety of degree $g \geq 2$ and level $n \geq 3$ defined over the cyclotomic field $k = \mathbb{Q}(\mu_n)$. Then none of $E_k(A_{g,n})$, $\mathrm{Out}_{G_k} \pi_1(\overline{A}_{g,n})$, $\bar{E}_k(A_{g,n})$ is isomorphic to $\mathrm{Aut}_k(A_{g,n})$.

For 'hyperbolists', $A_{g,n}$ is a very "pleasant" object; the associated manifold $A_{g,n}(\mathbb{C})$ is the quotient of the Siegel upper half space by a torsion-free discrete subgroup $\Gamma_g(n) = \{A \in \mathrm{Sp}(2g,\mathbb{Z}) \mid A \equiv 1_{2g} \bmod n\} \subset \mathrm{Sp}(2g,\mathbb{R})$; hence it is a locally symmetric space having negative curvature. In particular, it is $K(\Gamma_g(n),1)$. Moreover, $A_{g,n}(\mathbb{C})$ is a hyperbolic complex manifold in the sense of Kobayashi [Ko], and is a variety of log general type in the sense of Iitaka [Ii] (cf. [Mu] §4). Therefore, $\mathrm{Aut}_{\mathbb{C}}(A_{g,n})$ is a finite group. By rigidity theorems for locally symmetric spaces, $\mathrm{Aut}_{\mathbb{C}}(A_{g,n})$ amounts to 'a half' of the finite group $\mathrm{Out}\Gamma_g(n)$ ([Mo], [Ma]; [No]). One encounters $A_{g,n}$ as a typical example of a hyperbolic variety in text books of hyperbolic geometry.

Meanwhile, for 'anabelianists', two "unpleasant" phenomena occur in $A_{g,n}$. Although $\pi_1(A_{g,n}(\mathbb{C})) = \Gamma_g(n)$ has trivial center and is residually finite, its profinite completion $\pi_1(\overline{A}_{g,n}) = \hat{\Gamma}_g(n)$ has a big center when $g > 1$. In fact, since $\mathrm{Sp}(2g,\mathbb{Z})$ $(g > 1)$ has the congruence subgroup property (Bass-Lazard-Serre[BLS], Mennicke[Me]),

$$(3.1) \qquad \hat{\Gamma}_g(n) = \prod_{p \nmid n} \mathrm{Sp}_{2g}(\mathbb{Z}_p) \times \prod_{p \mid n} \{A \in \mathrm{Sp}_{2g}(\mathbb{Z}_p) \mid A \equiv 1_{2g} \bmod n\}.$$

So, the center Z of $\pi_1(\overline{A}_{g,n})$ is an infinite group that corresponds via (3.1) to the infinite product $\prod_p' \{\pm 1\}$ of the center $\{\pm 1\}$ of $\mathrm{Sp}_{2g}(\mathbb{Z}_p)$, where p runs over all primes $p \nmid n$, with an addition of $p = 2$ when $2 \parallel n$. On the other hand, k has infinitely many quadratic extensions. Therefore, the cohomology group $H^1_{cont}(G_k, Z)$ is infinite. Therefore, by (2.2), $E_k(A_{g,n})$ is also infinite and hence *cannot* approximate the finite group $\mathrm{Aut}(A_{g,n})$.

Remark. Existence of non-trivial torsions in $\hat{\Gamma}_g(n)$ $(n \geq 3)$ implies that $\mathrm{Sp}(2g,\mathbb{Z})$ is *not* a good group in the sense of Serre [Se1][3]. This is also rele-

---

[3] A discrete group is called *good* if the Galois cohomology of its profinite completion coincides with the corresponding (discrete) group cohomology. The cohomological dimension of $\hat{\Gamma}_g(n)$ is infinite due to the raised torsion, while that of the $K(\Gamma_g(n), 1)$-space $A_{g,n}(\mathbb{C})$ (for $n \geq 3$) is bounded ([T2]).

vant to the goodness condition appearing in the profinite Gottlieb theorem [N3] 1.3.

On the other hand, the image of the exterior Galois representation φ : $G_k \to \text{Out}\pi_1(\overline{A}_{g,n})$ is an *infinite abelian* group. To see this, let $K_{g,N}$ be the function field of $A_{g,N}$ over $\mathbb{Q}(\mu_N)$ ($N \geq 1$), and let L_g denote the composite of $K_{g,N}$ for all $N \geq 1$. Then, according to Shimura ([Sh1] Th.3 (brief account), [Sh2] Th.7.2 (details)), L_g is a Galois extension of $K_{g,1}$ with exact constant field $\mathbb{Q}(\mu_\infty)$, and moreover, there is an equivalence of two short exact sequences of profinite groups

$$(3.2) \quad 1 \to \text{Gal}(L_g/K_{g,1}(\mu_\infty)) \to \text{Gal}(L_g/K_{g,1}) \to \text{Gal}(\mathbb{Q}(\mu_\infty)/\mathbb{Q}) \to 1$$
$$\downarrow\wr \qquad\qquad \downarrow\wr \qquad\qquad \downarrow\wr\chi$$
$$(3.3) \quad 1 \to \text{Sp}(2g,\hat{\mathbb{Z}})/\{\pm 1\} \quad\to\quad \text{GSp}(2g,\hat{\mathbb{Z}})/\{\pm 1\} \xrightarrow{\nu} \hat{\mathbb{Z}}^\times \to \quad 1.$$

Here, χ is the cyclotomic character, GSp is the group of (symplectic) similitudes, and ν is the multiplier. Now let $n \geq 3$ and N run over all multiples of n. Then $\overline{A}_{g,N}$ is the finite etale covering of $\overline{A}_{g,n}$ that corresponds to the open normal subgroup $\hat{\Gamma}_g(N)$ of $\pi_1(\overline{A}_{g,n}) \cong \hat{\Gamma}_g(n)$, and by the congruence subgroup property, every finite etale covering of $\overline{A}_{g,n}$ is a subcovering of some $\overline{A}_{g,N}$. Therefore, by the above result of Shimura, the exterior action of $G_k = G_{\mathbb{Q}(\mu_n)}$ on $\pi_1(\overline{A}_{g,n})$ factors through $\text{Gal}(\mathbb{Q}(\mu_\infty)/\mathbb{Q}(\mu_n))$ and is given by the exterior action of

$$\{a \in \hat{\mathbb{Z}}^\times; a \equiv 1(\text{mod } n)\} = \prod_{p\nmid n}\mathbb{Z}_p^\times \times \prod_{p|n}\{a \in \mathbb{Z}_p^\times; a \equiv 1(\text{mod } n)\}$$

on $\hat{\Gamma}_g(n)$ (see (3.1)) *via* the "1_{2g} (mod n) part" of (3.3). But the kernel of the exterior action of $\mathbb{Z}_p^\times$ on $\text{Sp}(2g,\mathbb{Z}_p)/\{\pm I\}$ is exactly $(\mathbb{Z}_p^\times)^2$, because the centralizer of $\text{Sp}(2g,\mathbb{Z}_p)/\{\pm I\}$ in $\text{GSp}(2g,\mathbb{Z}_p)/\{\pm I\}$ consists only of scalar matrices $a_p \cdot I_{2g}(a_p \in \mathbb{Z}_p^\times)$ (whose multiplier being a_p^2). Therefore, $\varphi(G_k)$ contains an infinite abelian group

$$\prod_{p\nmid n}(\mathbb{Z}_p^\times/(\mathbb{Z}_p^\times)^2).$$

Now since $\varphi(G_k)$ is abelian, $\text{Out}_{G_k}(\pi_1(\overline{A}_{g,n}))$ contains $\varphi(G_k)$ itself which is infinite. Therefore, $\text{Aut}_k A_{g,n}$ cannot be isomorphic to $\text{Out}_{G_k}(\pi_1(\overline{A}_{g,n}))$.

It remains to examine whether the group $\bar{E}_k(A_{g,n})$ happens to approximate $\text{Aut}_k(A_{g,n})$ or not. But this is again negative: In fact, $\bar{E}_k(A_{g,n})$ still contains the exterior Galois image $\varphi(G_k)$, because of the following

Lemma. *Suppose that the projection* $p_{X/k} : \pi_1(X) \to G_k$ *has a splitting homomorphism* $s : G_k \to \pi_1(X)$ *such that the induced lift* $\tilde{\varphi} : G_k \to \mathrm{Aut}\pi_1(\overline{X})$ *of* $\varphi_{X/k} : G_k \to \mathrm{Out}\pi_1(\overline{X})$ *via* s *has an abelian image. Then,* $\varphi_{X/k}(G_k)$ *is contained in* $\bar{E}_k(X)$.

When $X = A_{g,n}$, $k = \mathbb{Q}(\mu_n)$, the assumption of the Lemma is satisfied by taking the preimages of the matrices $diag(a, \ldots, a, 1 \ldots, 1)$ for $s(G_k)$.

Proof of the lemma. It is not difficult to see from the assumption that, for any fixed $\sigma_0 \in G_k$, the map

$$xs(\sigma) \mapsto s(\sigma_0)xs(\sigma_0)^{-1}s(\sigma) \qquad (x \in \pi_1(\overline{X}), \ \sigma \in G_k)$$

gives a group automorphism of $\pi_1(X) = \pi_1(\overline{X}) \rtimes s(G_k)$. This provides us with a desired preimage of $\varphi(\sigma_0)$ in $E_k(X)$. $\diamond$

Example (H). Let F be a totally real number field of finite degree g over $\mathbb{Q}$, $\mathcal{O}_F$ be the ring of integers of F, and for each positive integer n, let $\Delta_F(n)$ be the Hilbert modular group of level n;

$$\Delta_F(n) = \{A \in SL(2, \mathcal{O}_F); A \equiv 1_2 (\mathrm{mod}\, n)\}.$$

Then the group $\Delta_F(n)$ acts on the product $\mathcal{H}^g$ of g copies of the complex upper half plane $\mathcal{H}$, in the usual manner, and if $n \geq 3$, $\Delta_F(n)$ is torsion-free. In this case, the quotient $\Delta_F(n)\backslash\mathcal{H}^g$ is also known to be Kobayashi-hyperbolic and of log-general type.

The congruence subgroup property for $SL(2, \mathcal{O}_F)$ is also valid if $g > 1$ [Se2]. Moreover, by Shimura ([Sh2] Th.7.2), $\Delta_F(n)\backslash\mathcal{H}^g$ has a standard model $A_{F,n}$ over $\mathbb{Q}(\mu_n)$ (not $F(\mu_n)$), and if $K_{F,n}$ denotes its function field, and $L_F = \bigcup_n K_{F,n}$, then there is again an equivalence of two short exact sequences:

$$1 \to \mathrm{Gal}(L_F/K_{F,1}(\mu_\infty)) \to \mathrm{Gal}(L_F/K_{F,1}) \to \mathrm{Gal}(\mathbb{Q}(\mu_\infty)/\mathbb{Q}) \to 1$$
$$\downarrow \wr \qquad\qquad\qquad \downarrow \wr \qquad\qquad\qquad \downarrow \wr \chi$$
$$1 \to \mathrm{SL}(2, \hat{\mathcal{O}}_F)/\{\pm 1\} \to \mathrm{GL}(2, \hat{\mathcal{O}}_F)|_{\det \in \hat{\mathbb{Z}}^\times}/\{\pm 1\} \overset{\det}{\to} \hat{\mathbb{Z}}^\times \to 1.$$

Using those primes p that decompose completely in F, we see easily that the image of $G_{\mathbb{Q}(\mu_n)}$ in $\mathrm{Out}\pi_1(\overline{A}_{F,n})$ is again an infinite abelian group.

Now it follows by the arguments parallel to those used in the Siegel modular case, that $A_{F,n}$ also fails the anabelianity test of §2.

§4. Braid configuration spaces

In this section, we consider braid configuration spaces of hyperbolic curves as typical candidates for higher dimensional anabelian varieties. Let $r \geq 1$ and C be a hyperbolic curve over a number field k. The r-dimensional (pure) braid configuration space $C^{(r)}$ is defined to be the product of r copies of C minus all the weak diagonals:

$$C^{(r)} := \{(x_1, \ldots, x_r) \in C^r \mid x_i \neq x_j \ (i \neq j)\}.$$

For this type of variety, $\pi_1(\overline{C}^{(r)})$ is a successive extension of free profinite groups (and a profinite surface group), and hence has trivial center (cf. e.g. [N3] 1.2). Therefore, three groups considered in the previous section coincide for braid configuration spaces: $E_k(C^{(r)}) = \bar{E}_k(C^{(r)}) = \mathrm{Out}_{G_k} \pi_1(\overline{C}^{(r)})$. We may also consider the 'pro-l version' of these groups for any fixed prime l, by replacing $\pi_1(\overline{C}^{(r)})$ by its maximal pro-l quotient $\pi_1^{pro-l}(\overline{C}^{(r)})$ and by using a naturally induced exact sequence from (2.1) for $X = C^{(r)}$:

$$1 \to \pi_1^{pro-l}(\overline{C}^{(r)}) \to \pi_1^{(l)}(C^{(r)}) \to G_k \to 1.$$

In the similar way, $E_k^{(l)}(C^{(r)})$, $\bar{E}_k^{(l)}(C^{(r)})$, $\mathrm{Out}_{G_k} \pi_1^{pro-l}(\overline{C}^{(r)})$ are defined and again these three groups are isomorphic. Thus, our anabelianity test (§2) for $C^{(r)}$ is reduced to the following

Question. Does $\mathrm{Out}_{G_k} \pi_1(\overline{C}^{(r)})$ or $\mathrm{Out}_{G_k} \pi_1^{pro-l}(\overline{C}^{(r)})$ recover $Aut_k(C^{(r)})$?

Before discussing the above question, we shall summarize recent achievements by Tamagawa and Mochizuki for Grothendieck's *fundamental conjecture of anabelian geometry* ([G1][G2]). Let C, C' be hyperbolic curves over k. Then,

Theorem A. (Tamagawa [T1]: affine case, Mochizuki [M1]: proper case) *The natural mapping*

$$\mathrm{Isom}_k(C, C') \to \mathrm{Isom}_{G_k}(\pi_1(C), \pi_1(C'))/\mathrm{Int}\pi_1(\overline{C}')$$

is a bijection.

Theorem B. (Mochizuki [M2]) *The natural mapping*

$$\mathrm{Isom}_k(C, C') \to \mathrm{Isom}_{G_k}(\pi_1^{(l)}(C), \pi_1^{(l)}(C'))/\mathrm{Int}\pi_1^{pro-l}(\overline{C}')$$

is a bijection.

Corollary AB. $\operatorname{Aut}_k(C) \cong \operatorname{Out}_{G_k} \pi_1(\overline{C}) \cong \operatorname{Out}_{G_k} \pi_1^{pro-l}(\overline{C})$.

Notes: The above works by Tamagawa and Mochizuki include much more essential ingredients beyond the number-basefield case. In effect, Tamagawa [T1] established new aspects of the finite-basefield case and Mochizuki [M2] introduced new ideas for the p-adic-basefield case. The statements of Theorems A, B can be divided into the following two forms respectively.

$$(A): \begin{cases} \text{(Equiv)}: & \text{If } \pi_1(C) \cong \pi_1(C') \text{ over } G_k, \text{ then } C \cong C' \text{ over } k. \\ \text{(Aut)}: & \operatorname{Aut}_k(C) \cong \operatorname{Out}_{G_k} \pi_1(\overline{C}). \end{cases}$$

$$(B): \begin{cases} \text{(Equiv}_l): & \text{If } \pi_1^{(l)}(C) \cong \pi_1^{(l)}(C') \text{ over } G_k, \text{ then } C \cong C' \text{ over } k. \\ \text{(Aut}_l): & \operatorname{Aut}_k(C) \cong \operatorname{Out}_{G_k} \pi_1^{pro-l}(\overline{C}). \end{cases}$$

Prior to [T1], [M1-2], some special cases had been studied by the second named author and H.Tsunogai for the number-basefield case of (Equiv) and (Aut$_l$). See ([AI] [N] $\sim$) [N1-7] [NT] (cf. [V]).

For the above Question, we have two kinds of results as follows.

Theorem C. ([T1] + [N3])

   $\operatorname{Aut}_k(C^{(r)}) \cong \operatorname{Out}_{G_k} \pi_1(\overline{C}^{(r)})$ for $C = \mathbb{P}^1 - \{0, 1, \infty\}$, $r \geq 1$.

In fact, prior to [T1], the problem for general r had been reduced to the case of $r = 1$ in [N3]. This, combined with [T1], settles Theorem C. In the case of $C = \mathbb{P}^1 - \{0, 1, \infty\}$, $C^{(r)}$ can be regarded as the moduli space $M_{0,n}$ of the n-pointed projective lines with $n = r + 3$, whose automorphism group is isomorphic to S_n, the symmetric group of degree n, for $n \geq 5$. Applying Theorem C and the triviality[4] of $\operatorname{Out} G_{\mathbb{Q}}$ to a group theoretical lemma ([N3] 1.6.2), we obtain a Galois analog of Ivanov's rigidity theorem ([Iv])[5]:

Corollary C. $\operatorname{Out} \pi_1(M_{0,n}/\mathbb{Q}) \cong S_n$ $(n \geq 5)$, $(\cong S_3 \ (n = 4))$.

For general hyperbolic curves C, we also have

Theorem D. ([M2] + [NTa])

   $\operatorname{Aut}_k(C^{(r)}) \cong \operatorname{Out}_{G_k} \pi_1^{pro-l}(\overline{C}^{(r)})$ for $r \geq 1$.

In fact, in a joint article with Takao [NTa], we had shown the following sequence of injective homomorphisms:

$$\operatorname{Aut}_k C \times S_r \hookrightarrow \operatorname{Aut}_k C^{(r)} \hookrightarrow \operatorname{Out}_{G_k} \pi_1^{pro-l}(\overline{C}^{(r)}) \hookrightarrow \operatorname{Out}_{G_k} \pi_1^{pro-l}(\overline{C}) \times S_r$$

---

[4] Neukirch $\sim$ Komatsu, Ikeda, Iwasawa, Uchida, cf. [Ne].

[5] Ivanov's rigidity asserts that the outer automorphism group of $\pi_1(M_{0,n}(\mathbb{C}))$ is an extension of $\{\pm 1\}$ by S_n for $n \geq 5$ ([Iv]). This result is generalized to the surface mapping class groups ([Iv] [Mc]). Our Corollary C particularly indicates that Galois compatibility condition drives out 'anti-holomorphic' self-equivalences from $\operatorname{Out} \pi_1$.

for any hyperbolic curve C of non-exceptional type. (For exceptional hyperbolic curves whose geometric types are $\mathbb{P}^1 - \{0, 1, \infty\}$ or one-point punctured elliptic curves, we need some modifications to the above sequence.) This, combined with [M2], settles Theorem D. Curiously, the following purely geometric statement follows immediately from this combination.

Corollary D. $\mathrm{Aut}_k(C^{(r)}) \cong \mathrm{Aut}_k(C) \times S_r$, *unless* $\overline{C}$ *is isomorphic to* $\mathbb{P}^1 - \{0, 1, \infty\}$ *or an elliptic curve minus one point. In the latter of the exceptional cases, the statement holds if* S_r *is replaced by* S_{r+1}.

These results apparently suggest that the Galois fundamental groups of braid configuration varieties[6] differ from those of Hilbert/Siegel modular varieties in group-theoretical nature. We observe, especially, the following distinguished anabelian (=far from abelian) properties of the former fundamental groups:

(E1) Every open subgroup of the geometric profinite fundamental group has trivial center (cf. [N3] 1.3, [N4]).

(E2) The Galois image in the outer automorphism group of the geometric profinite fundamental group has only finitely many centralizing elements.

Problem. *Prove these two properties (E1-2) for the 'Galois-Teichmüller modular group'* $\pi_1(M_{g,n}/\mathbb{Q})$ *([G1][O2]).*

Problem. *Find new examples of algebraic varieties possessing the properties (E1-2).*

---

[6] Some analogous statements also hold for products of configuration varieties of hyperbolic curves as results of Theorems A, B, C, D.

References

[AGV] M.Artin, A.Grothendieck and J.L.Verdier, Théorie des topos et cohomologie étale de schémas (SGA4-3) Lecture Notes in Math. **305**,
Springer-Verlag, 1973.

[AI]   G.Anderson, Y.Ihara, Pro-l branched coverings of $\mathbb{P}^1$ and higher cir-
cular l-units, Part 1, *Ann. of Math.* **128** (1988), 271–293, Part 2,
Intern. J. Math. **1** (1990), 119–148.

[BLS] H.Bass, M.Lazard, J.-P.Serre, Solution of the congruence subgroup
problem for SL_n ($n \geq 3$) and Sp_{2n} ($n \geq 2$), *Publ. Math. I.H.E.S.*
33 (1967), 59–137.

[De]   P.Deligne, Le groupe fondamental de la droite projective moins trois
points, *The Galois Group over Q*, Y.Ihara, K.Ribet, J-P. Serre, eds.,
Springer, 1989, 79–297.

[G1]   A.Grothendieck, Esquisse d'un Programme, 1984 (this volume).

[G2]   A.Grothendieck, Letter to G.Faltings, June 1983 (this volume).

[H]   D.Harbater, Fundamental groups of curves in characteristic p, Proc.
ICM, Zürich, 1994, 654–666.

[Ih1] Y.Ihara, Arithmetic analogues of braid groups and Galois representations, *Contemp. Math.* **78** (1988), 245–257.

[Ih2] ——, Braids, Galois groups and some arithmetic functions, Proc.
ICM, Kyoto, 1990, 99–120.

[Ii]   S.Iitaka, Logarithmic forms of algebraic varieties, *J. Fac. Sci. Univ.
Tokyo* **23** (1976), 525–544.

[Iv]   N.V.Ivanov, Automorphisms of Teichmüller modular groups, Springer
Lecture Notes in Math. **1346** (1988), 199–270.

[Ko]   S.Kobayashi, *Hyperbolic manifolds and holomorphic mappings*, Mar-
cel Dekker, Inc., New York, 1970.

[Mc]   J.D.McCarthy, Automorphisms of surface mapping class groups – A
recent theorem of N.Ivanov, *Invent. math.* **84** (1986), 49–71.

[Ma]   G.A.Margulis, *Discrete subgroups of semisimple Lie groups*, Springer-
Verlag, 1991.

[Me]   J.Mennicke, Zur Theorie der Siegelschen Modulgruppe, *Math. Ann.*
159 (1965), 115–129.

[M1]   S.Mochizuki, The profinite Grothendieck conjecture for closed hyperbolic curves over number fields, *J. Math. Sci. Univ. Tokyo*, to
appear.

[M2] S.Mochizuki, The local pro-p Grothendieck conjecture for hyperbolic curves, Preprint 1995 (RIMS-1045).

[Mo] G.D.Mostow, Strong rigidity of locally symmetric spaces, Annals of Math. Studies **78**, Princeton Univ. Press, 1973.

[Mu] D.Mumford, Hirzebruch's proportionality theorem in the non-compact case, *Invent. math.* **42** (1977), 239–272.

[N] H.Nakamura, Rigidity of the arithmetic fundamental group of a punctured projective line, *J. reine angew. Math.* **405** (1990), 117–130.

[N1] ——, Galois rigidity of the etale fundamental groups of punctured projective lines, *J. reine angew. Math.* **411** (1990), 205–216.

[N2] ——, On Galois automorphisms of the fundamental group of the projective line minus three points, *Math. Z.* **206** (1991), 617–622.

[N3] ——, Galois rigidity of pure sphere braid groups and profinite calculus, *J. Math. Sci., Univ. Tokyo* **1** (1994), 71–136.

[N4] ——, Galois rigidity of algebraic mappings into some hyperbolic varieties, *Intern. J. Math.* **4** (1993), 421–438.

[N5] ——, On exterior Galois representations associated with open elliptic curves, *J. Math. Sci. Univ. Tokyo* **2** (1995), 197–231.

[N6] ——, Coupling of universal monodromy representations of Galois–Teichmüller modular groups, *Math. Ann.* **304** (1996), 99–119.

[N7] ——, Galois rigidity of profinite fundamental groups Sugaku (in Japanese) **47** (1995), 1–17, English translation to appear in *Sugaku Exposition* (AMS).

[NTa] ——, N.Takao, Galois rigidity of pro-l pure braid groups of algebraic curves, *Trans. Amer. Math. Soc.*, to appear.

[NT] ——, H.Tsunogai, Some finiteness theorems on Galois centralizers in pro-l mapping class groups, *J. reine angew. Math.* **441** (1993), 115–144.

[Ne] J. Neukirch, Über die absoluten Galoisgruppen algebraischer Zahlkörper, *Astérisque* **41/42** (1977), 67–79.

[No] J.Noguchi, Moduli spaces of holomorphic mappings into hyperbolically imbedded complex spaces and locally symmetric spaces, *Invent. math.* **93** (1988), 15–34.

[O1] T.Oda, Galois actions on the nilpotent completion of the fundamental group of an algebraic curve, in *Advances in Number Theory*, F.Q.Gouvêa, N.Yui eds. Clarendon Press, Oxford, 1993, 213–232.

[O2] T.Oda, Etale homotopy type of the moduli spaces of algebraic curves, preprint 1990 (this volume).

[P]   F.Pop, On Grothendieck's conjecture of birational anabelian geome-
        try, *Ann. of Math.* **138** (1994), 145–182, Part 2, *Ann. of Math.*, to
        appear.

[Se1]   J.P.Serre, *Cohomologie Galoisienne*, Lecture Notes in Math. **5**, Sprin-
        ger-Verlag, Berlin Heidelberg New York, 1973.

[Se2]   J.P.Serre, Le problème des groupes de congruence pour SL_2, *Ann.
        of Math.* **92** (1970), 489–527.

[Sh1]   G. Shimura, On modular correspondences for $Sp(N, \mathbb{Z})$ and their
        congruence relations, *Proc Nat. Acad. Sci. U.S.A.* **49-6** (1963),
        824–828.

[Sh2]   G. Shimura, On the field of definition for a field of automorphic
        functions II, *Ann. of Math.* **81** (1965), 124–165.

[T1]   A.Tamagawa, The Grothendieck conjecture for affine curves, *Com-
        positio Math.*, to appear.

[T2]   A.Tamagawa, personal communication to Nakamura, July, 1993.

[V]   V.A.Voevodsky, Galois representations connected with hyperbolic
        curves, *Cep. Matem.* (in Russian) **55** (1991), 1331–1342, English
        translation in *Math. USSR Izv.* **39** (1992), 1281–1291.

Research Institute for Mathematical Sciences, Kyoto University

Dept. of Math. Sci., Univ. of Tokyo, and Institute for Advanced Study

The Fundamental Groups at Infinity of the Moduli Spaces of Curves

Pierre Lochak

Abstract

In this note we explicit and prove some assertions contained in the *Esquisse*, concerning the fundamental groups "at infinity" (see precise definitions below) of the fine moduli spaces of non singular pointed curves over the complex numbers. This is achieved essentially by connecting these assertions to some known results, obtained by topological methods. We also discuss related assertions for the coarse moduli spaces.

§1. Introduction

We shall here be mainly concerned with one page (p.7) of the *Esquisse*, dealing with the fundamental group of the moduli spaces of pointed curves in an analytic context (i.e. over $\mathbb{C}$). Let us first quote the key sentence (starting at the bottom of p.6): "Ce principe de construction de la tour de Teichmüller n'est pas démontré à l'heure actuelle – mais je n'ai aucun doute qu'il ne soit valable. Il résulterait [...] d'une propriété extrêmement plausible des multiplicités modulaires ouvertes $M_{g,\nu}$, dans le contexte analytique complexe, à savoir que pour une dimension modulaire $N \geq 3$, le groupe fondamental de $M_{g,\nu}$ (i.e. le groupe de Teichmüller habituel $T_{g,\nu}$) est isomorphe au 'groupe fondamental à l'infini' i.e. celui d'un 'voisinage tubulaire de l'infini' ".

Let us first recall a few standard definitions and introduce some notation (which differs slightly from Grothendieck's). But first a warning to algebraic-geometers; here we shall actually be concerned only with the *analytic* part of the theory ("la théorie transcendante" as Grothendieck puts it on p.7 of the *Esquisse*), and indeed mostly with topological or at most real analytic properties. In particular, we shall have at our disposal the Teichmüller spaces, which are typically analytic objects. We shall accordingly recall mainly the analytic definitions and only quickly mention (with references) the connection with algebraic geometry, postponing some remarks until §5. For background information on the analytic treatment, we refer to the several books which are available, in particular [IT] and [T].

So let $S_{g,n}$ be *the* topological surface of genus g with n *numbered* marked points $(p_1, \ldots, p_n)$; we fix an orientation and a differentiable structure and there is one and only one class of such data, up to diffeomorphism. Next, a Riemann surface X of type (g,n) is a compact Riemann surface of genus

g with n numbered marked points, $(x_1, \ldots, x_n)$. Finally, a marked surface of type (g, n) is defined as a pair (X, f) where X is a Riemann surface of type (g, n) and f is a (marking) diffeomorphism $f : S_{g,n} \to X$ such that $f(p_i) = x_i$ for $i = 1, \ldots, n$. One introduces an equivalence relation on the set of marked surfaces by declaring (X, f) and (X', f') equivalent if there exists a diffeomorphism h of S onto itself, *isotopic to the identity map*, preserving the marked points, and such that the map $f' \circ h \circ f^{-1}$ is a biholomorphic map from X to X'.

From now on, we drop the subscript (g, n) when the context is clear; also, in this note, we shall always deal with surfaces of hyperbolic type or, equivalently, of strictly negative Euler characteristic; that is, we assume throughout that $2g - 2 + n > 0$.

The Teichmüller space $\mathcal{T}(S_{g,n}) = \mathcal{T}_{g,n} = \mathcal{T}$ parametrizes the equivalence classes of *marked* surfaces of type (g, n). The point is that it is not hard to coordinatize this set of classes, using for instance the so-called Fenchel-Nielsen coordinates, and to endow $\mathcal{T}$ with a real analytic structure; it is a classical result (essentially due to Teichmüller, and the main motivation for introducing *marked* surfaces) that $\mathcal{T}_{g,n}$ is a contractible space of real dimension $6g - 6 + 2n$. It can actually also be given a *complex* analytic structure, making it into a contractible analytic manifold of complex dimension $3g - 3 + n$ (see e.g. [IT]).

The moduli space $\mathcal{M}_{g,n}$ is obtained by "forgetting the marking": namely let $\mathrm{Diff}^+(S)$ be the group of orientation preserving diffeomorphisms of the surface $S = S_{g,n}$ fixing the points p_i, and let $\mathrm{Diff}_0(S)$ be the connected component of the identity map, i.e. the subgroup of such diffeomorphisms isotopic to the identity. Let $\Gamma = \Gamma_{g,n} = \pi_0(\mathrm{Diff}^+(S)) = \mathrm{Diff}^+(S)/\mathrm{Diff}_0(S)$ be the group of the connected components of $\mathrm{Diff}^+(S)$; the group structure is inherited from that of $\mathrm{Diff}^+(S)$. This group $\Gamma_{g,n}$ is the one which Grothendieck denotes $T_{g,\nu}$ ($\nu = n$) in the quotation from the *Esquisse* given above. This mapping class group or Teichmüller modular group Γ acts on $\mathcal{T}$: an element $h \in \Gamma$ sends $(X, f) \in \mathcal{T}$ to $(X, f \circ h)$; here we do not notationally distinguish a diffeomorphism of S and its image in Γ, because there is no need to. The moduli space $\mathcal{M} = \mathcal{M}_{g,n}$ is defined as the quotient $\mathcal{T}/\Gamma$. We denote by p the quotient map $\mathcal{T} \to \mathcal{M}$.

It is easy to see that this really amounts to "forgetting the marking". Indeed, say that two (*un*marked) Riemann surfaces X and X' (of type (g, n)) are equivalent, simply if they are isomorphic as marked Riemann surfaces, i.e. if it exists a biholomorphic map ϕ between X and X' respecting the marked points. If X and X' come from marked surfaces (X, f) and (X', f'), the map ϕ will give rise to a diffeomorphism $h = f'^{-1} \circ \phi \circ f$ from S to itself, but it will not in general be isotopic to the identity; the quotient map

p thus appears as a "forgetful" (with respect to marking) surjective map between $\mathcal{T}$ and $\mathcal{M}$.

Let us add a word about marked points. In the above we have used "marked points" which are considered as part of the (differentiable or Riemann) surface. We could have used "punctures" instead, obtained by deleting the points from the surfaces. In some sense, it is more natural to use marked points when dealing with conformal structures, and punctures when dealing with hyperbolic structures. We shall sometimes not make the choice explicit, and just speak of surfaces – or other objects – "of type (g, n)". The point is that we always consider surfaces *without* boundary, which from the point of view of topology means that the Dehn twists around simple closed curves encircling the removed points vanish. Also, we have assumed above that the points are numbered and that the various maps do not permute them; in §4, we shall need the extension to the case where punctures can be permuted, and we therefore include a short discussion of this point at the beginning of §4.

Returning to the main stream, an important (not so easy; see e.g. [IT]) result asserts that the action of Γ on $\mathcal{T}$ is proper and discontinuous; in particular, the stabilizers have finite order: they correspond in fact to the automorphism groups of the underlying Riemann surfaces, which are well-known to be finite in the hyperbolic case $2g + n \geq 3$. The space $\mathcal{M}$ can now be considered as a – possibly singular – variety given as the quotient space $\mathcal{T}/\Gamma$. From the point of view of algebraic geometry, $\mathcal{M}$ viewed this way solves the *coarse* moduli problem for complex non singular algebraic curves of type (g, n) (alias smooth Riemann surfaces of type (g, n)). For a general discussion of the moduli problem, we refer to §5.1 of [MFK]; a coarse moduli scheme is defined in §5.2 of that book (Definition 5.6) in terms of representability of a functor. Roughly speaking, $\mathcal{M}$ meets the requirements because its points (recall we are working over $\mathbb{C}$ and "point" means "geometric point") are in one-to-one correspondence with the equivalence classes of curves, as briefly noted above (this is (i) in Definition 5.6 of [MFK]); the universality requirement ((ii) of Definition 5.6) can also be shown to hold true.

It is crucial to note however that in the quotation from the *Esquisse* given above, Grothendieck has in mind the *fine* moduli spaces. Roughly speaking, from the point of view of topology or analysis, it amounts to viewing the spaces $\mathcal{M}_{g,n}$ as orbifolds, retaining the information encoded in the finite groups defined at every point of $\mathcal{M}$; that is, to a point of $\mathcal{M}$ representing the Riemann surface X, one associates its finite group of analytic automorphisms $Aut(X)$. We postpone to the beginning of §3 a more detailed discussion of this important point, stressing that all in all we shall only make use of the definition-proposition 3 (in §3) which the reader may take as a working definition if she or he wishes to. We shall need nothing from Thurston's theory of orbifolds and shall only briefly sketch the connection with algebraic geometry i.e. with "multiplicities" (to use Grothendieck's word) or "stacks". Also, all the fundamental groups we consider are topological, not algebraic, i.e. they are discrete and not profinite groups.

So in this introduction, we just note that the $\mathcal{M}_{g,n}$'s can be regarded as open (noncompact) orbifolds and that it is natural to look at their fundamental groups in the orbifold category. On the other hand, define the fundamental group "at infinity" of a space M to be the inverse limit of the fundamental groups of the subspaces $M \setminus K$ where K runs over compact sets of M (see §2 for details). With these definitions in mind, Grothendieck's assertion (see the quotation from the *Esquisse* given above, and the more precise statement (*) in §3 following definition-proposition 3) makes sense in the analytic setting; this will be elaborated further in §3. Before that, we discuss in §2 some geometric properties of the moduli and Teichmüller spaces which are essential for understanding and proving Grothendieck's assertion. We have also included for completeness a fourth section which deals with the fundamental groups of the coarse moduli spaces, that is, as topological spaces, forgetting the orbifold structure. Finally, the last section consists of a short informal discussion of the context of Grothendieck's assertion and possible consequences.

In closing this introduction, it is a pleasure to thank X.Buff, A.Douady, I.Faucheux, J.Fehrenbach and L.Schneps for numerous discussions on this and related topics. I also wish to warmly thank G.Maltsiniotis, H.Nakamura and F.Oort for carefully reading a first version of the manuscript and for suggesting changes and improvements. In particular, H.Nakamura pointed out the paper of D.Patterson ([P]) to me.

§2. Some geometry at infinity of the moduli spaces

We fix a type (g, n) of hyperbolic surfaces (i.e. we assume that $2g+n > 2$) and let $\mathcal{T}$ and $\mathcal{M} = \mathcal{T}/\Gamma$ be the associated Teichmüller and moduli spaces as above, with projection $p : \mathcal{T} \to \mathcal{M}$. A point of $\mathcal{M}$ corresponds to a Riemann surface X which is canonically endowed with a Poincaré metric of constant curvature -1. For a given closed curve γ on X, we let $\ell(\gamma)$ denote the length of the unique geodesic curve which is freely homotopic to γ. It exists and is simple if γ is simple. Now, for $\varepsilon > 0$, we define $\mathcal{M}^{\varepsilon}$ as the set of Riemann surfaces X such that there exists on X a simple closed curve γ with $\ell(\gamma) < \varepsilon$; we define $\mathcal{T}^{\varepsilon} = p^{-1}(\mathcal{M}^{\varepsilon})$. These are open sets in $\mathcal{M}$ and $\mathcal{T}$ respectively.

Note that $\mathcal{M}^{\varepsilon}$ is indeed a neighbourhood of infinity in $\mathcal{M}$ (see below for more details) whereas the analog is not true for $\mathcal{T}^\varepsilon$, which for any ε extends "well inside" $\mathcal{T}$; the part at infinity of the Teichmüller space $\mathcal{T}$ will not play any role here. We also recall that on $\mathcal{T}$, but *not* on $\mathcal{M}$, there is a well-defined *length function*. Precisely, we may consider C, a simple closed curve on the reference surface S, or rather a free homotopy class of closed curves, containing a simple representative (we shall sometimes simply write "a curve C"). Then for any $(X, f) \in \mathcal{T}$, we use the marking f to determine the homotopy class $f_*(C)$ on the Riemann surface X, and define $\ell_C(X, f) = \ell(f_*(C))$. With this in mind one may redefine $\mathcal{T}^\varepsilon$ as the set of $(X, f) \in \mathcal{T}$ such that $\ell_C(X, f) < \varepsilon$ for *some* simple closed curve C, and then $\mathcal{M}^\varepsilon = p(\mathcal{T}^\varepsilon)$.

We now recall some facts from hyperbolic geometry, which *in fine* (see corollary below) will enable us to relate the fundamental group of $\mathcal{M}$ at infinity to the fundamental group of $\mathcal{M}^\varepsilon$ for ε small enough. First $\mathcal{M} \setminus \mathcal{M}^\varepsilon$ is compact for any ε, and the family $(\mathcal{M} \setminus \mathcal{M}^\varepsilon)_{\varepsilon > 0}$ is *cofinal* in the partially ordered (by inclusion) family of all compact subsets of $\mathcal{M}$. Indeed, a classical theorem of Mumford (cf. e.g. [T], Lemma 3.2.2 and Appendix C) asserts that the lengths of the simple closed geodesics of a compact family K of Riemann surfaces are bounded from below, which precisely means that $K \subset \mathcal{M} \setminus \mathcal{M}^\varepsilon$ for some $\varepsilon > 0$.

Before we formally define the fundamental group of a space M "at infinity", we briefly discuss the problem of base points in this setting. We assume that M is a countable union of compact sets; the compact subsets K of M, partially ordered by inclusion, define an inverse sytem of sets: if $K \subset K'$, we simply consider the inclusion $M \setminus K' \subset M \setminus K$. A *base point at infinity* (denoted $*$) is given by an open part $U \subset M$ such that for any compact set K, there exists a compact set K' with $K \subset K'$ and $U \setminus K'$ non empty and *simply* connected. Let now π_1 denote the fundamental group, either topological, or as an orbifold, for the time being; the two cases for $M = \mathcal{M}$ will be discussed in the next two sections. We set:

Definition. Let M be a space which is a countable union of compact sets and assume there exists a base point at infinity $*$ for M, defined by an open set U. We define the *fundamental group at infinity of M, based at $*$* as:

$$\pi_1^\infty(M, *) = \varprojlim \pi_1(M \setminus K, U \setminus K),$$

where the inverse limit is over the cofinal family of compact subsets K of M such that $U \setminus K$ is simply connected, partially ordered by inclusion and using the natural induced maps on the fundamental groups.

We want of course to apply this to the case $M = \mathcal{M}$, which necessitates the construction of – at least one – base point at infinity. We shall be rather sketchy at this point as in this note we just need the existence of such an object: in order to construct it, one can start from a pants decomposition (maximal multicurve) of the topological surface S, which defines a system of Fenchel-Nielsen coordinates on the Teichmüller space $\mathcal{T}$. These coordinates can be written as $(r_i, t_i) \in \mathbb{R}^{+*} \times \mathbb{R}$, $i = 1, \ldots, 3g - 3 + n$, where the pairs are indexed by the simple closed curves appearing in the decomposition; the r_i's denote the lengths of the curves, and the t_i's are the associated twist parameters. Now, if we consider the region defined by $t_i \notin 1 + 2\mathbb{Z}$ for all i, and let $U \subset \mathcal{M}$ be the projection of this region to $\mathcal{M}$, U defines a base point at infinity $*$. This being said, we shall now for simplicity drop the mention of base points (finite as well as at infinity) from the notation. The above should make what we write clear.

Now, from the discussion at the beginning of this section, we conclude that in both cases (topological and orbifold fundamental groups), $\pi_1^\infty(\mathcal{M}) = \lim \pi_1(\mathcal{M}^\varepsilon)$ as $\varepsilon \to 0$, where there are natural maps $\pi_1(\mathcal{M}^\varepsilon) \to \pi_1(\mathcal{M}^{\varepsilon'})$ when $0 < \varepsilon < \varepsilon'$, induced by the inclusion $\mathcal{M}^\varepsilon \subset \mathcal{M}^{\varepsilon'}$. Another important phenomenon then comes in; namely there exists an absolute constant μ (actually $\mu = \ln(1 + \sqrt{2})$ will do and is optimal) such that on any hyperbolic Riemann surface, two simple closed geodesics of lengths $< \mu$ do not intersect. This is a straightforward consequence of the so-called "collar lemma" (cf. [T], Lemma 3.2.1 and Appendix D) and $\mathcal{M}^\mu$ (resp. $\mathcal{T}^\mu$) can be called the *thin* part of the moduli (resp. Teichmüller) space. Using this, we can now state and prove (essentially following [Ha] §3.2) a proposition which is the first important step in the justification of Grothendieck's assertion.

Proposition 1. *Let ε and ε' be given, with $0 < \varepsilon < \varepsilon' < \frac{\mu}{3}$; then:*

i) $\mathcal{T} \setminus \overline{\mathcal{T}^\varepsilon} \simeq \mathcal{T} \setminus \overline{\mathcal{T}^{\varepsilon'}} \simeq \mathcal{T}$, where "$\simeq$" means "diffeomorphic to"; moreover the diffeomorphisms can be chosen equivariant with respect to the action of the Teichmüller modular group Γ;

ii) $\mathcal{T}^{\varepsilon'} \simeq \mathcal{T}^\varepsilon$, and the implied diffeomorphism can be chosen Γ-equivariant;

iii) $\mathcal{T}^{\varepsilon'}$ is a strong deformation retract of $\partial \overline{\mathcal{T}}^\varepsilon$; in particular, these two spaces have the same homotopy type. Again the retraction can be chosen Γ-equivariant.

Note that in i), which will not be used in the sequel, the assertions for $\mathcal{T}^\varepsilon$ and $\mathcal{T}^{\varepsilon'}$ are not visibly related and we simply have to prove the property for any $\varepsilon < \frac{\mu}{3}$. Here $\overline{\mathcal{T}}^\varepsilon$ denotes the closure of $\mathcal{T}^\varepsilon$, which consists of the marked surfaces (X, f) with $\ell_{C_0}(X, f) \leq \varepsilon$ for some curve C_0. The open set $\mathcal{T} \setminus \overline{\mathcal{T}}^\varepsilon$ thus consists of the marked surfaces such that $\ell_C(X, f) > \varepsilon$ for all curves C. The fact that this set is open, or equivalently that the description of $\overline{\mathcal{T}}^\varepsilon$ given above is correct, that is describes indeed the closure of $\mathcal{T}^\varepsilon$, is a consequence of the discreteness of the length spectrum of any Riemann surface: when verifying the above conditions, one has in fact to deal with a *finite* number of short curves only, which implies that the condition is indeed open, since for a given curve C, the length function ℓ_C is continuous on $\mathcal{T}$.

As for iii), note that $\mathcal{T}^\varepsilon \subset \mathcal{T}^{\varepsilon'}$ and that $\partial\overline{\mathcal{T}}^\varepsilon$ denotes the boundary of $\overline{\mathcal{T}}^\varepsilon$, comprising the marked surfaces such that $\ell_C(X, f) \geq \varepsilon$ for all curves C and $\ell_{C_0}(X, f) = \varepsilon$ for at least one curve C_0. The complement $K^\varepsilon = \mathcal{T} \setminus \mathcal{T}^\varepsilon$ is a compact set and $\partial\overline{\mathcal{T}}^\varepsilon = \partial K^\varepsilon$. The image $p(\partial\overline{\mathcal{T}}^\varepsilon)$ of this set in $\mathcal{M}$ is simply the set of Riemann surfaces which have all their simple closed geodesics of length $\geq \varepsilon$, with at least one of length exactly ε .

Proposition 1 is an easy consequence of the following lemma, due to S.Wolpert (see [Ha], Lemma 3.7).

Lemma 2. (S. Wolpert) *Let* $\phi : [0, \infty) \to [0, 1]$ *be a* C^∞ *function on the positive real axis, with support in* $[0, \mu)$. *Then there exists a* Γ-*equivariant vector field* $W = W_\phi$ *on* $\mathcal{T}$ *with the following property: let* Φ^t *be the flow generated by* W, *and* C *be any curve on* S, *defining a length function* ℓ_C *on* $\mathcal{T}$; *then one has:*

$$\frac{d}{dt}(\Phi^t \circ \ell_C) = \phi \circ \ell_C.$$

Here the surprise, if any, consists in the possibility of dealing with all the (classes of simple closed) curves simultaneously; it is indeed easy, using Fenchel-Nielsen coordinates, to produce a vector field with the above property for the curves of a given "pants decomposition" of the surface S, but here one achieves much more. This statement reflects the convexity properties of the length fonctions; the proof is not difficult, given some geometric work of S.Wolpert. Note that W actually vanishes outside the thin part $\mathcal{T}^\mu$ of $\mathcal{T}$. The equivariance of W means, as usual, that for any $\gamma \in \Gamma$, one has $W \circ \gamma = \gamma_* W$; thus W descends to a vector field on $\mathcal{M}$, vanishing outside the thin part $\mathcal{M}^\mu$ lying "at infinity".

Let us now turn to the proof of proposition 1, granting lemma 2. To prove i) for $\mathcal{T}^\varepsilon$, we choose (as in [Ha]) a function ϕ which is decreasing, and such that $\phi = 1$ on $[0, \varepsilon]$ and $\phi = 0$ on $[2\varepsilon, \infty)$. Consider the vector field $W = W_\phi$ whose existence is asserted by lemma 2 and the corresponding time ε diffeomorphism Φ^ε. It is easy to see that Φ^ε maps $\mathcal{T}$ diffeomorphically onto $\mathcal{T} \setminus \overline{\mathcal{T}}^\varepsilon$. Since W actually vanishes outside $\mathcal{T}^{2\varepsilon}$, the set $\mathcal{T} \setminus \overline{\mathcal{T}}^{2\varepsilon}$ is kept fixed pointwise by the diffeomorphism.

In order to prove ii), we choose another function ϕ. The support of ϕ is now contained in $[0, 2\varepsilon']$ and we require that $\phi(0) = 0$ and $\phi = 1$ over the interval $[\varepsilon, \varepsilon']$. The corresponding time $(\varepsilon' - \varepsilon)$ diffeomorphism, i.e. $\Phi^{\varepsilon'-\varepsilon}$, then maps $\mathcal{T}^\varepsilon$ onto $\mathcal{T}^{\varepsilon'}$.

To prove iii), we construct the required retraction Ψ as follows: on $\overline{\mathcal{T}}^\varepsilon$ we use the same function ϕ as in i), only with ε replaced with ε', and the corresponding vector field W and flow Φ^t; we flow the points forward, until they cross the boundary $\partial\overline{\mathcal{T}}^\varepsilon$. That is, if $\tau = (X, f) \in \overline{\mathcal{T}}^\varepsilon$, we set $\Psi(\tau) = \Phi^{t_*}(\tau)$ where $t_* \geq 0$ is the first (and actually only) instant t when $\Phi^t(\tau) \in \partial\overline{\mathcal{T}}^\varepsilon$. It is easy to see from i) that t_* exists, with $t_* \leq \varepsilon'$. Now, on $\mathcal{T}^{\varepsilon'} \setminus \mathcal{T}^\varepsilon$ we use the same function ϕ as in ii) and define the retraction by flowing the points backward until they reach $\partial\overline{\mathcal{T}}^\varepsilon$. This completes the proof of proposition 1.

We record for future use a corollary to proposition 1 (more precisely assertions ii) and iii)), which comes from the fact that ε and ε' are arbitrary, subject only to $0 < \varepsilon < \varepsilon' < \frac{\mu}{3}$.

Corollary. *For any ε with $0 < \varepsilon < \frac{\mu}{3}$, we have $\pi_1^\infty(\mathcal{M}) = \pi_1(\mathcal{M}^\varepsilon)$, and $\mathcal{T}^\varepsilon$ has the homotopy type of $\partial\overline{\mathcal{T}}^\varepsilon$.*

§3. The orbifold fundamental groups of the moduli spaces

In this section, we prove Grothendieck's assertion, using known topological results. We postpone to §5 below a short discussion of the context. Here, as mentioned in §1, we have to deal with the *fine* moduli spaces $\mathcal{M}_{g,n}$ and we proceed to say a few words about it; there are however several versions of the theory, with a more or less abstract and algebraic flavour. Again, all we need here is the (essentially topological) definition-proposition 3 below and the coming discussion has been added only for the sake of completeness.

The basic "defect" of the *coarse* moduli spaces is that they do not carry a *universal family*. Using the language of analytic geometry, recall that a family Z of curves of type (g, n) parametrized by an analytic space S is given by a smooth and proper map $Z \to S$ whose fibers (i.e. the preimages of –geometric– points) are curves of type (g, n). By definition, such a family $\mathcal{C}_{g,n}$, parametrized by $\mathcal{M}_{g,n}$, is universal if any family $Z \to S$ can be viewed (in a unique way) as the pull-back of $\mathcal{C}_{g,n} \to \mathcal{M}_{g,n}$ via a map $\phi : S \to \mathcal{M}_{g,n}$. Such a universal family (or "universal curve") does *not* exist in the category of analytic spaces, but it does exist in the category of –analytic– orbifolds and this is why these are important in this contexts. The nonexistence of a universal family in the context of analytic spaces is due to the existence of nontrivial automorphisms of some curves (or Riemann surfaces). And, as noted very briefly in §1, orbifolds are made to keep track of these automorphisms. More precisely, an orbifold is given by the data of suitably compatible charts such that in each chart the orbifold is modelled on a quotient D/G where D is an open set (a polydisk) of $\mathbb{C}^d$ and G is a finite group acting on it; it is important that the various groups G are considered as part of the data. One may endow the spaces $\mathcal{M}_{g,n}$ with orbifold structures, where the groups are the automorphism groups of the underlying (isomorphism classes of) Riemann surfaces. This is carried out in [Mu, §2] (in the case $n = 0$). From the analytic viewpoint, these are the objects we actually have to deal with. Concerning the fundamental group of an orbifold, again, we shall only make use of the definition-proposition 3 below; we refer the reader to [HQ] for a concrete topological approach in terms of loops and equivariant homotopies. In this section we always denote the orbifold fundamental group simply as "π_1".

We note that there is an algebraic way to deal with the above, namely using algebraic "stacks", which were introduced in [DM, §§ 4,5] in this context; the fact that they have thus defined a nice fine moduli space is given in their proposition §5.1, which says that $\mathcal{M}_{g,0}$ (they deal with the case $n = 0$ only), which is a priori defined via a representability property, is in effect a separated algebraic stack of finite type over $\text{Spec}(\mathbb{Z})$. One then proceeds to develop the theory of the algebraic fundamental groups of stacks, which is the subject matter of T.Oda's contribution to this volume, to which we refer the interested reader (who will also find there a quick introduction to stacks in general).

As for now, we return to more topological matters and state a simple

Definition-Proposition 3. *Let T be a simply connected non singular manifold on which the discrete group G acts properly discontinuously. Then the quotient space $M = T/G$ has a natural orbifold structure and the orbifold fundamental group of M (with respect to any base point) is isomorphic to G: $\pi_1(M, *) \simeq G$.*

This applies to the case when $T = \mathcal{T}$, $G = \Gamma$ and $M = \mathcal{M}$, once one knows that the action of Γ is proper and discontinuous, a classical result. Grothendieck's assertion now reads:

$$\pi_1^\infty(\mathcal{M}_{g,n}) = \pi_1(\mathcal{M}_{g,n}) = \Gamma_{g,n} \text{ if and only if } d = 3g - 3 + n > 2. \qquad (*)$$

Now, the action of Γ on $\mathcal{T}$ obviously restricts to a still proper and discontinous action on $\mathcal{T}^\varepsilon$, simply because this is an open domain of $\mathcal{T}$, invariant under the action of Γ. Moreover, one has $\mathcal{M}^\varepsilon = \mathcal{T}^\varepsilon/\Gamma$ by definition. Picking any ε with $0 < \varepsilon < \frac{\mu}{3}$, and refering to the first assertion of the corollary in the last section, we see that, according to definition-proposition 3 above, assertion $(*)$ is implied by (actually equivalent to) the following result:

Proposition 4. *For $\varepsilon > 0$ small enough (actually for $\varepsilon < \frac{\mu}{3}$), $\mathcal{T}_{g,n}^\varepsilon$ is simply connected if and only if $d = 3g - 3 + n > 2$.*

The next reduction comes by using the second assertion of the corollary, which asserts that $\mathcal{T}^\varepsilon$ has the same homotopy type as $\partial\overline{\mathcal{T}}^\varepsilon$. So in order to prove the result, one needs only determine the homotopy type of this last space.

In order to state the next result, one has to introduce complexes of curves, which have played a prominent role in the recent topological investigations of the Teichmüller modular groups (see [Ha], [I] and references therein). In fact, we need only define the complex $\mathcal{Z} = \mathcal{Z}_{g,n}$ which was originally introduced by Harvey in his short seminal paper ([H]). It is a simplicial complex whose dimension k simplexes are defined to be the isotopy classes of families $\{C_0, C_1, \ldots, C_k\}$ of disjoint simple closed curves on the reference topological surface $S = S_{g,n}$. Vertices (zero simplexes) are given by (the class of) one such curve and the complex has dimension $3g - 4 + n$, a simplex of maximal dimension being associated to a "pants decomposition" of S (a maximal multicurve), comprising $3g - 3 + n$ curves and defining a system of Fenchel-Nielsen coordinates on $\mathcal{T}$. To finish with the definition of $\mathcal{Z}_{g,n}$ one has to describe the face relations, and this is done quite naturally, by defining the k-simplex associated to the family $\{C_0', C_1', \ldots, C_j'\}$ to be a face of the one associated to $\{C_0, C_1, \ldots, C_k\}$ ($j \leq k$) if the (unordered) family $\{C_r'\}$ is a subset of $\{C_s\}$ (as isotopy classes of curves). The relevance of this object is obvious from the following

Lemma 5. *$\partial\overline{\mathcal{T}}_{g,n}^\varepsilon$ has the homotopy type of $\mathcal{Z}_{g,n}$.*

This is actually an easy lemma, which is proven by describing $\partial\overline{\mathcal{T}}_{g,n}^\varepsilon$ "explicitly" (see [Ha]); tracing through the definitions, it emerges that this is built up from $\mathcal{Z}$ by replacing cells by Teichmüller spaces of the right dimensions or the products of such spaces with Euclidean spaces. In both cases, these pieces are contractible and the homotopy type reduces to that of the complex $\mathcal{Z}$; we refer again to [Ha] for a brief account, as well as to the references there (and to [I]) for more details. This description also leads to the construction of a bordification of the Teichmüller space $\mathcal{T}$, which was the initial motivation of Harvey, when introducing $\mathcal{Z}$.

Lemma 5 reduces the proof of proposition 4 (hence also of (∗)) to investigating the homotopy type of $\mathcal{Z}$. This is settled by the following

Theorem 6. *The complex $\mathcal{Z}_{g,n}$ is homotopically equivalent to a wedge of k-spheres, where $k = 2g - 2$ for $g > 0$, $n = 0$, whereas $k = 2g - 3 + n$ for $g > 0$, $n > 0$, and $k = n - 4$ when $g = 0$ ($n \geq 4$).*

This finishes the proof of (∗), since in all cases, $k \geq 2$ if and only if $d = 3g-3+n > 2$, implying that $\mathcal{Z}_{g,n}$ is simply connected. The above theorem is stated in this form in [Ha, §4], with a sketch of proof. One can find a variety of such statements in the literature as it gradually emerged, starting with [H] and [HT], that many properties of the Teichmüller modular groups $\Gamma_{g,n}$ can be traced to homotopical properties of complexes of curves. Various such complexes have been introduced, and many variants and connections are recorded in [I], including the connection between $\mathcal{Z}$, as introduced in [H], and the – at first sight less natural – complex which enabled Hatcher and Thurston to prove (in [HT]) that the $\Gamma_{g,n}$'s are finitely presented (see also [W], [G]).

§4. The topological fundamental groups of the moduli spaces

In this section, we shall be concerned with the fundamental groups of the *coarse* moduli spaces, for which we retain the notation $\mathcal{M}_{g,n}$, "forgetting" however about the orbifold structure. These – discrete – fundamental groups will be denoted as "π_1^{top}". We have added the discussion below partly in order to emphasize the contrast with the previous section. Results are not quite complete and we shall have to leave some assertions as they are, hoping that they sound plausible enough and that some readers might want to try and provide the missing proofs (assuming they do not already exist in the literature, which we cannot of course guarantee).

Actually, for reasons which will appear clearly below, in this section, we have to include the possibility of permuting the marked points, and must accordingly slightly redefine the moduli spaces and their fundamental groups (the Teichmüller spaces remain unchanged). The permutation group $\mathcal{S}_n$ acts on the moduli space $\mathcal{M}_{g,n}$; actually the action is faithful except when $(g,n) = (0,4)$ (in which case it factors through $\mathcal{S}_3$), and $\mathcal{S}_n = Aut(\mathcal{M}_{g,n})$, the group of analytic automorphisms of the space $\mathcal{M}_{g,n}$, except again when $(g,n) = (0,4)$ (since $Aut(\mathcal{M}_{0,4}) = \mathcal{S}_3$). So we now consider the quotient space $\mathcal{M}_{g,[n]} = \mathcal{M}_{g,n}/\mathcal{S}_n$. Another way to put it is that $\mathcal{M}_{g,[n]}$ is obtained as a quotient of $\mathcal{T}_{g,n}$ by allowing maps which fix the set of marked points but can permute them; the notation $[n]$ is intended to suggest that the marked points are then considered setwise, not individually. The Teichmüller modular groups are modified accordingly, introducing $\Gamma_{g,[n]}$ as the orbifold fundamental group of $\mathcal{M}_{g,[n]}$, with the tautological exact sequence:

$$1 \to \Gamma_{g,n} \to \Gamma_{g,[n]} \to \mathcal{S}_n \to 1.$$

As a matter of terminology, the group $\Gamma_{g,[n]}$ is often called the *full* mapping class group ($\Gamma_{g,n}$ being the *pure* subgroup) and classifies the connected components of the group of orientable diffeomorphisms of the surface $S_{g,n}$, preserving the marked points setwise. Note the obvious equalities: $\Gamma_{g,[0]} = \Gamma_{g,0} = \Gamma_g$ and $\Gamma_{g,[1]} = \Gamma_{g,1}$. Here again, we shall sometimes drop the subscript $(g,[n])$ from the notation.

The starting point for investigating the fundamental groups of the coarse moduli spaces is a simple and useful result, contained in [A], which is to be contrasted with definition-proposition 3.

Proposition 7. *Let T be a simply connected non singular manifold on which the discrete group G acts properly discontinuously. Let G^f be the subgroup of G generated by those elements in G which have fixed points (when acting on T). Then the fundamental group of the (possibly singular) quotient space $M = T/G$ is isomorphic to G/G^f, i.e. $\pi_1^{top}(M,*) \simeq G/G^f$.*

Note that the subgroup G^f is normal, because for any $(g,g') \in G$ and $\tau \in T$, one has $g'gg'^{-1}(g'\tau) = g'\tau$ if $g\tau = \tau$. In the case we are interested in, namely again $T = \mathcal{T}$, $G = \Gamma$ and $M = \mathcal{M}$, the subgroup G^f coincides with the group generated by the elements of finite order in Γ. This comes from a classical result (due to Nielsen), which says that any element of finite order in Γ has a fixed point in $\mathcal{T}$. So in our case, $G^f = \Gamma^t$, the subgroup of the Teichmüller modular group Γ generated by the elements of finite order.

We thus obtain the equality: $\pi_1^{top}(\mathcal{M}_{g,[n]},*) \simeq \Gamma_{g,[n]}/\Gamma_{g,[n]}^t$, where $*$ denotes any given base point. This last group was computed by D.Patterson in [P] in the general case, with the same purpose as ours. The final result reads:

Theorem 8. *The coarse moduli spaces $\mathcal{M}_{g,[n]}$ are simply connected, i.e. we have $\pi_1^{top}(\mathcal{M}_{g,[n]},*) = \{1\}$, except when $g = 2$ and $n = 4 \mod 5$, in which cases $\pi_1^{top}(\mathcal{M}_{2,5k+4},*) = \mathbb{Z}/5\mathbb{Z}$.*

The result for $n = 0$ and any genus had been previously obtained by C.MacLachlan in [M], again as a consequence of proposition 7 and in conjunction with a note of J.Birman ([B]) in which she discusses certain explicit generators for the groups Γ_g. The existence of a series of exceptional cases in theorem 8 looks odd and we have no general explanation to offer for this phenomenon, which perhaps deserves further investigation. Note that it is at this point that one has to use the *full* modular groups, and consequently to allow permutation of the marked points in the definition of the moduli spaces. Indeed, whereas $\Gamma_{0,[n]}$ is generated by its torsion elements, $\Gamma_{0,n}$ is torsion free.

We turn briefly to the topological fundamental groups at infinity of the coarse moduli spaces. The definition of §2 is still valid in this setting, except that, for consistency, we have to use here the unfortunately heavy notation "$\pi_1^{\infty,top}$". We shall see that the equality of the fundamental group and the fundamental group at infinity for spaces of dimensions $d > 2$, is intimately connected with the distribution of Riemann surfaces with nontrivial symmetry groups, a theme which also appears in the *Esquisse*, as Grothendieck suggests to use these exceptional surfaces as base points for the fundamental Teichmüller groupoid. In order to investigate these topological fundamental groups at infinity, we start again from the corollary in §2. That is, we still have $\pi_1^{\infty,top}(\mathcal{M}) = \pi_1^{top}(\mathcal{M}^\varepsilon)$ for ε small enough ($\varepsilon < \frac{\mu}{3}$). Moreover, we may again use proposition 4 in conjunction with proposition 7. We conclude that for $d > 2$ (and only then), the thin part $\mathcal{T}_{g,n}^\varepsilon$ of Teichmüller space is simply connected (recall that $\mathcal{T}_{g,[n]}^\varepsilon = \mathcal{T}_{g,n}^\varepsilon$) and $\pi_1^{top}(\mathcal{M}_{g,[n]}^\varepsilon)$ is then determined by the – proper and discontinous – action of $\Gamma_{g,[n]}$ on this space.

We thus find that for $d > 2$, we have $\pi_1^{\infty,top}(\mathcal{M}_{g,[n]}) = \Gamma_{g,[n]}/\Gamma_{g,[n]}^f$, where $\Gamma_{g,[n]}^f$ is the subgroup of $\Gamma_{g,[n]}$ generated by the elements which have a fixed point when acting on $\mathcal{T}_{g,n}^\varepsilon$ ($\varepsilon < \frac{\mu}{3}$). We shall now elucidate the nature of this subgroup, making it plausible that it is indeed independent of ε and actually equal to $\Gamma_{g,[n]}^t$. So we state as a

Plausible assertion. *One has* $\pi_1^{\infty,top}(\mathcal{M}_{g,[n]}) = \pi_1^{top}(\mathcal{M}_{g,[n]})$ *if (and only if)* $d = 3g - 3 + n > 2$.

Again, compatible base points and base point are infinity (as explained in §2) are implied. Thus $\mathcal{M}_{g,[n]}$ would be "simply connected at infinity", except along the mysterious exceptional series which appears in theorem 8.

This assertion is of course a strengthening of theorem 8 since the torsion elements of $\Gamma_{g,n}$ are exactly those which have a fixed point when acting on the whole Teichmüller space $\mathcal{T}_{g,n}$. Also, in theorem 8, the cases of dimensions 1 and 2 ("les deux premiers étages" of the *Esquisse*) can be dealt with "by inspection". The relationship between the fixed points of the action of the Teichmüller modular groups and the Riemann surfaces with non trivial symmetries is classical and has been developped by several authors. Here we just recall a few relevant facts, refering to [GH] for more details and references. Fix some type (g,n) of surfaces, and let $H \subset \mathrm{Diff}^+(S)$ ($S = S_{g,n}$ etc.) be a finite group of of diffeomorphisms of the reference surface, such that no two disctinct elements are isotopic (permutations of marked points allowed). We also view H as a finite subgroup of the mapping class group Γ, by considering the classes of the elements of H modulo isotopy. Define now $\mathcal{T}(H)$ as the set of marked surfaces (X, f) such that there is a subgroup H_X of the group of the – conformal – automorphisms of X which is conjugate to H via the marking f; that is, $H_X \subset Aut(X)$ satisfies $H_X = f \circ H \circ f^{-1}$, where the elements of H_X are viewed as diffeomorphisms (and X as a topological surface). We shall identify H_X with H, via f. Finally, we let $\mathcal{M}(H) = p(\mathcal{T}(H))$ where p denotes as usual the projection from $\mathcal{T}$ onto $\mathcal{M}$. It is natural to call $\mathcal{M}(H)$ the set of – isomorphism classes of – Riemann surfaces with H-symmetry. By definition, if H and S are fixed as above, and if $X \in \mathcal{M}(H)$, there is a subgroup $H_X \subset Aut(X)$, isomorphic to H, and such that for any marking g, i.e. any diffeomorphism $g : S \to X$, the finite group $f^{-1} \circ H_X \circ f$ is *conjugate to* H in $\mathrm{Diff}^+(S)$. Warning: it may be that (X, g) is not in $\mathcal{T}(H)$, because the modular group Γ permutes its finite subgroups by conjugation: if $(X, f) \in \mathcal{T}(H)$ for some $H \subset \mathrm{Diff}^+(S)$, one has $(X, g) \in \mathcal{T}(H)$ if and only if $f^{-1} \circ g$ normalizes H in Γ.

There is another viewpoint on $\mathcal{T}(H)$, the equivalence being the subject matter of classical theorems (cf. Theorem A in [GH]). Namely, start from any finite subgroup of the mapping class group Γ, and let it act on $\mathcal{T}$ via the natural action of Γ: the fixed point set is precisely $\mathcal{T}(H)$. Moreover, by the positive answer to the Nielsen realization problem, given by Kerchkhoff (in [K]), $\mathcal{T}(H)$ defined this way is not empty, whatever the finite group $H \subset \Gamma$ (at least in the compact case, i.e. when $n = 0$). In particular, we can always start from such a subgroup and lift it to $\mathrm{Diff}^+(S)$ if necessary (as in the first description of $\mathcal{T}(H)$), because given an element (X, f) of $\mathcal{T}(H)$, the subgroup H_X of $Aut(X)$ will define such a lift; here one uses an old result, due to Hurwitz, which says that a nontrivial automorphism of a Riemann surface is not isotopic to identity. This sets up the connection between surfaces with symmetries and fixed points of the action of Γ on $\mathcal{T}$.

The space $\mathcal{T}(H)$ parametrizes the marked surfaces (X, f) with H-symmetry; now we may view H as a subgroup of $\mathrm{Diff}^+(S)$ and look at the quotient surface $R = S/H$, the projection map $S \to R$ being ramified at some points $p_i \in R, i = 1, \ldots, r$. We also introduce R^*, which is obtained by puncturing R at the points p_i, and we let (γ, ν) denote the type of the surface R^*. For any (X, f), the marking f induces a diffeomorphism from R^* to Y^*, obtained analogously to R^*, by puncturing $Y = X/H$ at the ramification points of the cover $X \to Y$. This being said, it should not come as a surprise that this construction induces a biholomorphic equivalence between $\mathcal{T}(H)$ and the Teichmüller space $\mathcal{T}_{\gamma,\nu}$ of surfaces of type (γ, ν) (cf. Theorem B in [GH]). We shall call a finite subgroup H of Γ *maximal* if $\mathcal{T}(H)$ (as well as $\mathcal{M}(H)$) has dimension 0, which by the above is tantamount to saying that R^* is a thrice punctured sphere. We shall say that $\mathcal{M}(H)$ *extends to infinity* if $\mathcal{M}(H) \cap \mathcal{M}^\varepsilon$ is not empty, for any $\varepsilon > 0$. We hope the following statements will now sound natural (perhaps tantalizing) to the reader:

i) For any (g, n), and any *non* maximal finite subgroup $H \subset \Gamma_{g,[n]}$, the subvariety $\mathcal{M}_{g,[n]}(H)$ extends to infinity;

ii) If $d = 3g - 3 + n > 2$, $\Gamma^t_{g,[n]}$ is actually generated by torsion elements which generate *non* maximal – cyclic – subgroups.

By the above, these two statements would imply the "plausible assertion". Statement ii) is purely group theoretic and could hopefully be extracted from a careful reading of [P]. Statement i) is geometric and is interesting in its own right. It can be rephrased in terms of the Teichmüller space $\mathcal{T}(H)$ and the thin part $\mathcal{T}^\epsilon$ and can be further reduced as follows: going to infinity in the moduli spaces is done by pinching simple closed geodesics of the corresponding Riemann surfaces; so, all one has to do in order to prove i) is to construct such a geodesic, in the situation at hand. More precisely, let $X \in \mathcal{M}(H)$ be a surface with H-symmetry, with $H_X \subset Aut(X)$ the corresponding symmetry group; set again $Y = X/H_X$. Thinking in hyperbolic terms, X is endowed with its canonical Poincaré metric μ (constant curvature -1), which descends to a metric $\bar\mu$ on Y (since the elements of H_X act isometrically on X), which is however singular at the ramification points of the covering map $X \to Y$. It does not of course coincide with the Poincaré metric of the punctured surface Y^*. Now, assertion i) above would be a direct consequence of the following:

iii) Under the above assumptions (in particular H non maximal), there exists on Y a simple closed geodesic for the quotient metric $\bar\mu$, which does not contain a ramification point of the covering map $X \to Y$.

We shall end at this point our discussion of the fundamental groups at infinity of the coarse moduli spaces, hoping to have somehow convinced the reader that the analogs of Grothendieck's assertion in this situation also give rise to interesting problems, that moreover seem to be "within reach".

§5. Informal discussion

This volume is mainly intended as a guide to certain themes which appear more or less explicitly in the *Esquisse*. We thus feel it may be useful to devote this last section to some remarks which may help the "nonexpert" reader put the above in a more general context. The groups $\Gamma_{g,n}$ appear in two different contexts, namely topology, where they go under the name "mapping class groups" ($\Gamma_{g,n} = \pi_0(\text{Diff}^+(S_{g,n}))$) and analytic or algebraic geometry where they are often called "Teichmüller modular groups" ($\Gamma_{g,n} = \pi_1(\mathcal{M}_{g,n})$). It is of course easy to recognize that these define one and the same family of groups, but the flavour and the methods of investigation remain quite distinct. There are many reasons why these groups have attracted a lot of attention since almost a century ago, which can often be traced to the fact that they embody an alternative (w.r.t. to the classical arithmetic groups) generalization of $SL(2, \mathbb{Z})$ (cf. *Esquisse*, p.6, the second underlined sentence). This is perhaps exemplified essentially by the $n = 0$ cases, while the braid groups of the sphere ($g = 0$) and their manifold generalizations trace a somewhat different story.

These groups $\Gamma_{g,n}$ have thus been the subject matter of a long series of investigations, conducted mainly via topological methods (see [B 1,2]). In particular, it took a lot of efforts to prove that they are finitely generated and a lot more time to show that they are actually finitely presented. This was achieved essentially by looking at their natural faithful action on the complexes of curves ([H] and [HT]), eventually leading to an explicit presentation ([W], [G]) and making it possible to extract information on their cohomology. Since the Teichmüller spaces are contractible, the rational group cohomology of the $\Gamma_{g,n}$ coincides with the rational cohomology of the spaces $\mathcal{M}_{g,n}$, an identity which is often exploited. In any case, the starting point of these – comparatively recent – studies often resides in statements which sound very much like theorem 6 in §3. Now, Grothendieck is of course on the algebraic-geometric side and he extracted a statement (formally (*) in §3) which, as we have seen, can be proved by "translating" it into topological terms. In fact, Grothendieck, being apparently largely unaware of the topological ideas which were being developed almost simultaneously, had an insight which, among other things, would eventually lead to a proof that the $\Gamma_{g,n}$'s are finitely presented (see below for more details), a result people had been after for half a century.

This statement about the fundamental groups at infinity of the moduli spaces should of course not be taken as an isolated assertion, as we shall now try to make more precise. Let us say right away that it is not clear to the author what exactly could be done on this particular subject along the lines suggested by Grothendieck, which has not already been achieved by topological methods. But it is certainly a beautiful and compelling vision, which in any case deserves to be explored. We first note that Grothendieck does suggest a way of proving assertion (*) by algebraic-geometric methods (p.7), and it would be interesting to pursue this suggestion. Then, this assertion should be seen as the first (or say the general) step of a "dévissage", justifying the important "two level reconstruction principle", stated on p.6 (first underlined sentence). Indeed this "example" was apparently the main incentive that lead Grothendieck to developing an algebraic theory of "stratified structures". This is explained in section 5 of the *Esquisse* (in particular on pp.35-36), and note 7 (p.56, related to p.34, bottom) is also quite suggestive in that respect. Two of the main tools appearing are the notion of "tubular neighbourhood" in an algebraic context and Van Kampen theorem with several base points (cf. bottom of p.5), i.e. in a groupoid version. We proceed to make this more precise in the context of moduli spaces, recalling on the way some "well-known"' facts and references about the divisors at infinity of their stable completions, which are actually useful to keep in mind when reading the *Esquisse*.

The moduli space $\mathcal{M}_{g,n}$ can be compactified by adding the moduli points of "stable curves", here Riemann surfaces with nodes (ordinary double points, when viewed as algebraic curves). This compactification $\overline{\mathcal{M}}_{g,n}$ was first studied by Deligne and Mumford in a more general setting, and the structure of the divisor at infinity $\mathcal{D}_{g,n} = \overline{\mathcal{M}}_{g,n} \setminus \mathcal{M}_{g,n}$ was studied by Knudsen in [Kn]. It is made of finitely many irreducible components, which may be enumerated as follows: Let $S_{g,n}$ be again the reference topological surface of type (g,n) and consider the simple closed curves on $S_{g,n}$, up to diffeomorphism of $S_{g,n}$. There are finitely many such classes; restricting from now on to the case $n = 0$ for simplicity, they are as follows. First all non disconnecting curves (i.e. the curves γ such that S_g remains connected after cutting along γ) are equivalent; then if a curve disconnects S_g into two surfaces S_1 and S_2 of type $(g_1, 1)$ and $(g_2, 1)$ respectively, with $g_1 + g_2 = g$, the class of the curve depends only on the unordered pair (g_1, g_2). A component of the divisor at infinity $\mathcal{D}_g$ is associated to pinching along such a simple closed curve, again up to diffeomorphism of S_g (since we are dealing with the moduli space, surfaces are unmarked, and only these classes make sense anyway). So by the above there are $1 + [g/2]$ such irreducible components (where $[a]$ denotes the integer part of a). Moreover, each component is itself "almost" isomorphic to a moduli space of lower dimension, or to a product of two such spaces. More precisely, assuming again that $n = 0$, one sees that the nondisconnecting curves give rise to a component of type $\mathcal{M}_{g-1,2}$ (or rather the compactification of this space), whereas disconnecting curves are associated to the $[g/2]$ direct products $\mathcal{M}_{k,1} \times \mathcal{M}_{g-k,1}$ (or the product of the compactifications). Knudsen proved among other things that the natural "clutching morphism" (glueing the two surfaces "along the nodes") $\mathcal{M}_{k,1} \times \mathcal{M}_{g-k,1} \to \mathcal{M}_g$, is a closed immersion if $k \neq g - k$, whereas the map $\mathcal{M}_{g-1,2} \to \mathcal{M}_g$ is at most finite.

So, the complete space $\overline{\mathcal{M}}_{g,n}$ is indeed a beautiful stratified structure, with open highest dimensional stratum the open moduli space $\mathcal{M}_{g,n}$ and essentially copies of lower dimensional such spaces in the divisor at infinity $\mathcal{D}_{g,n} = \overline{\mathcal{M}}_{g,n} \setminus \mathcal{M}_{g,n}$. The lowest (zero) dimensional strata are obtained by pinching *all* the curves of a "pants decomposition" of the reference surface $S_{g,n}$, again up to diffeomorphisms; there are finitely many such decompositions, each consisting of $3g - 3 + n$ non intersecting simple curves (the cells of maximal dimension in the complex $\mathcal{Z}_{g,n}$ of §3), which in principle can be combinatorially enumerated. Finally, all the intersections are "as good as possible". More precisely, all intersecting pairs of irreducible components of the divisor at infinity have normal crossings: this reflects the simple fact that they correspond to Riemann surfaces with nodes at different places, and the intersections to surfaces with two nodes. Again, the intersections are essentially moduli spaces of lower dimensions and products thereof. The types can be determined simply by looking at the topology. So, the global picture, which however ignores some technical difficulties (as mentioned above, some maps may not be closed immersions), is that of an intersection graph for $\mathcal{D}_{g,n}$, whose structure can be combinatorially determined, with the vertices corresponding to the irreducible components and the edges to their intersections (including self-intersections); all the spaces appearing are – completed – moduli spaces or products of two copies of such spaces.

Returning to the quotation we started this note with ($Esquisse$ p.7), we see that in dimensions strictly larger than 2, the – orbifold – fundamental group of $\mathcal{M}_{g,n}$ is indeed roughly the same as that of a tubular neighbourhood of $\mathcal{D}_{g,n}$ (where this divisor is of course not included), namely $\mathcal{M}_{g,n}^{\varepsilon}$ for ε small enough. Computing such a fundamental group is the subject matter of the Grothendieck-Murre theory, expounded in [GM], where however it is required that the irreducible components of the divisor with normal crossings be non singular, which here is not always the case. Recently, Nakamura was able to apply part of this theory in this setting, in order to obtain (in [N]) some arithmetic results on the Galois action on the –profinite– Teichmüller modular groups. Returning briefly to geometry, we note that the fundamental group of the tubular neighbourhood is generated by two kinds of loops: First those which come from $\mathcal{D}_{g,n}$; more precisely, one can shift the generators of the fundamental groups of the components of $\mathcal{D}_{g,n}$ off into the non singular part $\mathcal{M}_{g,n}$. Then one has to add the loops which arise when "going around" the components of $\mathcal{D}_{g,n}$ and correspond to Dehn twists performed along the curves which are pinched when "tending to infinity". This is made precise in [GM], again under some smoothness assumptions. Putting these data together in order to express the fundamental group $\Gamma_{g,n}$ is apparently what it means to apply Van Kampen theorem, with carefully selected base points. Since again all the spaces involved are moduli spaces, this procedure should be viewed as one step of a "dévissage", here a descending induction on the maximal dimension of the spaces involved, which ends when one reaches dimension 2 and the fundamental (dimensions 1 and 2) pieces of the "lego".

References.

[A] M.A.Armstrong, The fundamental group of the orbit space of a discontinuous group, $Proc.\ Cambridge\ Phil.\ Soc.$ **64**, 1968, 299-301.

[B] J.S.Birman, Automorphisms of the fundamental group of a closed orientable two manifold, $Proc.\ Amer.\ Soc.$ **21**, 1969, 351-354.

[B1]   J.S.Birman, *Braids, Links, and Mapping Class Groups*, Annals of Math. Studies **82**, Princeton Univ. Press, 1974.

[B2]   J.S.Birman, Mapping class groups of surfaces, in *Braids*, Contemporary Math. **78**, AMS Publ., 1988, 13-43.

[DM]   P.Deligne and D.Mumford, The irreducibility of the space of curves of given genus, *Publ. Math. IHES* **36**, 1969, 75-109.

[G]   S.Gervais, Presentation and central extensions of mapping class groups, to appear in *Trans. of the AMS*.

[GH]   G.González-Díez and W.J.Harvey, Moduli of Riemann surfaces with symmetry, in *Discrete groups and geometry*, London Math. Soc. LNS **173**, W.J.Harvey and C.MacLachlan eds., Cambridge Univ. Press, 1992, 75-93.

[GM]   A.Grothendieck and J.-P.Murre, *The tame fundamental group of a formal neighborhood of a divisor with normal crossings on a scheme*, Springer LN **208**, Springer Verlag, 1971.

[H]   W.J.Harvey, Boundary structure for the modular group, in *Riemann surfaces and related topics*, I.Kra and B.Maskit eds., Annals of Maths Studies **97**, Princeton Univ. Press, 1980, 245-251.

[Ha]   J.L.Harer, The cohomology of the moduli space of curves, in *Theory of moduli*, Springer LN **1337**, E.Sernesi ed., 1988, 138-218.

[HQ]   A.Haefliger and Quach Ngoc Du, Une présentation du groupe fondamental d'une orbifold, in *Structure transverse des feuilletages, Astérisque* **116**, 1984, 98-107.

[HT]   A.Hatcher and W.Thurston, A presentation for the Mapping Class Group of a closed orientable surface, *Topology* **19**, 1980, 221-237.

[I]   N.V.Ivanov, Complexes of curves and the Teichmüller modular group, *Russian Math. Surveys* **42**, 1987, 55-107.

[IT]   Y.Imayoshi and M.Taniguchi, *An introduction to Teichmüller spaces*, Springer-Verlag, 1992.

[K]   S.P.Kerckhoff, The Nielsen realization problem, *Annals of Math.* **117**, 1983, 235-285.

[Kn]   F.F.Knudsen, The projectivity of the moduli space of stable curves II: The stacks $M_{g,n}$, *Math. Scand.* **52**, 1983, 161-199.

[M]   C.MacLachlan, Modulus space is simply connected, *Proc. Amer. Soc.* **29**, 1971, 85-86.

[Mu]   D.Mumford, Towards an enumerative geometry of the moduli space of curves, in *Arithmetic and Geometry* (dedicated to I.Shafarevitch), volume II, Progress in Math. **36**, Birkhäuser Verlag, 1983, 271-328.

[MFK]   D.Mumford, J.Fogarty and F.Kirwan, *Geometric invariant theory*, third enlarged edition, Ergeb. Math. **34**, Springer Verlag, 1994.

[N]   H.Nakamura, Coupling of the universal monodromy representations of Galois-Teichmüller modular groups, *Math. Ann.* **304**, 1996, 99-119.

[P]   D.B.Patterson, The fundamental group of the modulus space, *Michigan Math. J.* **26**, 1979, 213-223.

[T]   A.J.Tromba, *Teichmüller Theory in Riemannian Geometry*, Lectures in Mathematics ETH Zürich, Birkhäuser Verlag, 1992.

[W]   B.Wajnryb, A simple presentation for the Mapping Class Group of an orientable surface, *Israël J. Math.* **45**, 1983, 157-174.

URA 762 du CNRS, Ecole Normale Supérieure, 45 rue d'Ulm, 75230 Paris Cedex 05

Galois representations in the
profinite Teichmüller modular groups

Hiroaki Nakamura[*]

§1. Introduction

Let $M_{g,1}$ be the moduli stack over $\mathbb{Q}$ of the one point marked smooth projective curves of genus $g \geq 1$. Then, the Galois-Teichmüller modular group '$\pi_1(M_{g,1})$' is a group extension of $G_{\mathbb{Q}} = \mathrm{Gal}(\overline{\mathbb{Q}}/\mathbb{Q})$ by the profinite completion of the mapping class group Γ_g^1 of a 1-pointed genus g surface (cf. [Oda]):

(1.1) $$1 \to \hat{\Gamma}_g^1 \to \pi_1(M_{g,1}) \to G_{\mathbb{Q}} \to 1.$$

It is well known that Γ_g^1 has a finite number of generators $a_1, \ldots, a_{2g}, d$ which are Dehn twists along simple closed curves $\alpha_1, \ldots, \alpha_{2g}, \delta$ in the following figure respectively (Lickorish [L], Humphries [Hu]).

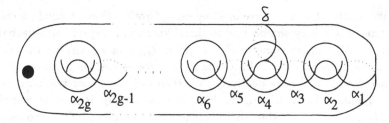

In this note, we prove the following

Theorem A. *There exists a good splitting homomorphism* $s : G_{\mathbb{Q}} \to \pi_1(M_{g,1})$ *of (1.1) such that the conjugate action* $* \mapsto s(\sigma) * s(\sigma)^{-1}$ $(\sigma \in G_{\mathbb{Q}})$ *transforms the twist generators* $a_1, \ldots, a_{2g}, d$ *of* $\hat{\Gamma}_g^1$ *as follows.*

$$\begin{cases} \sigma(d) &= d^{\chi(\sigma)}, \\ \sigma(a_i) &= \mathfrak{f}_\sigma(y_i, a_i^2)^{-1} a_i^{\chi(\sigma)} \mathfrak{f}_\sigma(y_i, a_i^2) \end{cases} \quad (1 \leq i \leq 2g).$$

Here, $\chi : G_{\mathbb{Q}} \to \hat{\mathbb{Z}}^\times$ *is the cyclotomic character acting on the roots of unity,*

$$y_1 = 1, \quad y_i = a_{i-1} \cdots a_1 \cdot a_1 \cdots a_{i-1} \quad (2 \leq i \leq 2g),$$

[*] The author was partially supported by Yoshida Foundation for sciences and technology.

and $\mathfrak{f}_\sigma(X,Y)$ is a unique "pro-word" in X,Y defined as an element of the commutator subgroup of the free profinite group $\pi_1(\mathbb{P}^1_{\overline{\mathbb{Q}}}-\{0,1,\infty\},\overrightarrow{01})$ on the standard loops X,Y turning around the punctures $0,1$ respectively, on which $\sigma \in G_{\mathbb{Q}}$ acts as $\sigma(X) = X^{\chi(\sigma)}$, $\sigma(Y) = \mathfrak{f}_\sigma(X,Y)^{-1}Y^{\chi(\sigma)}\mathfrak{f}_\sigma(X,Y)$.

Our splitting homomorphism s in Theorem A is provided by a sophisticated use of Deligne's notion of "tangential base points" ([De]). In fact, we construct two tangential base points lying on the hyperelliptic locus $\mathcal{H}_{g,1}$ of $M_{g,1}$. One is the image of a tangential base point on the "braid configuration space", originated from Drinfeld [Dr], Ihara-Matsumoto [IM], which a priori succeeds to a Galois action of desired form on $a_1,\dots,a_{2g}$. The other is the one induced from a certain maximally degenerate hyperelliptic curve over $\mathbb{Q}[[q]]$ whose special fibre is in the form where the simple closed curves $\delta_{\pm i}$ $(1 \leq i \leq g)$, ϵ_j $(1 \leq j \leq g-1)$ indicated below vanish:

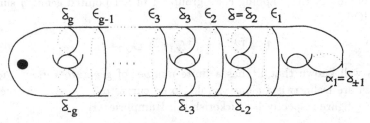

We construct such a curve in §3 by using Grothendieck's formal patching technique in a very similar way to Ihara-Nakamura [IN] (cf. also Harbater [Ha]). At the second tangential base point, $G_{\mathbb{Q}}$ acts a priori via the cyclotomic character on the Dehn twists $d_{\pm i}$, e_j corresponding to those simple closed curves $\delta_{\pm i}$, ϵ_j respectively. We then estimate these two tangential base points in the local neighborhood of a maximally degenerate ∞-point of $M_{0,2g+2}$, and show that the differences of the corresponding two Galois actions preserve the $d_{\pm i}, e_j$'s respectively. From this we conclude that the desired Galois action is obtained from the first tangential base point (twisted by a dummy 1-cocycle on the "hyperelliptic involution".) Especially, our proof shows more information on our Galois action:

Theorem A'. *The Galois action given in Theorem A on $\hat{\Gamma}^1_g$ transforms the Dehn twists $d_{\pm *}, e_*$'s (corresponding to $\delta_{\pm *}, \epsilon_*$'s respectively) by the cyclotomic character:*

$$\sigma(d_i) = d_i^{\chi(\sigma)}, \quad \sigma(e_j) = e_j^{\chi(\sigma)} \quad (1 \leq |i| \leq g, 1 \leq j \leq g-1; \ \sigma \in G_{\mathbb{Q}}).$$

In [Dr], Drinfeld introduced what is called the Grothendieck-Teichmüller group $\widehat{GT}$, into which, thanks to Belyi [Be], $G_{\mathbb{Q}}$ is embedded by the parameters $(\chi(\sigma), \mathfrak{f}_\sigma)$ of Theorem A (cf. [Ih2], [N, Appendix]). The group Γ_g^1 is known to be finitely presented (A.Hatcher-W.Thurston), and the relations for the Humphries' generators are listed in Wajnryb [W]. Only by using defining relations of $\widehat{GT}$, one can check directly that our $G_{\mathbb{Q}}$-action of Theorem A preserves almost all Wajnryb's relations except for the lantern relation (due to M.Dehn, D.Johnson). These calculations are relevant to the problem of approximating $G_{\mathbb{Q}}$ by $\widehat{GT}$ which was taken up also by L.Schneps, P.Lochak at the Luminy conference in several contexts. See the article [S] by L. Schneps in this volume for various background materials on $\widehat{GT}$.

At the same conference, M.Matsumoto posed a remarkable approach to genus 3 case from his E_7-singularity viewpoint, which motivated the author to make the present work. Matsumoto also worked out his resultant article [M2] soon in which another type of tangential base point on M_3 is displayed in connection with the Artin group of type E_7.

§2. Hyperelliptic locus

Let $\mathbb{A}_v^{2g+1}$ be the $(2g+1)$-dimensional affine space over $\mathbb{Q}$ with coordinates $v = (v_1, \ldots, v_{2g+1})$ and $\Delta = \bigcup_{i \neq j} \Delta_{ij}$ be the weak diagonal divisor on it, where $\Delta_{ij} = \{v \mid v_i = v_j\}$. The symmetric group S_{2g+1} acts naturally on $\mathbb{A}_v^{2g+1} - \Delta$, and its quotient variety is in the form of $\mathbb{A}_u^{2g+1}$ minus the discriminant locus D. The points $u = (u_1, \ldots, u_{2g+1}) \in \mathbb{A}_u^{2g+1}$ are identified with the monic polynomials $f_u(x) = x^{2g+1} + u_1 x^{2g} + \cdots + u_{2g+1}$, and $u \notin D$ if and only if the equation $f_u(x) = 0$ has only simple zeros. We have then a family of hyperelliptic curves $\{y^2 = f_u(x)\}_u$ over $\mathbb{A}_u^{2g+1} - D$ each fibre of which has ∞ as a specially attached point. Thus, there exists a representing morphism from $\mathbb{A}_u^{2g+1} - D$ to the hyperelliptic locus $\mathcal{H}_{g,1}$ of the moduli stack $M_{g,1}$ whose point represents, by definition, a hyperelliptic curve Y with one marked point fixed by the hyperelliptic involution. Every such $[Y] \in \mathcal{H}_{g,1}$ can be realized as a double cover of $\mathbb{P}^1$ with $2g + 2$ branch points (one of which is distinguished from others as the point ∞) so that there exists a natural morphism $\mathcal{H}_{g,1} \to M_{0,2g+2}/S_{2g+1}$, where $M_{0,2g+2}$ is the moduli of (ordered) $2g + 2$-pointed projective lines and S_{2g+1} is the automorphism group of $M_{0,2g+2}$ "fixing the $(2g+2)$-nd marking point ∞". We also have an obvious morphism $\mathbb{A}_v^{2g+1} \setminus \Delta \to M_{0,2g+2}$ mapping v to the class of $(\mathbb{P}^1; v_1, \ldots, v_{2g+1}, \infty)$ so as to fit into the commutative diagram:

$$
\begin{array}{ccc}
\mathbb{A}_v^{2g+1} \setminus \Delta & \longrightarrow & M_{0,2g+2} \\
\downarrow & & \downarrow \\
\mathbb{A}_u^{2g+1} \setminus D & \to \mathcal{H}_{g,1} \to & M_{0,2g+2}/S_{2g+1}.
\end{array}
$$

(2.1)

Now, the geometric fundamental group of $\mathbb{A}_v^{2g+1} \setminus D$ is the profinite braid group $\hat{B}_{2g+1}$ with standard generators $\sigma_1, \ldots, \sigma_{2g}$ and relations $\sigma_i \sigma_j = \sigma_j \sigma_i$ ($|i - j| \geq 2$), $\sigma_i \sigma_{i+1} \sigma_i = \sigma_{i+1} \sigma_i \sigma_{i+1}$ ($i = 1, \ldots, 2g - 1$), and its center is a free procyclic subgroup generated by $w_{2g+1} = (\sigma_1 \cdots \sigma_{2g})^{2g+1}$. The lower horizontal arrows of the above diagram induce projections of $\hat{B}_{2g+1}$ leading to

$$\pi_1(M_{0,2g+2}/S_{2g+1} \otimes \overline{\mathbb{Q}}) \cong \hat{B}_{2g+1}/\langle w_{2g+1} \rangle.$$

Moreover, the natural homomorphism $\pi_1(\mathcal{H}_{g,1}) \to \pi_1(M_{g,1})$ maps σ_i to a_i for $i = 1, \ldots, 2g$ (cf. [BH]). In $\hat{B}_{2g+1}$, we have a distinguished commutative subgroup generated by $y_i = \sigma_{i-1} \cdots \sigma_1 \cdot \sigma_1 \cdots \sigma_{i-1}$ ($2 \leq i \leq 2g + 1$). When mapped into $\pi_1(M_{g,1})$, these y_i ($2 \leq i \leq 2g$) coincide with those of Theorem A, while $w_{2g+1} = y_{2g+1} \cdots y_2$ gives a topological mapping class of a "hyperelliptic involution".

§3. Hyperelliptic stable curve

In this section, we shall construct a certain hyperelliptic curve over $\mathbb{Q}[[q]]$ with a special type of maximal degeneration. Our construction process goes on exactly parallel to that of Ihara-Nakamura [IN] §2, with an additional care to the hyperelliptic involution making the curve be a double-cover of a degenerate projective line (cf. also [Ha]). In [IN], we showed an explicit method for constructing a curve over $\mathbb{Q}[[q_1, \ldots, q_{m'}]]$ from a maximally degenerate stable marked curve – "$\mathbb{P}^1_{01\infty}$-diagram" – over $\mathbb{Q}$ and its "distinguished coordinates" of the irreducible components. In this note, we present a variant of this method by introducing a certain $\mathbb{P}^1_{0\pm1\infty}$-diagram Y^0 appearing as a double cover of a standard $\mathbb{P}^1_{01\infty}$-tree X^0. This variant is useful when extending the natural involution on Y^0 to that on the deformed family over $\mathbb{Q}[[q]]$.

Now, let us start from the definition of X^0. It is a connected stable curve over $\mathbb{Q}$ consisting of rational irreducible components X^0_λ ($\lambda \in \Lambda$), ordinary double points P^0_μ ($\mu \in M$) and marking points Q^0_ν ($\nu \in N$) such that

$$(3.1) \quad \Lambda = \{\lambda_1, \ldots, \lambda_{2g}\}, \quad M = \{\mu_1, \ldots, \mu_{2g-1}\}, \quad N = \{\nu_1, \ldots, \nu_{2g+2}\},$$

and the incidence relations are given by

$$(3.2) \quad \begin{cases} \mu_i/\lambda_i, \ \mu_i/\lambda_{i+1} \ (1 \leq i \leq 2g-1), \\ \nu_1, \nu_2/\lambda_1, \ \nu_i/\lambda_{i-1} \ (3 \leq i \leq 2g), \ \nu_{2g+1}, \nu_{2g+2}/\lambda_{2g}, \end{cases}$$

where μ/λ (resp. ν/λ) means that P^0_μ (resp. Q^0_ν) lies on X^0_λ. The dual graph of X^0 (with "legs" corresponding to Q^0_ν) is as follows.

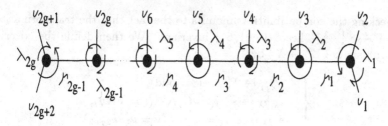

For each incidence pair μ/λ, ν/λ, we introduce distinguished coordinates $t_{\mu/\lambda}$ (resp. $t_{\nu/\lambda}$) of X_λ^0 which has value 0 at P_μ^0 (resp. Q_ν^0) and $1,\infty$ at the other distinguished points (i.e., double/marking points) on X_λ^0. Regarding the above figure as a plane tree, we introduce such coordinates in the way that the values at the distinguished points on each X_λ^0 are anticlockwise arranged in the same cyclic order as $0, 1, \infty$ except for

$$\left\{ \begin{array}{l} t_{\mu_{2i}/\lambda_{2i+1}}(Q_{\nu_{2i+2}}^0) = \infty \quad (1 \le i \le g - 1), \\ t_{\nu_1/\lambda_1}(Q_{\nu_2}^0) = t_{\nu_{2g+1}/\lambda_{2g}}(Q_{\nu_{2g+2}}^0) = \infty. \end{array} \right.$$

Next, we construct a double cover Y^0 over X^0 also as a connected stable curve. Its irreducible components Y_λ^0 ($\lambda \in \Lambda$) are again all rational components, and the marking points R_ν^0 ($\nu \in N$) lie on them in the same incidence relations ν/λ as in (3.2) above. But the double points $\{P_\kappa^0\}$ ($\kappa \in K$) on Y^0 are more complicated. The index set K is taken to be

$$\{\kappa_i ; |i| \le 2g + 1, \text{ odd}\} \cup \{\kappa_i ; 2 \le i \le 2g - 2, \text{ even}\}$$

and the incidence relations are given by $\kappa_j/\lambda_{|j|}, \lambda_{|j|+1}$ for all j.

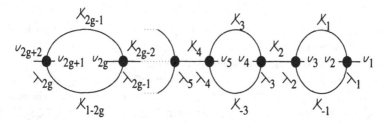

Since each Y_λ^0 has 4 distinguished points, we need to impose some condition on the relative locations of them on each component. This is done by introducing distinguished coordinates $s_{\kappa/\lambda}$, $s_{\nu/\lambda}$ in compatible ways so that their values at the distinguished points are $\{0, \pm 1, \infty\}$. We define them by $s_{\kappa/\lambda}(P_\kappa^0) = 0$, $s_{\nu/\lambda}(R_\nu^0) = 0$ and

$$s_{\kappa_i/\lambda}(P_{\kappa_{-i}/\lambda}^0) = \infty, \quad s_{\kappa_i/\lambda_{|i|}}(R_{\nu_{|i|+1}}^0) = s_{\kappa_i/\lambda_{|i|+1}}(R_{\nu_{|i|+2}}^0) = -1 \quad (i = \text{odd}),$$

$$s_{\kappa_i/\lambda}(P_{\kappa_{\text{odd}>0}}^0) = 1, \quad s_{\kappa_i/\lambda}(P_{\kappa_{\text{odd}<0}}^0) = -1 \quad (i = \text{even}),$$

$$s_{\nu_i/\lambda}(P_{\kappa_{\text{odd}>0}}^0) = 1, \quad s_{\nu_i/\lambda}(P_{\kappa_{\text{odd}<0}}^0) = -1 \quad (1 \le i \le 2g + 2).$$

Checking the compatibilities amounts to the fact that the transformations $s \mapsto -s$, $\frac{1}{s}$, $\frac{1-s}{1+s}$ keep $\{0, \pm 1, \infty\}$ invariant. We then define the covering morphism $\varpi : Y^0 \to X^0$ by

(3.3)
$$\begin{cases} P^0_{\kappa_{\pm i}} \mapsto P^0_{\mu_{|i|}} & (i = \text{odd}), \\ s_{\kappa_i/\lambda} \mapsto s^2_{\kappa_i/\lambda} = t_{\mu_i/\lambda} & (i = \text{even}), \\ s_{\nu/\lambda} \mapsto s^2_{\nu/\lambda} = t_{\nu/\lambda} & (\nu = \nu_1, \nu_{2g+2}). \end{cases}$$

Note that $\varpi : Y_0 \to X_0$ is ramified at all the Q^0_ν's and the $P^0_{\mu_{\text{even}}}$'s so that each component of Y^0 is a double cover of the corresponding component of X^0 ramified over exactly two points.

Let us then deform Y^0 to a 1-parameter family $Y/\mathbb{Q}[[q]]$ of hyperelliptic curves, by Grothendieck's formal patching technique ([G] EGA III Sect. 5.4; cf. also [DR], [Ha], [IN]). We prepare the following $\mathbb{Q}[[q]]$-algebras as parts of Y:

(a1) $$A_\kappa = \mathbb{Q}[s, s', \frac{1}{1 \pm s}, \frac{1}{1 \pm s'}][[q]]/(ss' - q) \qquad (\kappa \in K),$$
$$s = s_{\kappa/\lambda}, \; s' = s_{\kappa/\lambda'} \; (\lambda \neq \lambda'),$$

(a2) $$A_\nu = \mathbb{Q}[s, \frac{1}{1 \pm s}][[q]] \qquad (\nu \in N),$$
$$s = s_{\nu/\lambda},$$

(a3) $$A_\lambda = \mathbb{Q}[s, \frac{1}{s}, \frac{1}{1 \pm s}][[q]] \qquad (\lambda \in \Lambda),$$
$$s = s_{\nu_{i+1}/\lambda_i}.$$

Then, since $A_\kappa[\frac{1}{s}]/q^N \cong \mathbb{Q}[s, \frac{1}{s}, \frac{1}{1 \pm s}][[q]]/q^N$ etc., the first two kinds of spectrums $\text{Spec}(A_\kappa/q^N)$ ($\kappa \in K$) and $\text{Spec}(A_\nu/q^N)$ ($\nu \in N$) are glued together by identifying their open parts with $\text{Spec}(A_\lambda/q^N)$ ($\lambda \in \Lambda$) along the diagram Y^0 so as to produce a scheme $\mathfrak{Y}^N$ over $\mathbb{Q}[q]/q^N$ ($N \geq 1$). The resulting sequence $Y^0 = \mathfrak{Y}^1 \subset \mathfrak{Y}^2 \subset \cdots$ over artinian schemes are compatible to form a proper regular formal scheme $\mathfrak{Y}$ over $\text{Spf}\,\mathbb{Q}[[q]]$. We denote the algebraization of $\mathfrak{Y}$ by $Y/\mathbb{Q}[[q]]$, and identify its special fibre with Y^0 in the obvious manner.

Observe that in each step of the above process, we have an involution on $\mathfrak{Y}_N$ interchanging local data compatibly as

$$s \leftrightarrow -s \quad \text{in } A_{\kappa_i}(i : \text{even}), A_\nu, A_\lambda$$

$$A_{\kappa_i} \leftrightarrow A_{\kappa_{-i}}; \; s_{\kappa_i/\lambda} \leftrightarrow s_{\kappa_{-i}/\lambda}, \; s_{\kappa_i/\lambda'} \leftrightarrow s_{\kappa_{-i}/\lambda'} \; (i : \text{odd}).$$

These involutions on $\mathfrak{Y}^N$ ($N \geq 1$) define an involution on $Y/\mathbb{Q}[[q]]$ extending the covering transformation of Y^0/X^0. Moreover, each marking point R^0_ν

has natural extensions $R_\nu^N \in \mathfrak{Y}^N(\mathbb{Q}[q]/q^N)$ and hence $R_\nu \in Y(\mathbb{Q}[[q]])$ fixed under the respective involutions. In particular, the generic fibre Y_η is a complete smooth curve over $\mathbb{Q}((q))$ with $2g + 2$ fixed $\mathbb{Q}((q))$-points under an involution, hence is a hyperelliptic curve of the form

$$y^2 = (x - v_1(q)) \cdots (x - v_{2g+1}(q)),$$

where $v_i(q) \in \mathbb{Q}((q))$ corresponds to the branch at $R_{\nu_i}^0$ $(i = 1, \ldots, 2g + 1)$. These coordinates $v(q) = (v_1(q), \ldots, v_{2g+1}(q))$ give a $\mathbb{Q}((q))$-valued point of $\mathbb{A}_v^{2g+1} \setminus \Delta$. We have thus obtained a tangential base point $\vec{v}$ on $\mathbb{A}_v^{2g+1} \setminus \Delta$ induced from the $\mathbb{Q}((q))$-rational point $v(q)$.

§4. Tangential base points

In the previous section, we constructed a deformation of a double cover Y^0 over X^0 using a single deformation parameter q to control all parts of the deformation procedure. In this section, we shall consider another direct construction of an explicit deformation of X^0 — a chain of $\mathbb{P}^1$'s — by allowing each singular point to deform independently by its own deformation parameter q_i. This construction provides a universal deformation of X^0 which will be related with the former deformation later in (4.3).

What we wish to be concerned with here is a standard tangential base point $\vec{b}$ on $\mathbb{A}_v^{2g+1} \setminus \Delta$ having the following two properties (4.1) and (4.2).

(4.1) $\vec{b}$ induces a sectional homomorphism $s_{\vec{b}} : G_\mathbb{Q} \to \pi_1(\mathbb{A}_u^{2g+1} \setminus D)$ such that the conjugate action by $s_{\vec{b}}(\sigma)$ $(\sigma \in G_\mathbb{Q})$ on the standard generators $\sigma_1, \ldots, \sigma_{2g} \in \hat{B}_{2g+1}$ is given by

$$s_{\vec{b}}(\sigma)\, \sigma_i\, s_{\vec{b}}(\sigma)^{-1} = \mathfrak{f}_\sigma(y_i, \sigma_i^2)^{-1} \sigma_i^{\chi(\sigma)} \mathfrak{f}_\sigma(y_i, \sigma_i^2) \qquad (1 \leq i \leq 2g),$$

where $y_1 = 1$, $y_i = \sigma_{i-1} \cdots \sigma_1 \cdot \sigma_1 \cdots \sigma_{i-1}$ $(2 \leq i \leq 2g)$.

(4.2)   In $M_{0,2g+2}$, the image of $\vec{b}$ coincides with the tangential base point coming from a 1-parameter family X_b over $\mathbb{Q}[[q]]$ of deformation of X^0 constructed explicitly, as in [IN] §2, from a system of distinguished coordinates $\{r_{\mu/\lambda}\}_{\mu/\lambda}$ such that the values at the points $Q_{\nu_2}^0, \ldots, Q_{\nu_{2g+1}}^0$ are always 1 and with $\{r_{\nu/\lambda} := t_{\nu/\lambda}\}_{\nu/\lambda}$. (In [IN], we called $\{r_{\mu/\lambda}\}_{\mu/\lambda}$ a tangential structure on the '$\mathbb{P}^1_{01\infty}$-diagram' X^0.)

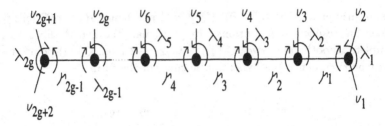

This kind of (tangential) base point was suggested by Drinfeld [Dr] after interpreting Grothendieck [G3], whose Galois property (4.1) was established by Ihara-Matsumoto [IM] in detail. For $\vec{b}$ satisfying both (4.1) and (4.2), one may employ the image, via a natural open immersion $M_{0,2g+4} \hookrightarrow \mathbb{A}^{2g+1} \setminus \Delta$, of the tangential base point in $M_{0,2g+4}$ constructed from the similar tangential-structured $(2g+4)$-pointed $\mathbb{P}^1_{01\infty}$-tree as in Ihara-Nakamura [IN]. Here, however, we shall look at a way to attach the property (4.2) to the tangential base point of Ihara-Matsumoto [IM], by introducing a canonical coordinate system of [IN] §2 on the formal neighborhood of the locus of X^0 in the moduli stack $\mathfrak{M}_{0,2g+2}$ of the stable $(2g+2)$-pointed $\mathbb{P}^1$-trees. Namely, gluing the following $\mathbb{Q}[[q_1, \ldots, q_{2g-1}]]$-algebras

(b1) $B_{\mu_i} = \mathbb{Q}[r, r', \dfrac{1}{1-r}, \dfrac{1}{1-r'}][[q_1, \ldots, q_{2g-1}]]/(rr' - q_i),$

         $r = r_{\mu_i/\lambda_i}, \ r' = r_{\mu_i/\lambda_{i+1}} \ (1 \le i \le 2g-1),$

(b2) $B_{\nu_i} = \mathbb{Q}[r, \dfrac{1}{1-r}][[q_1, \ldots, q_{2g-1}]],$

         $r = r_{\nu_i/\lambda} \ (1 \le i \le 2g+2),$

(b3) $B_{\lambda_i} = \mathbb{Q}[r, \dfrac{1}{r}, \dfrac{1}{1-r}][[q_1, \ldots, q_{2g-1}]],$

         $r = r_{\mu_j/\lambda_i} \ (1 \le j \le i \le 2g)$

along X^0, and applying Grothendieck's formal geometry, we obtain a sequence $X^0 = \mathfrak{X}^1 \subset \mathfrak{X}^2 \subset \cdots$ over the sequence of artinian schemes $\{\operatorname{Spec} \mathbb{Q}[[q_1, \ldots, q_{2g-1}]]/\mathfrak{q}^N\}_{N \ge 1}$, where $\mathfrak{q} = (q_1, \ldots, q_{2g-1})$, and hence a universal deformation $\tilde{X} \to \operatorname{Spec} \mathbb{Q}[[q_1, \ldots, q_{2g-1}]]$ of $X^0/\mathbb{Q}$. The representing morphism for this $\tilde{X}$ gives a local coordinate system of the locus of X^0 in $\mathfrak{M}_{0,2g+2}$. Our 1-parameter family X_b over $\mathbb{Q}[[q]]$ (4.2) is the pull back of $\tilde{X}$ by the diagonal specialization $\mathbb{Q}[[q_1, \ldots, q_{2g-1}]] \to \mathbb{Q}[[q]] \ (\forall q_i \mapsto q)$. Meanwhile, by simple calculations, we see that the local universal family $\tilde{X}$ generically parameterizes $(2g+2)$-pointed projective lines $(\mathbb{P}^1; Q_1, \ldots, Q_{2g+2})$ with $Q_1 = 0$, $Q_{2g+1} = 1$, $Q_{2g+2} = \infty$ and $Q_i = q_{i-1} \cdots q_{2g-1} \ (i = 2, \ldots, 2g)$. From this and the relations $Q_i = (v_i - v_1)/(v_{2g+1} - v_1)$, we can conclude that our q_i coincides with Ihara-Matsumoto's "t_i" (cf. [IM] p.179).

Let us compare the images of $\vec{v}$ and $\vec{b}$ on $M_{0,2g+2}$. Recall that $\vec{v}$ corresponds to the 1-parameter family $X_v/\mathbb{Q}[[q]]$ obtained from the sequence $\{\mathfrak{X}_v^N\}_{N\geq 1}$, where $\mathfrak{X}_v^N$ is the quotient of $\mathfrak{Y}^N$ by the hyperelliptic involution. Since this is also a deformation of X^0, there is a specialization homomorphism representing $X_v/\mathbb{Q}[[q]]$ in the form of

$$\mathbb{Q}[[q_1,\ldots,q_{2g-1}]] \longrightarrow \mathbb{Q}[[q]]$$
$$q_i \longmapsto f_i(q) \qquad (1 \leq i \leq 2g-1),$$

where $f_i(q)$ is a power series with $f_i(0) = 0$.

Lemma (4.3) *(i)* $f_i(q) = q^2 + \{higher\ terms\}$ *(i =even).*
(ii) $f_i(q) = 16q + \{higher\ terms\}$ *(i =odd).*

Proof. Let $\varpi : Y^0 \to X^0$ be the double covering morphism constructed in §3, and let $U_\mu^0 \subset X^0$ denote the affine open $\mathrm{Spec}\,(B_\mu/\mathbf{q})$. Then, $\mathfrak{X}_v^N|U_\mu^0$ is the spectrum of the ring of invariant functions on $\varpi^{-1}(U_\mu^0)$ mod q^N under the hyperelliptic involution.

(i) When i is even, this ring is $\mathbb{Q}[t,t',\frac{1}{1-t},\frac{1}{1-t'}][[q]]/(tt'-q^2,q^N)$ by (3.3), where $t = t_{\mu_i/\lambda_i}$, $t' = t_{\mu_i/\lambda_{i+1}}$. By assumption, this ring has to be isomorphic to $\mathbb{Q}[r,r',\frac{1}{1-r},\frac{1}{1-r'}][[q]]/(rr'-f_i(q),q^N)$ $(r = r_{\mu_i/\lambda_i}, r' = r_{\mu_i/\lambda_{i+1}})$ via some variable transformations of the form $r \equiv \frac{t}{t-1}, r' \equiv \frac{t'}{t'-1}$ mod q. Observing this isomorphism localized at (t,t'), we get $f_i(q) = q^2 + O(q^3)$.

(ii) When i is odd, we may employ a more a posteriori argument. On U_μ^0, the sequence $\{\mathfrak{X}_v^N|U_\mu^0\}_{N\geq 1}$ coincides with that induced from the Tate elliptic curve of level 2 ([DR]) modulo $\{\pm 1\}$. In this case, the Legendre function $\lambda(q) = 16q+\cdots$ $(q = e^{\pi\sqrt{-1}\tau})$ uniformizing $\mathbb{P}^1-\{0,1,\infty\}$ measures the difference between $\{\mathfrak{X}_v^N|U_\mu^0\}_{N\geq 1}$ and $\{\mathfrak{X}_b^N|U_\mu^0\}_{N\geq 1}$ (cf. [N3] §4). Since different values of $f_i'(0)$ give different deformation rings of $\mathbb{Q}[[t,t']]/(tt')$ over $\mathbb{Q}[q]/q^N$ $(N \geq 2)$ near P_μ^0, we conclude that 16 is the exact value. $\diamond$

§5. End of the proof

The fundamental group of the local neighborhood $\mathrm{Spec}\,\mathbb{Q}[[\mathbf{q}]]$ (where $\mathbf{q} = (q_1,\ldots,q_{2g-1})$) within $M_{0,2g+2}$ can be identified with $\mathrm{Aut}(\overline{\mathbb{Q}}\{\{\mathbf{q}\}\}/\mathbb{Q}[[\mathbf{q}]])$, where $\overline{\mathbb{Q}}\{\{\mathbf{q}\}\}$ is the union of the rings $k[[q_1^{1/n},\ldots,q_{2g-1}^{1/n}]]$ $(n \geq 1, [k : \mathbb{Q}] < \infty)$. It has an abelian normal subgroup $\hat{\mathbb{Z}}(1)^{2g-1}$ with independent generators $w_2,\ldots,w_{2g}$, where $w_{i+1} : q_i^{1/n} \mapsto q_i^{1/n}\zeta_n^{-1}$ $(\zeta_n = e^{2\pi\sqrt{-1}/n})$, $q_j^{1/n} \mapsto q_j^{1/n}$ $(j \neq i)$, and fits into the following exact sequence:

(5.1) $1 \to \hat{\mathbb{Z}}(1)^{2g-1} \to \mathrm{Aut}(\overline{\mathbb{Q}}\{\{\mathbf{q}\}\}/\mathbb{Q}[[\mathbf{q}]]) \to G_\mathbb{Q} \to 1.$

The image of w_i via the natural map $\mathrm{Aut}(\overline{\mathbb{Q}}\{\{\mathbf{q}\}\}/\mathbb{Q}[[\mathbf{q}]]) \to \pi_1(M_{0,2g+2},\vec{b})$ corresponds to the monodromy around the singular divisor '$q_{i-1} = 0$' ($2 \leq i \leq 2g$). This is the Dehn twist along a simple closed curve ω_i on the $(2g+2)$-pointed sphere pinching $P^0_{\mu_{i-1}} \in X^0$ (indicated below), and comes from $w'_i = y_2 y_3 \cdots y_i = (\sigma_1 \cdots \sigma_{i-1})^i \in \hat{B}_{2g+1}$.

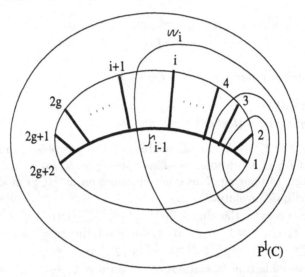

Our two tangential base points $\vec{b}$ and $\vec{v}$ give different splitting sections s_b, s_v of (5.1) respectively. The $s_b(\sigma)$ ($\sigma \in G_{\mathbb{Q}}$) transforms each Puiseux power series $\sum_{\alpha \in \mathbb{Q}^{2g-1}} a_\alpha \mathbf{q}^\alpha$ to $\sum_{\alpha \in \mathbb{Q}^{2g-1}} \sigma(a_\alpha) \mathbf{q}^\alpha$. On the other hand, we can perceive the action by $s_v(\sigma)$ to be the coefficientwise Galois action on the Puiseux power series after specialized via $q_i^{1/n} \to f_i(q)^{1/n}$ ($n \geq 1$): The specialization process via (4.3) becomes 'power-compatible' after setting $1^{1/n} = \zeta_n$, $2^{1/n} \in \mathbb{R}_{>0}$ (this corresponds to a choice of 'natural' chemin connecting $\vec{v}$ and $\vec{b}$.) Then, for each $\alpha = (\alpha_i) \in \mathbb{Q}^{2g-1}$, $f(\mathbf{q}^\alpha) = \prod_i f_i(q)^{\alpha_i}$ makes sense in $\overline{\mathbb{Q}}\{\{q\}\}$, and $s_v(\sigma)$ transforms it into $e^{2\pi\sqrt{-1}(4\rho_2(\sigma)\sum_{i:\mathrm{odd}} \alpha_i)} f(\mathbf{q}^\alpha)$, where $\rho_2 : G_{\mathbb{Q}} \to \hat{\mathbb{Z}}(1)$ is the Kummer 1-cocycle with $2^{1/n(\sigma-1)} = \zeta_n^{\rho_2(\sigma)}$. Comparing these two operations on Puiseux series, we obtain:

$$s_b(\sigma) = \prod_{\substack{j=2 \\ \mathrm{even}}}^{2g} w_j^{4\rho_2(\sigma)} s_v(\sigma).$$

Here, it is noteworthy that, although the i-th component of the "tangent vector" $\vec{v}$ vanishes via f_i for i even, its non-trivial principal term ('q^2', in this case) still works well in carrying Galois properties from the tangential base point $\vec{v}$. The author is indebted to Prof. Deligne for this crucial remark.

Then, let us be back to the diagram (2.1), and let s'_b, s'_v be the splitting homomorphisms of the surjection $\pi_1(\mathcal{H}_{g,1}) \to G_\mathbb{Q}$ coming down from the tangential base points $\vec{b}$, $\vec{v}$ on $A_v^{2g+1} \setminus \Delta$. Considering the above relation in $\pi_1(M_{0,2g+2}/S_{2g+1})$ and lifting it back to $\pi_1(A_u^{2g+1} \setminus D)$ (2.1), we see

$$s'_b(\sigma) = \prod_{\substack{j=2 \\ \text{even}}}^{2g} (w'_j)^{4\rho_2(\sigma)} \cdot w_{2g+1}^{c_\sigma} \cdot s'_v(\sigma)$$

for some 1-cocycle $c : G_\mathbb{Q} \to \hat{\mathbb{Z}}(1)$ ($\sigma \mapsto c_\sigma$). Let s'_b, s'_v also denote the induced sectional homomorphisms $G_\mathbb{Q} \to \pi_1(M_{g,1})$ from (2.1) and $\mathcal{H}_{g,1} \hookrightarrow M_{g,1}$. Then the conjugate actions by $s'_b(\sigma)$ ($\sigma \in G_\mathbb{Q}$) on $a_1, \cdots, a_{2g}$ are described just as direct images of (4.1):

$$s'_b(\sigma) a_i s'_b(\sigma)^{-1} = \mathfrak{f}_\sigma(y_i, a_i^2)^{-1} a_i^{\chi(\sigma)} \mathfrak{f}_\sigma(y_i, a_i^2) \qquad (1 \leq i \leq 2g).$$

On the other hand, the conjugate actions by $s'_v(\sigma)$ ($\sigma \in G_\mathbb{Q}$) on $d_{\pm *}, e_*$'s are a priori via the cyclotomic character. The reason is that our $Y^0/\mathbb{Q}$ lies over a representative point of a maximally degenerate locus in the moduli stack $\mathfrak{M}_{g,1}$ of the 1-pointed stable curves of genus g, whose local neighborhood within $M_{g,1}$ has the geometric fundamental group $\hat{\mathbb{Z}}(1)^{3g-2}$ with the commutative $3g - 2$ generators $d_{\pm *}, e_*$'s.

We define then the sectional homomorphism $s : G_\mathbb{Q} \to \pi_1(M_{g,1})$ of Theorem A by

$$s(\sigma) := w_{2g+1}^{-c_\sigma} s'_b(\sigma) \qquad (\sigma \in G_\mathbb{Q}).$$

Since $w_{2g+1}(=$"hyperelliptic involution") commutes with $a_1, \ldots, a_{2g}$, the conjugate action by $s(\sigma)$ on the a_*'s are in the same way as that by $s'_b(\sigma)$, hence in the desired way. As for $d_{\pm *}, e_*$'s, notice that $\{w'_{\text{even}}\}$ and $\{d_{\pm *}, e_*\}$ commute elementwise with each other. Then, we see that $s(\sigma)$ operates on the $d_{\pm *}, e_*$'s in the same way as $s'_v(\sigma)$ by conjugation, i.e., via the cyclotomic character. Thus, Theorems A and A' are both settled.

§6. Complementary notes

This section describes complementary remarks to the results of this note, whose details will be included in a forthcoming paper [N3]. Let $\mathfrak{M}_{g,n}$ denote the stack$_{/\mathbb{Q}}$ of the ordered n-pointed stable curves of genus g, and $M_{g,n} \subset \mathfrak{M}_{g,n}$ its nonsingular locus (Deligne-Mumford [DM], Knudsen [K]). By using Grothendieck-Murre's theory [GM], one can observe behaviors of the fundamental group of the tubular neighborhood in $M_{g,n}$ of the divisor of the form $\mathfrak{M}_{g_1,n_1} \times \mathfrak{M}_{g_2,n_2}$ ($g = g_1 + g_2, n = n_1 + n_2 - 2$) inside $\pi_1(M_{g,n})$

(cf. [N2]). Roughly speaking, the "coupling device" considered in [N2] enables one to relate Galois-Teichmüller modular groups of different genera by "sewing up" two topological types of Riemann surfaces along boundaries.

By looking at the arguments of previous sections along the coupling of $\mathfrak{M}_{g-1,1} \times \mathfrak{M}_{1,2} \subset \mathfrak{M}_{g,1}$, we can see that the indeterminate parameter c_σ in Sect.5 is negligible. Thus, the $G_{\mathbb{Q}}$-action at the base point $\vec{b}$ is essentially the desired one. Meanwhile, the $G_{\mathbb{Q}}$-action at $\vec{v}$ differs from it by the factors $(w'_j)^{4\rho_2(\sigma)}$ ($j \geq 2$, even). Since $\rho_2(\sigma)$ is recovered from the ratio of the upper components of $\mathfrak{f}_\sigma(\left(\begin{smallmatrix}1&2\\0&1\end{smallmatrix}\right), \left(\begin{smallmatrix}1&0\\-2&1\end{smallmatrix}\right)) \in \mathrm{SL}_2(\hat{\mathbb{Z}})$ ([N3] §4), we may say that both $G_{\mathbb{Q}}$-actions on $\hat{\Gamma}_g^1$ can be written in terms of parameters $(\chi(\sigma), \mathfrak{f}_\sigma) \in \widehat{GT}$.

The natural forgetful map $M_{g,n} \to M_{g,0}$ obtained by forgetting the marking points induces the exact sequence

$$1 \to \hat{\Pi}_{g,0}^{(n)} \to \hat{\Gamma}_g^n \to \hat{\Gamma}_g^0 \to 1,$$

where $\hat{\Gamma}_g^n$ (resp. $\hat{\Pi}_{g,0}^{(n)}$) denotes the profinite completion of the mapping class group of an n-pointed genus g surface (resp. of the pure braid group with n-strings on a genus g surface). Note that our $G_{\mathbb{Q}}$-actions on $\hat{\Gamma}_g^1$ induce those on $\hat{\Gamma}_g^0$ by the above forgetful mapping with $n = 1$. Matsumoto [M] studied the Galois action on the profinite braid group for a fixed affine smooth curve, and decomposed it into the Galois actions on $\hat{B}_n$ and the π_1 of the curve. One can also consider his insight in our coupling context as follows. Let us introduce $M_{g,[n]} := M_{g,n}/S_n$, the moduli stack over $\mathbb{Q}$ obtained by letting the marking points unordered. Then the kernel $\hat{\Gamma}_g^{[n]}$ of $\pi_1(M_{g,[n]}) \to G_{\mathbb{Q}}$ includes $\hat{\Gamma}_g^n$ as an open subgroup, and is isomorphic to the profinite completion of the mapping class group of a closed surface of genus g preserving n points $Q_1, \ldots, Q_n$ as a set.

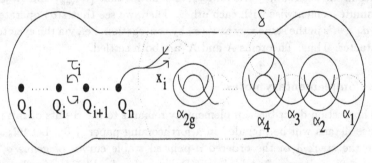

The group $\hat{\Gamma}_g^{[n]}$ has the following three types of generators: (1) $a_1, \ldots, a_{2g}$, d (Dehn twists); (2) $\tau_1, \ldots, \tau_{n-1}$ (braids); (3) $x_1, \ldots, x_{2g}$ (peripheral paths of Q_n around the handles).

By modifying the constructions of this note, one can get a tangential base point attached to the locus of the maximally degenerate marked stable curve whose dual graph (with legs) looks like the following picture, at which $\sigma \in G_{\mathbb{Q}}$ acts on $\hat{\Gamma}_g^{[n]}$ by

(1) $\sigma(a_i) = \mathfrak{f}_\sigma(y_i, a_i^2)^{-1} a_i^{\chi(\sigma)} \mathfrak{f}_\sigma(y_i, a_i^2)$, $\sigma(d) = d^{\chi(\sigma)}$ $(1 \leq i \leq 2g)$;

(2) $\sigma(\tau_j) = \mathfrak{f}_\sigma(\eta_j, \tau_j^2)^{-1} \tau_j^{\chi(\sigma)} \mathfrak{f}_\sigma(\eta_j, \tau_j^2)$ $(1 \leq j \leq n - 1)$,
        where $\eta_1 = 1$, $\eta_j = \tau_{j-1} \cdots \tau_1 \cdot \tau_1 \cdots \tau_{j-1}$ $(j \geq 2)$;

(3) $\sigma(x_i)$ $(1 \leq i \leq 2g)$ are described explicitly in terms of $\widehat{GT}$.

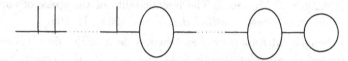

The third part action is, in effect, the main theme of [N3], where, based on [IN], established is a concrete procedure of computing the limit behaviors of exterior Galois representations when (marked) algebraic curves maximally degenerate to various types of marked stable curves consisting of 3-pointed projective lines. This procedure is, as shown in the author's talk at the Luminy conference, described in terms of a graph of profinite groups over the dual graph of the special fibre whose edge/vertex groups are products of free profinite groups of rank 1 or 2 with standard Galois actions.

Acknowledgments. The author would like to thank Prof. P.Lochak, L.Schneps for sharing enjoyable tests of some known relations of twist generators of mapping class groups by direct calculations after Th. A,A'. He is also very grateful to Prof. P.Deligne, V.G.Drinfeld for helpful suggestions on ideas of tangential base points and moduli theory (cf. Sect.4,5), and to Prof. D.Harbater, F.Pop for instructive discussions on patching techniques from different points of view. The result of the present note wouldn't have occurred to the author without his joint work with Prof. Y.Ihara [IN]. The author would like to express his sincere thanks to Prof. Y.Ihara also for several pieces of advice for improving the presentation of this note. He also should like to thank Prof. S.Morita for several helpful comments on literature and topological figures. In addition to them, his hearty thanks should also go to many Japanese colleagues on Galois-π_1-theory and related topologists together with foreign inverse-Galois researchers (cf. [F]), for their mutually related works helped his imagination into the formula of this note.

References

[Be]   G.V.Belyi, On galois extensions of a maximal cyclotomic field, *Math. U.S.S.R. Izv.* **14** (1980), 247–256.

[BH]   J.Birman, H.Hilden, Isotopies of homomorphisms of Riemann surfaces *Ann. of Math.* **97** (1973), 424–439.

[De]   P.Deligne, Le groupe fondamental de la droite projective moins trois points, in *The Galois Group over* $\mathbb{Q}$, eds. Y.Ihara, K.Ribet, J-P. Serre, Springer-Verlag, 1989, 79-297.

[DM]   P.Deligne, D.Mumford, The irreducibility of the space of curves of given genus, *Publ. Math. I.H.E.S.* **36** (1969), 75–109.

[DR]   P.Deligne, M.Rapoport, Les schémas de modules de courbes elliptiques, in *Modular functions of one variable II*, Lecture Notes in Math. **349**, Springer, Berlin Heidelberg New York , 1973.

[Dr]   V.G.Drinfeld, On quasitriangular quasi-Hopf algebras and a group closely connected with $\mathrm{Gal}(\overline{\mathbb{Q}}/\mathbb{Q})$, *Leningrad Math. J.* **2(4)** (1991), 829–860.

[F]   M.Fried et al. (eds.) *Recent Developments in the Inverse Galois Problem*, Contemp. Math. **186**, 1995.

[G]   A.Grothendieck, *Éléments de Géométrie Algébrique III*, Publ. Math. I.H.E.S. **11**, 1961.

[G2]   A.Grothendieck, *Revêtements Etales et Groupe Fondamental (SGA1)*, Lecture Notes in Math. **224**, Springer, Berlin Heidelberg New York, 1971.

[G3]   A.Grothendieck, *Esquisse d'un Programme*, 1984, (this volume).

[GM]   A.Grothendieck, J.P.Murre, *The tame fundamental group of a formal neighborhood of a divisor with normal crossings on a scheme*, Springer Lect. Notes in Math. **208**, Berlin, Heidelberg, New York, 1971.

[Ha]   D.Harbater, Formal patching and adding branch points *Amer. J. Math.* **115** (1993), 487–508.

[Hu]   S.Humphries, Generators for the mapping class group of an orientable surface, Lect. Notes in Math. **722** Springer, Berlin Heidelberg New York, 1979, 44–47.

[Ih]   Y.Ihara, Braids, Galois groups and some arithmetic functions Proc. ICM, Kyoto 1990, 99–120.

[Ih2] Y.Ihara, On the embedding of $Gal(\bar{Q}/Q)$ into $\widehat{GT}$ in *The Grothendieck theory of Dessins d'Enfants*, L.Schneps (ed.), London Math. Soc. Lect. Note Ser. **200**, Cambridge Univ. Press, 1994, 289-306.

[IM] Y.Ihara, M.Matsumoto, On Galois actions on profinite completions of braid groups, in [F], 173-200.

[IN] Y.Ihara, H.Nakamura, On deformation of maximally degenerate stable marked curves and Oda's problem, to appear in *J. reine angew. Math.*

[K] F.F.Knudsen, The projectivity of the moduli space of stable curves II: The stacks $M_{g,n}$, *Math. Scand* **52** (1983), 161–199.

[L] W.B.R.Lickorish, A finite set of generators for the homeotopy group of a 2-manifold, *Proc. Camb. Phil. Soc.* **60** (1964), 769–784.

[M] M.Matsumoto, On Galois representations on profinite braid groups of curves, *J. reine angew. Math.* **474** (1996), 169–219.

[M2] M.Matsumoto, Galois group G_Q, singularity E_7, and moduli M_3, *Geometric Galois Actions II*.

[N] H.Nakamura, Galois rigidity of pure sphere braid groups and profinite calculus, *J. Math. Sci., Univ. Tokyo* **1** (1994), 71–136.

[N2] ———, Coupling of universal monodromy representations of Galois-Teichmüller modular groups, *Math. Ann.* **304** (1996), 99–119.

[N3] ———, Limits of Galois representations in fundamental groups along maximal degeneration of marked curves I, II, in preparation.

[P] F.Pop, 1/2 Riemann existence theorem with Galois action, in *Algebra and number theory*, G.Frey, J.Ritter (eds.), de Gruyter, Berlin, 1994, 193-218.

[Oda] Takayuki Oda, Etale homotopy type of the moduli spaces of algebraic curves, preprint 1990, (this volume).

[S] L.Schneps, The Grothendieck-Teichmüller group $\widehat{GT}$: a survey, (this volume)

[W] B. Wajnryb, A simple presentation for the mapping class group of an orientable surface, *Israel J. Math.* **45** (1983), 157–174, Erratum (with J.Birman), *loc. cit.* **88** (1994), 425–427.

Dept. of Math. Sci., Univ. of Tokyo, and
Institute for Advanced Study
h-naka@ms.u-tokyo.ac.jp

Deux lettres sur la cohomologie non abélienne *

Jean-Pierre Serre

Chère Leila

Voici comment je vois ces questions de H^0 et de H^1.

Tout d'abord, une notation: soit $\mathcal{C}$ une catégorie de groupes finis satisfaisant à l'axiome suivant: si $G \in \mathcal{C}$, tout groupe isomorphe à un sous-groupe de G, à un quotient de G, ou à un produit $G \times \cdots \times G$, appartient à $\mathcal{C}$. Si $F = F(x,y)$ est le groupe libre de base $\{x,y\}$, je noterai $F_{\mathcal{C}}$ la limite projective des quotients de F qui appartiennent à $\mathcal{C}$ (c'est le complété de F pour la "$\mathcal{C}$-topologie", en un sens évident).

Si l'on prend pour $\mathcal{C}$ la catégorie de tous les groupes finis, on trouve pour $F_{\mathcal{C}}$ le groupe profini libre qui t'intéresse; mais j'ai envie de pouvoir prendre d'autres catégories, par exemple:

les groupes finis dont l'ordre ne fait intervenir qu'un ensemble donné de nombres premiers (par exemple 2 et 3, pour la suite); en particulier les p-groupes.

J'aurai besoin plus loin que $\mathcal{C}$ possède la propriété suivante, relative à un nombre premier p:

(E_p) – Si $1 \to N \to G \to H \to 1$ est une suite exacte de groupes finis telle que $H \in \mathcal{C}$ et que N soit un p-groupe (ou $N \in \mathcal{C}$ et H un p-groupe), alors on a $G \in \mathcal{C}$. (Bref, $\mathcal{C}$ est stable par extensions par les p-groupes.)

1. Action d'un élément s d'ordre 2.

Soit $S = \langle s \rangle$ un groupe d'ordre 2. On le fait agir sur F (et aussi sur $F_{\mathcal{C}}$) par $x,y \mapsto y,x$. On peut donc parler de $H^i(S, F_{\mathcal{C}})$ pour $i = 0, 1$.

Théorème 1 – *On suppose que $\mathcal{C}$ satisfait à l'axiome E_2 ci-dessus. On a alors $H^0(S, F_{\mathcal{C}}) = 0$ et $H^1(S, F_{\mathcal{C}}) = 0$.*

(Cela s'applique dans le cas où $\mathcal{C}$ est la catégorie de tous les groupes, et aussi dans le cas où $\mathcal{C}$ est la catégorie des 2-groupes. Cela ne s'applique pas au cas où $\mathcal{C}$ est la catégorie des p-groupes, $p \neq 2$, auquel cas on a d'ailleurs $H^0(S, F_{\mathcal{C}}) \neq 0$, comme on le voit facilement.)

---

* Editor's note: These letters are dated April 30,1995 and May 22,1995 respectively.

Démonstration de $H^0(S, F_C) = 0$.

Formons le produit semi-direct $S \ltimes F_C$; il contient s, x, y, et l'on a $s^2 = 1$, $sxs^{-1} = y$. Il s'agit de montrer que le centralisateur $Z(s)$ de s dans ce groupe est réduit à $\langle 1, s \rangle$. Or, si $G \in C$ est quotient de ce groupe, G est muni d'un couple s_G, x_G avec $s_G^2 = 1$. Si de plus, z est un élément de $Z(s)$ distinct de 1 et de s, alors z donne un élément z_G de G appartenant au centralisateur de s_G dans G; de plus on peut choisir G tel que z_G soit distinct de 1 et de s_G (et aussi que $s_G \neq 1$): cela se voit en prouvant que $S \ltimes F_C$ est limite projective de groupes finis appartenant à C, ce qui est facile.

On va tirer de là une contradiction, grâce au lemme suivant, que je démontrerai plus bas:

Lemme 0 – *Soient p un nombre premier, G un groupe fini et u un élément de G d'ordre p. Il existe une extension scindée de G:*

$$1 \to P \to G' \to G \to 1,$$

et un élément u' de G', d'ordre p, relevant u, tels que:

(a) *P est un p-groupe abélien élémentaire;*

(b) *l'image dans G du centralisateur $Z_{G'}(u')$ est contenue dans $\langle u \rangle$.*

Appliquons ce lemme au groupe G ci-dessus, et à l'élément $u = s_G$ (avec $p = 2$, bien sûr). On trouve un groupe G', et un élément u' de G' d'ordre 2 relevant u. Choisissons un relèvement x' de x_G, et posons $y' = u'x'u'^{-1}$. D'après (E_2), on a $G' \in C$. Il existe donc un unique morphisme $F_C \to G'$ qui applique x, y sur x', y', et ce morphisme se prolonge à $S \ltimes F_C$ en demandant que s aille sur u'. L'élément z de $Z(s)$ a une image z' dans G' qui appartient au centralisateur de u' dans G'. Mais, d'après la propriété (b) du lemme, l'image de z' dans G appartient au groupe $\langle s_G \rangle$; or cette image est égale à z_G. Contradiction.

Il me reste á démontrer le lemme. Ce n'est pas bien difficile. Appelons $\underline{1}$ le $\langle u \rangle$-module $\mathbf{Z}/p\mathbf{Z}$ (avec action triviale – aucune autre n'est possible...), et définissons P comme le G-module *induit* correspondant. On peut voir P, de façon plus concrète, comme le G-module des fonctions $f : G \to \mathbf{Z}/p\mathbf{Z}$ qui sont telles que $f(ux) = f(x)$ pour tout $x \in G$, le groupe G agissant par $(g.f)(x) = f(xg)$. Notons e l'élément de P qui vaut 1 sur $\langle u \rangle$ et 0 ailleurs. On a $g.e = e$ si et seulement si g appartient à $\langle u \rangle$. En particulier u fixe e. On va prendre pour G' le *produit semi-direct* $G' = P \rtimes G$. L'élément e s'identifie alors à un élément d'ordre p de P, qui est centralisé par les éléments de $\langle u \rangle$, mais n'est pas centralisé par les autres éléments de G. On choisit pour u' le produit $e.u$, et l'on vérifie tout de suite qu'il a les propriétés voulues.

Démonstration de $H^1(S, F_\mathcal{C}) = 0$.

J'utilise encore le produit semi-direct $S \ltimes F_\mathcal{C}$; c'est une extension (scindée)

$$1 \to F_\mathcal{C} \to S \ltimes F_\mathcal{C} \to S \to 1,$$

et l'on sait que l'on peut identifier les éléments de $H^1(S, F_\mathcal{C})$ aux classes de conjugaison (par $F_\mathcal{C}$) des différents scindages $S \to S \ltimes F_\mathcal{C}$. On veut montrer qu'il n'y a, à conjugaison près, que le scindage évident $S \to S \ltimes F_\mathcal{C}$. En termes plus concrets, il nous faut donc prouver que, si $s' \in S \ltimes F_\mathcal{C}$ est d'ordre 2, et a pour image s dans S (i.e. $s' = s.x$ avec $x \in F_\mathcal{C}$), alors s' est conjugué de s par un élément de $F_\mathcal{C}$. Supposons que ce ne soit pas le cas. Il y aurait alors un quotient fini G de $F_\mathcal{C}$ appartenant à $\mathcal{C}$, tel que les images s'_G et s_G de s' et s ne soient pas conjuguées dans G. On va utiliser le lemme suivant, que je démontrerai plus bas:

Lemme 1 – *Soient G un groupe fini, et $u, u_1, \ldots, u_k$ des éléments de G d'ordre p. On suppose que u n'est conjugué d'aucun élément de l'un des $\langle u_i \rangle$. Il existe alors une extension*

$$1 \to P \to G' \to G \to 1$$

ayant les propriétés suivantes:

(a) *P est un p-groupe abélien élémentaire*;

(b) *les u_i sont relevables en des éléments d'ordre p de G', mais u ne l'est pas.*

(Bien sûr, p est un nombre premier fixé.)

On applique le lemme à G, avec $p = 2$, $k = 1$, $u = s'_G$ et $u_1 = s_G$. On choisit un relèvement $x_{G'}$ de x_G dans G', ainsi qu'un relèvement $s_{G'}$ d'ordre 2 de s_G; on définit $y_{G'}$ comme le conjugué de $x_{G'}$ par $s_{G'}$. Comme précédemment, ces éléments définissent un morphisme $S \ltimes F_\mathcal{C} \to G'$. L'image de s' par ce morphisme est un élément d'ordre 2 de G' qui relève s'_G; cela contredit (b).

Il me reste à démontrer le lemme. On choisit le même groupe P que pour le lemme 0, à savoir le G-module induit du $\langle u \rangle$-module $\underline{1}$. D'après le "lemme de Shapiro", $H^2(G, P)$ s'identifie à $H^2(\langle u \rangle, \underline{1})$, qui est d'ordre p. On choisit alors pour G' une extension de G par P qui est *non scindée*. On vérifie (par des arguments du genre "induction-restriction" à la Mackey) que la restriction de cette extension à $\langle u \rangle$ (resp. à l'un des $\langle u_i \rangle$) est non triviale (resp. est triviale). C'est ce que l'on veut.

Remarque. Des arguments analogues s'appliquent dans la catégorie des groupes discrets (éventuellement infinis). On en déduit que $H^0(S, F) = 0$

et $H^1(S, F) = 0$. Ce n'est pas très intéressant, car le premier énoncé se démontre tout de suite en utilisant la structure des mots d'un groupe libre $F(x, y)$, et le second se prouve en appliquant un résultat général sur les sous-groupes finis des amalgames (cf. par exemple *Arbres, amalgames et* SL$_2$, p. 13 et p. 53).

2. Action d'un element t d'ordre 3.

Soit $T = \langle t \rangle$ un groupe d'ordre 3. On le fait agir sur F, et sur F_C, par $x, y, z \mapsto y, z, x$, où z est défini par la formule $xyz = 1$. D'où encore des $H^i(T, F_C)$ pour $i = 0, 1$; on va les déterminer:

Théorème 2 – *On suppose que C satisfait à l'axiome (E₃) du début. On a alors*
$H^0(T, F_C) = 0$, *et $H^1(T, F_C)$ a deux éléments* (qui seront explicités).

(Cela s'applique lorsque C est la catégorie de tous les groupes finis, ou bien celle des 3-groupes.)

Démonstration de $H^0(T, F_C) = 0$.

Ici encore, on va se servir du produit semi-direct $T \ltimes F_C$; il contient t, x, y, z avec $t^3 = 1$, $txt^{-1} = y$, $tyt^{-1} = z$, $tzt^{-1} = x$ et $xyz = 1$. Si l'on définit θ par $\theta = xt$, on a

$$\theta^3 = xtxtxt = x.txt^{-1}.t^2xt^{-2}.t^3 = xyz.t^3 = 1.$$

De plus, t et θ "reconstituent" x, y, z en définissant x par $x = \theta t^{-1}$, $y = txt^{-1}$ et $z = tyt^{-1}$. Il nous faut montrer que le centralisateur de T dans $T \ltimes F_C$ est réduit à T. Supposons donc que ce centralisateur possède un élément g qui ne soit pas dans T. Il existerait alors un quotient fini G de $T \ltimes F_C$ dans lequel t, θ, g donnent des éléments t_G, θ_G, g_G, avec g_G commutant à t_G, g_G non dans $\langle t_G \rangle$, et t_G et θ_G d'ordre 3. On applique le lemme 0, avec $p = 3$ et $u = t_G$. On obtient un groupe G', extension scindée de G par P, et un relèvement u' d'ordre 3 de u dont le centralisateur s'envoie dans $\langle u \rangle$. On relève alors θ en un élément θ' d'ordre 3 de G' (c'est possible vu que l'extension est scindée). Le couple u', θ' définit un morphisme $T \ltimes F_C \to G'$; d'où un élément $g_{G'}$ de G' qui appartient au centralisateur de u' et a pour image g_G dans G, ce qui contredit les hypothèses faites.

Détermination de $H^1(T, F_C)$.

La méthode est la même qu'au n° 1: on interprète les éléments de $H^1(T, F_C)$ comme classes de conjugaison de scindages $T \to T \ltimes F_C$. Or il y a *deux* tels scindages qui sont évidents:

le scindage trivial $t \mapsto t$;

le scindage $t \mapsto \theta = xt$;

(En termes de cocycles, le premier scindage correspond au cocycle $a : T \to F_C$ tel que $a(1) = a(t) = a(t^2) = 1$; le second correspond au cocycle b tel que $b(1) = 1$, $b(t) = x$, $b(t^2) = xy$.)

Ces deux scindages ne sont pas conjugués (cela se voit en rendant F_C abélien, par exemple). Il nous faut montrer qu'il n'y en a pas d'autre, à conjugaison près. S'il y en avait un autre, il existerait un élément g d'ordre 3 de $T \ltimes F_C$ qui ne serait contenu, à conjugaison près, ni dans $\langle t \rangle$, ni dans $\langle \theta \rangle$. On aurait une situation analogue dans un quotient fini G convenable, où cela donnerait des éléments d'ordre 3: g_G, t_G et θ_G. On applique alors le lemme 1 à G, avec $p = 3$, $k = 2$, $u = g_G$, $u_1 = t_G$, $u_2 = \theta_G$. On en déduit une extension $G' \to G$ où t_G et θ_G se relèvent en des éléments d'ordre 3, mais pas g_G. Le couple des relèvements de t_G et θ_G définit alors un morphisme $T \ltimes F_C \to G'$, et l'image par ce morphisme de g est un élément d'ordre 3 de G' relevant g_G: contradiction.

3. Action du groupe S_3.

Soit S_3 le groupe symétrique de trois lettres. Je note s un élément d'ordre 2 de ce groupe, et t un élément d'ordre 3. On fait agir S_3 sur F et F_C par

$$s : x, y, z \mapsto y^{-1}, x^{-1}, z^{-1} \quad \text{et} \quad t : x, y, z \mapsto y, z, x.$$

(Noter que l'action de s est différente de celle du n°1, mais qu'elle lui est isomorphe, comme on le voit en remarquant que s permute les générateurs X, Y de F donnés par $X = x$, $Y = y^{-1}$.)

Théorème 3 – *On suppose que C satisfait à (E$_2$) et (E$_3$). On a alors $H^0(S_3, F_C) = 0$ et $H^1(S_3, F)$ a deux éléments.*

L'assertion sur H^0 est triviale à partir du th.1 (ou du th.2: au choix!)

En ce qui concerne le H^1, on peut encore l'interpréter comme l'ensemble des classes de scindages $S_3 \to S_3 \ltimes F_C$. Or, à côté du scindage trivial $s, t \mapsto s, t$, il y a aussi:

$$s, t \mapsto \sigma, \theta \quad \text{où } \sigma = xys = sz \text{ et } \theta = xt \text{ comme au n°2}.$$

(Pour vérifier que ces formules donnent un scindage, on doit voir que $\sigma^2 = 1$ et $\sigma\theta = \theta^2\sigma$: petit calcul!) Ces scindages ne sont pas conjugués car leurs restrictions à $T = \langle t \rangle$ ne le sont pas. Il faut voir qu'il n'y a qu'eux, à conjugaison près. Or on connait déjà, grâce au th.2, leurs restrictions à T. On est ainsi ramené à traiter deux cas:

a) le scindage $S_3 \to S_3 \ltimes F_C$ a pour restriction à T le scindage trivial $t \mapsto t$. Soit alors s' l'image de s dans ce scindage; s' conjugue t en t^2. Comme s a la même propriété, le produit ss' centralise t; d'autre part, ce produit appartient à F_C. Or on sait (th.2) que le seul élément de F_C qui commute à t est l'élément 1. On a donc $ss' = 1$, i.e. $s' = s$, et le scindage considéré est le scindage trivial.

b) le scindage $S_3 \to S_3 \ltimes F_C$ a pour restriction à T le scindage $t \mapsto \theta$. Soit σ' l'image de s dans ce scindage. Les éléments σ' et σ conjuguent θ en θ^2. Leur produit $\sigma\sigma'$ centralise donc θ. Comme ce produit appartient à F_C, on est ainsi ramené à montrer que *le centralisateur de θ dans F_C est réduit à 1*. Je vais déduire ce résultat de celui déjà démontré pour t (th.2), en prouvant que l'automorphisme de F_C défini par θ est *conjugué* (par un automorphisme de F_C – et même de F) de celui défini par t. Pour cela, on considère les trois éléments X,Y,Z de F définis par

$$X = x^{-1}, \ Y = xy^{-1}x^{-1}, \ Z = xy.$$

On a $XYZ = 1$, et il est clair qu'il y a un automorphisme $f : F \to F$ qui transforme x,y,z en X,Y,Z. D'autre part, on vérifie que

$$\theta X \theta^{-1} = Y, \ \theta Y \theta^{-1} = Z, \ \theta Z \theta^{-1} = X.$$

D'où le fait que θ est conjugué de t par f. Cqfd.

Remarque finale. Il est clair que le genre d'argument que j'ai employé peut s'appliquer, plus généralement, pour démontrer que, dans certains groupes profinis, les éléments d'ordre fini sont "ceux qui sont évidents" (à conjugaison près), et que leurs centralisateurs sont "tout petits" – exactement comme dans le cas discret, pour les amalgames. Si cela t'intéresse, je peux essayer de fabriquer des énoncés plus précis.

Bien à toi

J-P. Serre

Chère Leila

Je voudrais te raconter une autre façon d'aborder les questions de centralisateurs, point fixes, etc. dont nous avons déjà parlé.

Soit G un groupe, et soient G_i une famille finie de sous-groupes finis de G. On va s'intéresser à la propriété suivante des G_i:

(*) Tout sous-groupe fini A de G est conjugué d'un sous-groupe de l'un des G_i. De plus, si $gG_ig^{-1} \cap G_j \neq 1$, on a $i = j$ et $g \in G_i$. (Note que cela entraîne que le centralisateur de G_i dans G est contenu dans G_i si $G_i \neq 1$ (ce que je supposerai dans la suite); même chose pour le normalisateur.

C'est une condition très forte. Elle est toutefois satisfaite dans quantité de cas intéressants. Par exemple:

– G est produit libre des G_i et d'un groupe sans torsion. Cela se démontre par des méthodes arboricoles.

Ce cas contient comme cas particulier celui que l'on rencontre dans les revêtements ramifiés de la droite projective, où G est engendré par des éléments z_j et x_i avec les relations $\prod z_j \prod x_i = 1$, les x_i étant d'ordre finis imposés e_i; bien sûr on prend pour G_i les $\langle x_i \rangle$.

– Même chose que précédemment, mais avec pour G le groupe universel pour les surfaces de genre g quelconque, avec ramification imposée.

(J'ai oublié de dire que, dans les deux cas, il vaut mieux imposer au revêtement universel d'être hyperbolique. Peu importe.)

Note que la propriété (*) s'applique aussi bien aux groupes discrets qu'aux groupes profinis. J'ai envie de donner un critère pour que (*) soit vrai dans le cadre profini :

Je pars d'un groupe discret G avec des G_i comme ci-dessus. Je fais une première hypothèse sur G:

(H) Il existe un entier N tel que, pour tout entier $q > N$, et tout G-module fini M, l'application de restriction $H^q(G, M) \to \prod H^q(G_i, M)$ soit un isomorphisme.

Editor's note: a complete exposition of the ideas and results described in this letter is given in the appendix by Claus Scheiderer to the article *A cohomological interpretation of the Grothendieck-Teichmüller group* by P. Lochak and L. Schneps, *Inv. Math.*

(On démontre que (H) est vrai dans les cas des groupes fondamentaux ci-dessus, avec $N = 1$ ou 2.)

Je me suis aperçu il y a longtemps que (H) *entraîne* (*), pour les groupes discrets. La démonstration a été publiée par Huebschmann, J. Pure Appl. Algebra **14** (1979), 137-143.

Soit maintenant $\hat{G}$ le complété profini de G (vis-à-vis de la catégorie de tous les groupes finis – on pourrait demander moins). Je vais faire l'hypothèse que G est "bon" au sens de *Cohomologie Galoisienne*, Chap. I, n° 2.6, exerc. 2, i.e. que les groupes de cohomologie de G et de $\hat{G}$ sont "les mêmes". La propriété (H) est alors vrai pour $\hat{G}$. Or je me suis aperçu (récemment, cette fois) que (H) entraîne aussi (*) dans le cas profini. D'où:

Théorème – *Si G est bon, et a la propriété (H), alors $\hat{G}$ jouit de la propriété* (*).

La condition "bon" n'est pas gênante pour les groupes qui t'intéressent. En effet, on démontre facilement qu'un groupe libre, un groupe de tresses, un groupe fondamental de surface compacte, sont "bons"; et si un groupe a un sous-groupe d'indice fini qui est bon, il est lui-même bon. Bref tous les groupes que tu rencontres sont bons. Quant à la propriété (H), elle est bien connue pour ce genre de groupe, et résulte de leur action, soit sur un arbre, soit sur le demi-plan de Poincaré (c'est une propriété liée au rang 1, si j'ose dire).

Cas particulier – Le complété profini d'un groupe du type envisagé au début a la propriété (*) (pour les sous-groupes évidents).

Tu n'auras aucun mal à déduire de là les résultats de ma lettre précédente ainsi que pas mal d'autres.

Bien à toi

Jean-Pierre

The Grothendieck-Teichmüller group $\widehat{GT}$: a survey

Leila Schneps

In this short survey, we recall the definition of the *profinite version of the Grothendieck-Teichmüller group* $\widehat{GT}$. The Grothendieck-Teichmüller group was originally defined by Drinfel'd in [D] (as it happens, not in the profinite case, but in the discrete, pro-ℓ and k-pro-unipotent cases; these definition are all analogous). We state most of the results on the profinite version $\widehat{GT}$ known to date, and give references to all the proofs. (It should be noted that there are several deep results and conjectures concerning only the pro-ℓ version of the Grothendieck-Teichmüller group; we do not discuss them here.) One of the most interesting aspects of $\widehat{GT}$ is the fact that it contains the absolute Galois group $\mathrm{Gal}(\overline{\mathbb{Q}}/\mathbb{Q})$ as a subgroup in a way compatible with natural actions of both groups on arithmetico-geometric objects such as certain covers defined over $\overline{\mathbb{Q}}$ and fundamental groups of varieties defined over $\mathbb{Q}$; the possibility that these two groups are actually isomorphic is an exciting one, suggesting a new way of considering $\mathrm{Gal}(\overline{\mathbb{Q}}/\mathbb{Q})$. It reflects Grothendieck's suggestion in the *Esquisse d'un Programme* of studying the combinatorial properties of the absolute Galois group $\mathrm{Gal}(\overline{\mathbb{Q}}/\mathbb{Q})$ by studying its natural action on what he calls the *Teichmüller tower*, described as follows:

Il s'agit du système de toutes les multiplicités $\mathcal{M}_{g,n}$ pour g, n variables, liées entre elles par un certain nombre d'opérations fondamentales (telles les opérations de 'bouchage de trous' i.e. de 'gommage' de points marqués, celle de 'recollement', et les opérations inverses), qui sont le reflet en géométrie algébrique absolue de caractéristique zéro (pour le moment) d'opérations géométriques familières du point de vue de la 'chirurgie' topologique ou conforme des surfaces. La principale raison sans doute de cette fascination, c'est que cette structure géométrique très riche sur le système des multiplicités modulaires 'ouvertes' $\mathcal{M}_{g,n}$ se reflète par une structure analogue sur les groupoïdes fondamentaux correspondants, les 'groupoïdes de Teichmüller' $\hat{T}_{g,n}$, et que ces opérations au niveau des $\hat{T}_{g,n}$ ont un caractère suffisamment intrinsèque pour que le groupe de Galois $\mathbb{T}$ de $\overline{\mathbb{Q}}/\mathbb{Q}$ opère sur toute cette 'tour' de groupoïdes de Teichmüller, en respectant toutes ces structures. Chose plus extraordinaire encore, cette opération est *fidèle* – à vrai dire, elle est fidèle déjà sur le premier 'etage' non trivial de cette tour, à savoir $\hat{T}_{0,4}$ – ce qui signifie aussi, essentiellement, que l'action extérieure de $\mathbb{T}$ sur le groupe fondamental $\hat{\pi}_{0,3}$ de la droite projective standard $\mathbb{P}^1$ sur $\mathbb{Q}$, privée des trois points 0, 1, ∞, est déjà fidèle. Ainsi **le groupe de Galois $\mathbb{T}$ se réalise comme un groupe d'automorphismes d'un groupe profini des plus concrets**, respectant d'ailleurs certaines structures essentielles de ce groupe. Il s'ensuit qu'un élément de $\mathbb{T}$ peut être 'parametré' (de diverses façons équivalentes d'ailleurs) par un élément convenable de ce groupe profini (un groupe profini libre à deux générateurs), ou par un système de tels éléments, ce ou ces éléments étant d'ailleurs soumis à certaines conditions simples, nécessaires (et sans doute non suffisantes) pour que ce ou ces éléments corresponde(nt) bien à un élément de Γ. Une des tâches les plus fascinantes ici, est justement d'appréhender des conditions nécessaires et suffisantes sur un automorphisme extérieur de $\hat{\pi}_{0,3}$, i.e. sur le ou les paramètres correspondants, pour qu'il provienne d'un élément de Γ – ce qui fournirait une description 'purement algébrique', en termes de groupes profinis et sans référence à la théorie de Galois des corps de nombres, du groupe de Galois $\Gamma = \text{Gal}(\overline{\mathbb{Q}}/\mathbb{Q})$!

Drinfel'd pointed out in [D] the connection between his construction of $\widehat{GT}$ and the kind of group Grothendieck is referring to; he suggests that if one were to construct a tower of Teichmüller fundamental groupoids with *tangential base points*, then the full automorphism group of this tower should be exactly $\widehat{GT}$. We quote the relevant passage here, with the following remarks about the notation: formulae (4.3), (4.4) and (4.10) correspond to the defining relations (I), (II) and (III) of $\widehat{GT}$ (see §1.1), and the group $\hat{\Gamma}$ denotes the profinite completion of the quotient of the Artin braid group B_3 by its center (again, see §1.1).

It is proposed in [G] to consider, for any g and n, the "Teichmüller groupoid" $T_{g,n}$, i.e. the fundamental groupoid of the module stack $\mathcal{M}_{g,n}$ of compact Riemann surfaces X of genus g with n distinguished points $x_1, \ldots, x_n$. The fundamental groupoid differs from the fundamental group in that we choose not one, but several distinguished points. In the present case it is convenient to choose the distinguished points "at infinity" (see §15 of [De]) in accordance with the methods of "maximal degeneration" of the set $(X, x_1, \ldots, x_n)$. Since degeneration of the set $(X, x_1, \ldots, x_n)$ results in decreasing g and n, the groupoids $T_{g,n}$ for different g and n are connected by certain homomorphisms. The collection of all $T_{g,n}$ and all such homomorphisms is called in [G] the Teichmüller tower. It is observed in [G] that there exists a natural homomorphism $\text{Gal}(\overline{\mathbb{Q}}/\mathbb{Q}) \to G$, where G is the group of automorphisms of the profinite analogue of the Teichmüller tower (in which $T_{g,n}$ is replaced by its pro-finite completion $\hat{T}_{g,n}$). It is also stated in [G], as a plausible conjecture, that $\hat{T}_{0,4}$ and $\hat{T}_{1,1}$ in a definite sense generate the whole tower $\{\hat{T}_{g,n}\}$ and that all relations between generators of the tower come from $\hat{T}_{0,4}$, $\hat{T}_{1,1}$, $\hat{T}_{0,5}$ and $\hat{T}_{1,2}$. This conjecture has been proved, apparently, in Appendix B of the physics paper [MS]. In any case, it is easily seen that $\hat{T}_{0,4}$ generates the subtower $\{\hat{T}_{0,n}\}$, and that all relations in $\{\hat{T}_{0,n}\}$ come from $\hat{T}_{0,4}$ and $\hat{T}_{0,5}$. It can be shown that $\widehat{GT}$ is the automorphism group of the tower $\{\hat{T}_{0,n}\}$. Indeed, an automorphism of this tower is uniquely determined by its action on $\hat{T}_{0,4}$, i.e. on $\hat{\Gamma}$. This action is described by an element (λ, f) satisfying (4.3) and (4.4), and (4.10) is necessary and sufficient for the automorphism of $\hat{T}_{0,4}$ to extend to one of $\hat{T}_{0,5}$. Grothendieck's conjecture implies that the group of automorphisms of the tower $\{\hat{T}_{g,n}\}$ that are compatible with the natural homomorphism $\hat{T}_{0,4} \to \hat{T}_{1,1}$ (to a quadruple of points on $\mathbb{P}^1$ is assigned the double covering of $\mathbb{P}^1$ ramified at these points) is also equal to $\widehat{GT}$.

The possibility that $\widehat{GT}$ may provide the full answer to the double question asked by Grothendieck – namely, what is the automorphism group of the Teichmüller tower and is it equal to $\mathrm{Gal}(\overline{\mathbb{Q}}/\mathbb{Q})$ – constitutes the main motivation behind the study of its properties enumerated in this survey.

§1. Group-theoretic properties of $\widehat{GT}$.

§1.1. Basic ingredients.
Let us recall the definitions of the Artin braid groups and the mapping class groups.

Definition. Let B_n denote the Artin braid group on n strands, generated by $\sigma_1, \ldots, \sigma_{n-1}$ with the braid relations $\sigma_i \sigma_{i+1} \sigma_i = \sigma_{i+1} \sigma_i \sigma_{i+1}$. Let $M(0,n)$ be the quotient of B_n by $\sigma_{n-1} \cdots \sigma_1^2 \cdots \sigma_{n-1} = 1$ and $(\sigma_1 \cdots \sigma_{n-1})^n = 1$. Let K_n and $K(0,n)$ be the *pure* subgroups of B_n and $M(0,n)$, i.e. kernels of the natural surjections of B_n and $M(0,n)$ onto S_n. Both K_n and $K(0,n)$ are generated by the elements $x_{ij} = \sigma_{j-1} \cdots \sigma_{i+1} \sigma_i^2 \sigma_{i+1}^{-1} \cdots \sigma_{j-1}^{-1}$ for $1 \le i < j \le n$. Note that $K(0,5)$ is generated by $x_{i,i+1}$ for $i \in \mathbb{Z}/5\mathbb{Z}$, with relations $x_{i,i+1}$ commutes with $x_{j,j+1}$ if $|i-j| \ge 2$ and $x_{51} x_{23}^{-1} x_{12} x_{34}^{-1} x_{23} x_{45}^{-1} x_{34} x_{51}^{-1} x_{45} x_{12}^{-1} = 1$.

For any discrete group G, let $\hat{G}$ denote its profinite completion and $\hat{G}'$ the derived group of $\hat{G}$. Let F_2 denote the free group on two generators x and y. Let θ be the automorphism of $\hat{F}_2$ defined by $\theta(x) = y$ and $\theta(y) = x$. Let ω be the automorphism of F_2 and $\hat{F}_2$ defined by $\omega(x) = y$ and $\omega(y) = (xy)^{-1}$. Let ρ be the automorphism of $K(0,5)$ and $\hat{K}(0,5)$ given by $\rho(x_{i,i+1}) = x_{i+3,i+4}$. If $\hat{K}(0,5)$ is considered as a subgroup of $\hat{M}(0,5)$, then ρ is just the restriction to $\hat{K}(0,5)$ of the inner automorphism $\mathrm{Inn}(c)$ where $c = (\sigma_4 \sigma_3 \sigma_2 \sigma_1)^3$ of $\hat{M}(0,5)$ (here inner automorphism means $\mathrm{Inn}(\alpha)(x) = \alpha^{-1} x \alpha$). Note that $\hat{M}(0,5)$ is generated by σ_1 and c since $c^{-1}\sigma_1 c = \sigma_4$, $c^{-2}\sigma_1 c^2 = \sigma_2$ and $c^{-4}\sigma_1 c^4 = \sigma_3$ (we also have $c^{-3}\sigma_1 c^3 = \sigma_{51}$). For all $f \in \hat{F}_2$, let $\tilde{f}$ denote the image of f in $\langle x_{12}, x_{23} \rangle$ under the isomorphism $\hat{F}_2 \overset{\sim}{\to} \langle x_{12}, x_{23} \rangle \subset \hat{K}(0,5)$ given by $x \mapsto x_{12}$ and $y \mapsto x_{23}$.

As a final point of notation, whenever we have a homomorphism $\phi : \hat{F}_2 \to G$ for a profinite group G, with $\phi(x) = a$ and $\phi(y) = b$, we write $\phi(f) = f(a, b)$ for all $f \in \hat{F}_2$. In particular, when ϕ is the identity we have $f = f(x, y)$

Definition. Let $\widehat{GT}_0$ be the set

$$\{(\lambda, f) \in \hat{\mathbb{Z}}^* \times \hat{F}_2' \,|\, (I) \quad \theta(f)f = 1;$$
$$(II) \quad \omega^2(fx^m)\omega(fx^m)fx^m = 1 \text{ where } m = (\lambda - 1)/2\}.$$

The couples (λ, f) in $\widehat{GT}_0$ induce endomorphisms of $\hat{F}_2$ via $x \mapsto x^\lambda$ and

$y \mapsto f^{-1}y^{\lambda}f$. Let us define a composition law on the elements of $\widehat{GT}_0$ by

$$(\lambda, f)(\mu, g) = (\lambda\mu, fF(g)),$$

where F is the endomorphism of $\hat{F}_2$ corresponding to the couple (λ, f). Let $\widehat{GT}_0$ denote the set of elements $(\lambda, f) \in \widehat{GT}_0$ which have a two-sided inverse in $\widehat{GT}_0$ under this law; in particular, they are automorphisms of $\hat{F}_2$. Let $\widehat{GT}$ denote the subset of elements $(\lambda, f) \in \widehat{GT}_0$ satisfying the further relation, which takes place in $\hat{K}(0,5)$:

$$(III) \quad \rho^4(\tilde{f})\rho^3(\tilde{f})\rho^2(\tilde{f})\rho(\tilde{f})\tilde{f} = 1.$$

§1.2. $\widehat{GT}$ is a group.

The proof that $\widehat{GT}_0$ (resp. $\widehat{GT}$) is a group, i.e. that if (λ, f) and $(\mu, g) \in \widehat{GT}_0$ (resp. $\widehat{GT}$) then $(\lambda\mu, fF(g)) \in \widehat{GT}_0$ (resp. $\widehat{GT}$) is not at all immediate. Drinfel'd obtained this as a consequence of the definition of $\widehat{GT}$ as a group of transformations of a certain category; this argument is given more simply in the profinite case in [IM, A.4]. Here we will describe a proof based only on group theory, citing [LS] for parts of the proof. The key to the proof is the fact that the two (resp. three) relations defining $\widehat{GT}_0$ (resp. $\widehat{GT}$) can be interpreted as the relations necessary to make elements of $\widehat{GT}_0$ (resp. $\widehat{GT}$) into automorphisms of $\hat{F}_2$ which extend to automorphisms of $\hat{B}_3$ (resp. $\hat{M}(0,5)$) in a certain way. Let us give the precise statement for $\widehat{GT}_0$ first.

Lemma 1. *Suppose* $(\lambda, f) \in \hat{\mathbb{Z}}^* \times \hat{F}_2'$ *gives an automorphism of* $\hat{F}_2$ *via* $x \mapsto x^{\lambda}$ *and* $y \mapsto f^{-1}y^{\lambda}f$. *Identify* $\hat{F}_2$ *with the pure subgroup of* $\hat{B}_3$ *by setting* $x = \sigma_1^2$ *and* $y = \sigma_2^2$. *Then* (λ, f) *extends to an automorphism of* $\hat{B}_3$ *via*

$$\sigma_1 \mapsto \sigma_1^{\lambda}, \quad \sigma_2 \mapsto f^{-1}\sigma_2^{\lambda}f$$

if and only if $(\lambda, f) \in \widehat{GT}_0$, *i.e. if and only if* (λ, f) *satisfies relations (I) and (II).*

The proof of this lemma consists in an easy direct computation of the constraints imposed on the couple (λ, f) by the relation $\sigma_1\sigma_2\sigma_1 = \sigma_2\sigma_1\sigma_2$ and the fact that $(\sigma_1\sigma_2)^3$ is central in $\hat{B}_3$. It is given on p. 335 of [LS].

Now, two pairs (λ, f) and (μ, g) in $\widehat{GT}_0$ give automorphisms of $\hat{B}_3$, so their composition $(\lambda\mu, fF(g))$ gives an automorphism of $\hat{B}_3$, so by lemma 1 it also lies in $\widehat{GT}_0$. An automorphism (λ, f) also has an inverse of the form (μ, g), so $\widehat{GT}_0$ is a group.

Let us give the analogous result for $\widehat{GT}$.

Lemma 2. *A pair $F = (\lambda, f) \in \hat{\mathbf{Z}}^* \times \hat{F}_2'$ gives an automorphism of $\hat{M}(0,5)$ via $\sigma_1 \mapsto \sigma_1^\lambda$ and $c \mapsto c\tilde{f}$ if and only if the couple lies in $\widehat{GT}$, i.e. satisfies (I), (II) and (III).*

Let us sketch the proof of this lemma. The first point is to notice that the subgroup $\langle \sigma_1, \sigma_2 \rangle \subset \hat{M}(0,5)$ is isomorphic to $\hat{B}_3$, and that this subgroup is preserved by the proposed action since $\sigma_2 = c^{-2}\sigma_1 c^2 \mapsto \tilde{f}^{-1}c^{-2}\sigma_1^\lambda c^2\tilde{f} = \tilde{f}^{-1}\sigma_2^\lambda\tilde{f}$ and $\tilde{f} \in \langle x_{12} = \sigma_1^2, x_{23} = \sigma_2^2 \rangle$. Thus, by lemma 1, $(\lambda, f) \in \widehat{GT}_0$, i.e. it satisfies relations (I) and (II).

Now, one direction of the proof is easy; if the proposed action of (λ, f) gives an automorphism of $\hat{M}(0,5)$, then in particular it respects the equation $c^5 = 1$, so $(c\tilde{f})^5 = 1$. But since $\rho = \text{Inn}(c)$, we have

$$(c\tilde{f})^5 = c^5\rho^4(\tilde{f})\rho^3(\tilde{f})\rho^2(\tilde{f})\rho(\tilde{f})\tilde{f} = 1,$$

which is exactly relation (III) defining $\widehat{GT}$.

The other direction is more complicated. To show that the action in the statement extends to an automorphism of $\hat{M}(0,5)$, it is necessary to show that it respects all the relations in a presentation of $\hat{M}(0,5)$. It is easiest to use the presentation with generators σ_1, σ_2, σ_3 and σ_4. To compute the action of (λ, f) on the σ_i, we use the fact that they are simply the conjugates of σ_1 by the powers of c; of course, we do not know that (λ, f) is an automorphism yet. But if we show that assuming relation (III), the map defined as follows on the σ_i is an automorphism, then we can compute the image of c directly and find that $c \mapsto c\tilde{f}$, so that also gives an automorphism.

$$\sigma_1 \mapsto \sigma_1^\lambda$$
$$\sigma_2 \mapsto f(x_{23}, x_{12})\sigma_2^\lambda f(x_{12}, x_{23})$$
$$\sigma_3 \mapsto f(x_{34}, x_{45})\sigma_3^\lambda f(x_{45}, x_{34})$$
$$\sigma_4 \mapsto \sigma_4^\lambda$$
$$\sigma_{51} \mapsto f(x_{23}, x_{12})f(x_{51}, x_{45})\sigma_{51}^\lambda f(x_{45}, x_{51})f(x_{12}, x_{23}).$$

These expressions were first computed by Nakamura (see the appendix of [N]) in a somewhat different (Galois) situation; he proves there that if this action gives an automorphism of $\hat{M}(0,5)$, then f satisfies relation (III). Conversely, assuming that f satisfies relation (III), we need to check that each of the relations between the σ_i are respected by this action. It turns out that all but one are respected simply by virtue of the fact that (λ, f) lies in $\widehat{GT}_0$; the last one is respected by the assumption that f satisfies (III). This computation is done explicitly in [LS], p. 340-342.

As in the case of $\widehat{GT}_0$, we immediately obtain the result we are seeking for:

Corollary. $\widehat{GT}$ *is a group.*

Symmetry. $\widehat{GT}$ is a natural group to look at also because of the following theorem, stating that it consists of *all* automorphisms of $\hat{K}(0,5)$ extending automorphisms of the subgroup $\hat{F}_2 = \langle x_{12}, x_{23} \rangle$ and satisfying certain simple symmetry conditions with respect to the actions of θ and ω on $\hat{F}_2$ and ρ on $\hat{K}(0,5)$, which is essentially equivalent to considering the *outer* actions of S_2 and S_3 on $\hat{F}_2$ and S_5 on $\hat{K}(0,5)$. This theorem is proved in the article [HS] in this volume. For a pair $(\lambda, f) \in \hat{\mathbb{Z}}^* \times \hat{F}'_2$, we write F for the associated endomorphism of $\hat{F}_2$.

Theorem. *(i) An invertible pair (λ, f) (for the usual composition law) satisfies relation (I) if and only if the images of F and θ commute in* $\mathrm{Out}(\hat{F}_2)$.

(ii) Given an invertible pair (λ, f) satisfying (I), it satisfies relation (II) if and only if the images of F and ω commute in $\mathrm{Out}(\hat{F}_2)$.

(iii) Let (λ, f) be an invertible pair satisfying (I) and (II), so in $\widehat{GT}_0$. Then (λ, f) lies in $\widehat{GT}$ if and only if there exists an automorphism F of $\hat{K}(0,5)$ extending that of $\hat{F}_2$ in the sense that $F(x_{12}) = x_{12}^\lambda$ and $F(x_{23}) = f(x_{23}, x_{12}) x_{23}^\lambda f(x_{12}, x_{23})$, such that F and ρ commute in $\mathrm{Out}(\hat{K}(0,5))$.

§1.3. Cohomological interpretation

Given the definition of $\widehat{GT}$, it is natural to ask oneself if it is possible to give a complete description of the elements $f \in \hat{F}'_2$ satisfying each of the three relations. This makes sense for relations (I) and (III), in which λ does not intervene; we let $\mathcal{E}_I$ and $\mathcal{E}_{III}$ denote the subsets of $\hat{F}_2$ of elements f satisfying relations (I) and (III) respectively. For relation (II), we let $\mathcal{E}_{II} = \{f \in \hat{F}'_2 \mid f \text{ satisfies relation (II) for some value of } \lambda\}$. In fact, it is not difficult to show that if $f \in \hat{F}'_2$ satisfies relation (II) for some λ, then it satisfies relation (II) also for $-\lambda$ but not for any other value (cf. [LS2], Lemma 9).

Let us give a complete description of the sets $\mathcal{E}_I$, $\mathcal{E}_{II}$ and $\mathcal{E}_{III}$. Consider for instance $\mathcal{E}_I$. θ is an automorphism of $\hat{F}_2$ of order 2 and the relation $\theta(f)f = 1$ is a cocycle relation for this automorphism. The set of classes of these cocycles up to coboundaries (i.e. elements $f = \theta(g)^{-1}g \in \hat{F}_2$, all of which satisfy $\theta(f)f = 1$) by the elements of the non-abelian cohomology set $H^1(\langle\theta\rangle, \hat{F}_2)$. Computation of this cohomology set shows that it is trivial (cf. [LS2]), so that we have

$$\mathcal{E}_I = \{\theta(g)^{-1}g \mid g \in \hat{F}_2\},$$

i.e. a complete list of the elements of $\mathcal{E}_I$. The full set of elements of $\mathcal{E}_{II}$ and $\mathcal{E}_{III}$ can be computed similarly. It is necessary to compute the non-abelian cohomology sets $H^1(\langle\omega\rangle, \hat{F}_2)$ and $H^1(\langle\rho\rangle, \hat{K}(0,5))$; then all elements of $\mathcal{E}_{II}$ (resp. $\mathcal{E}_{III}$) are given up to coboundaries $\omega(h)^{-1}h$ for $h \in \hat{F}_2$ (resp. $\rho(k)^{-1}k$ for $k \in \hat{K}(0,5)$) by representative cocycles of these cohomology sets. The result of these computations is given in the following theorem, the main result of [LS2].

Theorem. *Let* $(\lambda, f) \in \widehat{GT}$, *and let* $m = (\lambda - 1)/2$. *Then there exist elements* g *and* $h \in \hat{F}_2$ *and* $k \in \hat{K}(0,5)$ *such that we have the following equalities, of which the first two take place in* $\hat{F}_2$ *and the third in* $\hat{K}(0,5)$:

$$(I') \qquad f = \theta(g)^{-1}g$$

$$(II') \qquad fx^m = \begin{cases} \omega(h)^{-1}h & \text{if } \lambda \equiv 1 \bmod 3 \\ \omega(h)^{-1} xy\,h & \text{if } \lambda \equiv -1 \bmod 3 \end{cases}$$

$$(III') \qquad f(x_{12}, x_{23}) = \begin{cases} \rho(k)^{-1}k & \text{if } \lambda \equiv \pm 1 \bmod 5 \\ \rho(k)^{-1} x_{34}x_{51}^{-1}x_{45}x_{12}^{-1}\,k & \text{if } \lambda \equiv \pm 2 \bmod 5. \end{cases}$$

§1.4. Further group-theoretic remarks and questions.

Over the last two or three years, many people have asked me if $\widehat{GT}$ possesses certain properties which are known or suspected for the absolute Galois group $\mathrm{Gal}(\overline{\mathbb{Q}}/\mathbb{Q})$ (which injects into $\widehat{GT}$; see §3.1). Here are several of the more group-theoretic of these questions, some solved, some unsolved, and some only partially solved.

(1) Is the "complex conjugation" element $(-1, 1)$ self-centralizing in $\widehat{GT}$? This question was asked by Y. Ihara; its answer turns out to be yes. A computation reduces this result to showing that the only element of $\hat{F}_2$ which is fixed under the automorphism ι given by $\iota(x) = x^{-1}$ and $\iota(y) = y^{-1}$ is the trivial element, in other words that the centralizer of ι in the semi-direct product $\hat{F}_2 \rtimes \langle\iota\rangle$ is exactly $\langle\iota\rangle$; this can be shown in a variety of ways (cf. [LS2]), suggested to me by J-P. Serre.

(2) The derived subgroup $\widehat{GT}'$ of $\widehat{GT}$ is contained in the subgroup $\widehat{GT}^1$ of pairs $(\lambda, f) \in \widehat{GT}$ with $\lambda = 1$. Are these two subgroups equal? This question was also asked by Y. Ihara, and remains unsolved.

(3) Is $\widehat{GT}$ itself a profinite group? This question was asked and (almost immediately) answered by Florian Pop, who noted that indeed $\widehat{GT}$ is the inverse limit of its own images in the (finite) automorphism groups of the quotients $\hat{F}_2/N$, where N runs over the characteristic subgroups of finite

index of $\hat{F}_2$. There are other ways of obtaining $\widehat{GT}$ as an inverse limit, cf. [HS] in this volume.

(4) Is there any torsion in $\widehat{GT}$ apart from the elements of order 2 given by $(-1, 1)$ and its conjugates? After a short discussion with Florian Pop, we were able to prove the weak result that any such torsion elements become trivial in the *pro-nilpotent quotient* of $\widehat{GT}$; no stronger result seems to be known. Let us indicate the proof of this result by showing there are any elements of order 2 with $\lambda = 1$ become trivial in the pro-nilpotent quotient of $\widehat{GT}$ (note that any torsion element must have $\lambda = \pm 1$). Suppose there exists $(1, f) \in \widehat{GT}$ such that $(1, f)(1, f) = 1$. Then $f(x, y)f(x, f(y, x)yf(x, y)) = 1$. We know that $f \in \hat{F}_2'$. Considering this equation modulo the second commutator group $\hat{F}_2'' = [\hat{F}_2', \hat{F}_2']$, we see that modulo $\hat{F}_2''$, we have $f(x, y)^2 = 1$, so since there is no torsion in this group, we must have that $f(x, y) \in \hat{F}_2''$. Working modulo $\hat{F}_2'''$ and so on, we quickly find that f lies in the intersection of all the successive commutator subgroups of $\hat{F}_2$, i.e. that the image of $f(x, y)$ in the nilpotent completion $\hat{F}_2^{\text{nil}}$ is trivial.

(5) Is the outer automorphism group of $\widehat{GT}$ trivial? (F. Pop, absolutely unsolved).

(6) Is it possible to determine the finite quotients of $\widehat{GT}$? (Everyone connected with inverse Galois theory; unsolved except for the obvious remark that $(\lambda, f) \mapsto \lambda$ gives a surjection $\widehat{GT} \to \hat{\mathbb{Z}}^*$, and therefore all abelian groups occur, just as for $\text{Gal}(\overline{\mathbb{Q}}/\mathbb{Q})$.)

(7) Is the subgroup $\widehat{GT}^1$ of pairs (λ, f) with $\lambda = 1$ profinite free? (Everyone interested in the Shafarevich conjecture; unsolved.)

(8) Less ambitiously than in (7), is the cohomological dimension of $\widehat{GT}^1$ finite? (J-P. Serre, unsolved).

(9) Is it possible to determine the image of, say, the elements $f \in \hat{F}_2$ belonging to pairs $(\lambda, f) \in \widehat{GT}$, in a given finite quotient $\hat{F}_2/N$ of $\hat{F}_2$? (J. Oesterle was the first of several people to ask me this question. Unfortunately, we do not know how to compute this image algorithmically, though there is a non-algorithmic (i.e. ending after a finite number of steps but one doesn't know how many) and very unwieldy procedure for doing it, and also an easier approximative procedure, giving a set which may be too large (cf. [Harbater-Schneps] in this volume). The point is to determine the $\widehat{GT}$-orbits of covers of $\mathbb{P}^1 - \{0, 1, \infty\}$ and compare them with the $\text{Gal}(\overline{\mathbb{Q}}/\mathbb{Q})$-orbits.

(10) If $\beta : X \to \mathbb{P}^1\mathbb{C}$ is a Belyi cover, then an element of $\widehat{GT}$ sends it to another Belyi cover $\beta' : X' \to \mathbb{P}^1\mathbb{C}$. Given X, is the curve X' independent

of the choice of β? Certainly this is so whenever the element of $\widehat{GT}$ happens to lie in $\mathrm{Gal}(\overline{\mathbb{Q}}/\mathbb{Q})$.

(11) Is it possible to define an action of $\widehat{GT}$ on $\overline{\mathbb{Q}}$ using its action on covers – even an action weaker than an automorphism? For instance, if the answer to question (10) were yes, then letting X range over the elliptic curves, we would obtain a well-defined action of $\widehat{GT}$ on $\overline{\mathbb{Q}}$ by considering their j-invariants. But perhaps it is possible to define such an action (corresponding to the usual Galois action when restricted to the subgroup $\mathrm{Gal}(\overline{\mathbb{Q}}/\mathbb{Q})$ of $\widehat{GT}$) without answering question (10).

The answers to questions (1), (3), (4), (5), (8) and of course (10) and (11) are known for $\mathrm{Gal}(\overline{\mathbb{Q}}/\mathbb{Q})$.

§2. $\widehat{GT}$-actions

In this section we describe several related objects which are equipped with a $\widehat{GT}$-action (compatible with a natural $\mathrm{Gal}(\overline{\mathbb{Q}}/\mathbb{Q})$-action when $\mathrm{Gal}(\overline{\mathbb{Q}}/\mathbb{Q})$ is considered as a subgroup of $\widehat{GT}$, cf. §3.2).

To begin with, we should mention the first appearance of $\widehat{GT}$, or at least of its discrete, pro-ℓ and k-pro-unipotent versions. They were first defined by Drinfel'd in [D] as groups of transformations of the associativity and commutativity morphisms of a quasi-triangular quasi-Hopf algebra. In the same article, he showed that the k-pro-unipotent version of $\widehat{GT}$ is an automorphism group of the k-pro-unipotent completions $B_n(k)$ of the Artin braid groups B_n respecting the natural inclusion automorphisms $B_{n-1}(k) \hookrightarrow B_n(k)$ given by $\sigma_i \mapsto \sigma_i$ for $i = 1, \ldots, n-2$. Following ideas of Grothendieck in the *Esquisse d'un Programme*, he further indicated that using the relations between braided tensor category structures and the structure of the fundamental groupoid of the moduli spaces of Riemann surfaces based at "tangential base points", the profinite version of $\widehat{GT}$ should be the automorphism group of the tower of these fundamental groupoids linked by certain natural connecting homomorphisms. All the $\widehat{GT}$-actions described below are essentially concretizations of these ideas of Drinfel'd and of Grothendieck.

§2.1. $\widehat{GT}$-action on groups

(1) **Artin braid groups.** $\widehat{GT}$ acts on the profinite completions of the Artin braid groups B_n. Up to an easy adaptation to the profinite case (cf.

[IM, A.4] or [LS]), Drinfel'd gave the action on the $\hat{B}_n$ as

$$\sigma_i \mapsto f(\sigma_i^2, y_i)\sigma_i^\lambda f(y_i, \sigma_i^2),$$

where $y_i = \sigma_{i-1} \cdots \sigma_1 \cdot \sigma_1 \cdots \sigma_{i-1}$. The proof that this action really induces an automorphism follows, in his paper, from the role of $\widehat{GT}$ as transformations of a quasi-triangular quasi-Hopf algebra. A direct proof is given in [LS]; indeed the result proved there is actually stronger (see §2.3 below).

(2) Genus zero mapping class groups. The group $M(0, n)$ is the quotient of the group $\hat{B}_n$ by the two relations $(\sigma_{n-1} \cdots \sigma_1)^n = 1$ and $\sigma_{n-1} \cdots \sigma_1^2 \cdots \sigma_{n-1} = 1$. In order to show that the action of $\widehat{GT}$ on $\hat{B}_n$ given above passes to the quotient, it suffices to show that in fact the two elements $\omega_n = (\sigma_{n-1} \cdots \sigma_1)^n$ and $y_n = \sigma_{n-1} \cdots \sigma_1^2 \cdots \sigma_{n-1}$ are sent to ω_n^λ and y_n^λ by an element $(\lambda, f) \in \widehat{GT}$ acting on $\hat{B}_n$ as in (1); this is an easy computation (cf. [IM] or [LS]). Note that since the pure mapping class subgroups $K(0, n)$ are characteristic, $\widehat{GT}$ acts on them as well.

The reason it is interesting for us to have $\widehat{GT}$ act on the groups $\hat{K}(0, n)$ and $\hat{M}(0, n)$ is that these groups have geometric significance: they are the (algebraic) fundamental groups of the moduli spaces of Riemann spheres with n *ordered* resp. *unordered* distinct marked points. We denote these moduli spaces by $\mathcal{M}_{0,n}$ and $\mathcal{M}_{0,[n]}$ respectively. (There are many references for the definitions of these moduli spaces; see for instance [Lochak] in this volume.) This remark is developed further in the examination of $\widehat{GT}$-actions on groupoids below.

(2') Dessins d'enfants. Dessins d'enfants are the drawings on topological surfaces obtained by taking $\beta^{-1}([0, 1])$ where $\beta : X \to \mathbb{P}^1\mathbb{C}$ is a cover of Riemann surfaces ramified over 0, 1 and ∞ only. They are in bijection with the pairs (β, X) up to isomorphism, and therefore with the set of conjugacy classes of subgroups of finite index of $\hat{F}_2 = \pi_1(\mathbb{P}^1 - \{0, 1, \infty\})$. The link with the situation of (2) is given by the fact that $\mathbb{P}^1 - \{0, 1, \infty\}$ is isomorphic to the first non-trivial moduli space $\mathcal{M}_{0,4}$, i.e. $K(0, 4) \simeq F_2$. The two main problems of the theory of dessins d'enfants are the following: (i) given a dessin, i.e. a purely combinatorial object, find the equations for β and X explicitly; (ii) find a list of (combinatorial? topological? algebraic) invariants of dessins which completely identify their $\mathrm{Gal}(\overline{\mathbb{Q}}/\mathbb{Q})$-orbits. The second problem can be interestingly weakened from $\mathrm{Gal}(\overline{\mathbb{Q}}/\mathbb{Q})$ to $\widehat{GT}$, but it remains absolutely non-trivial.

(3) Fundamental groups of spheres with n removed points. For every $n \geq 5$, there is a homomorphism $\iota : K(0, n) \to K(0, n-1)$ obtained by removing the n-th strand of the braids; ι passes to the profinite completions.

It is easily checked by direct computation that the $\widehat{GT}$-action on the $\hat{K}(0,n)$ for $n \geq 4$ are compatible with this homomorphism. This remark has the following geometric interpretation. Let $S_{0,n-1}$ denote a topological sphere with $n-1$ punctures. For a suitable choice of base points, we have an exact sequence

$$1 \to \hat{\pi}_1(S_{0,n-1}) \to \hat{\pi}_1(\mathcal{M}_{0,n}) \to \hat{\pi}_1(\mathcal{M}_{0,n-1}) \to 1,$$

where the map between fundamental groups on moduli spaces comes from the map $\mathcal{M}_{0,n}$ is given by erasing the n-th point on every sphere. This exact sequence corresponds to considering the moduli space $\mathcal{M}_{0,n}$ as the *universal curve* over $\mathcal{M}_{0,n-1}$. The map from $\hat{\pi}_1(\mathcal{M}_{0,n}) \simeq \hat{K}(0,5)$ to $\pi_1(\mathcal{M}_{0,n-1}) \simeq \hat{K}(0,n-1)$ is exactly ι. Thus by the exact sequence we obtain a $\widehat{GT}$-action on the kernel of ι, namely on $\hat{\pi}_1(S_{0,n-1})$ (which is, of course, a free group of rank $n-2$). This $\widehat{GT}$-action can be explicitly computed. It is compatible with a (lifting of) the outer $\mathrm{Gal}(\overline{\mathbb{Q}}/\mathbb{Q})$-action on $\hat{\pi}_1(S_{0,n-1})$ associated to a certain choice of "tangential" sphere with n marked points. Indeed, using tangential base point techniques, Ihara and Nakamura were able to explicitly compute (cf. [IN] and further unpublished work) a $\widehat{GT}_0$-action on $\pi_1(S_{g,n-1})$ where $S_{g,n-1}$ denotes a topological surface of genus $g \geq 0$ with $n-1$ punctures; see the end of this article for the $\mathrm{Gal}(\overline{\mathbb{Q}}/\mathbb{Q})$-compatibility properties of this action.

(4) Higher genus mapping class groups. In the case $g > 0$, we know that there is a $\widehat{GT}$-action on the mapping class groups $\hat{M}(g,n)$ for the following pairs (g,n): $(1,1)$, $(1,2)$, $(2,0)$, and $(2,1)$. These actions are obtained quite simply by the well-known similarities between these groups and Artin braid groups or genus zero mapping class groups. Indeed, the group $M(1,1)$ is isomorphic to B_3 modulo its center (and to $\mathrm{PSL}_2(\mathbb{Z})$). Since the natural action of $\widehat{GT}$ (and even $\widehat{GT}_0$) on $\hat{B}_3$ passes to the quotient mod center, $\hat{M}(1,1)$ inherits it. The group $\hat{M}(1,2)$ is a quotient of a certain subgroup of $\hat{B}_5$ and again, it is easily confirmed that the $\widehat{GT}$-action passes to it. By the Birman-Hilden presentations of $M(2,1)$ and $M(2,0)$, one can show directly that $M(2,1)$ is isomorphic to the quotient of $\hat{B}_6$ by the single relation

$$\sigma_5\sigma_4\sigma_3\sigma_2\sigma_1^2\sigma_2\sigma_3\sigma_4\sigma_5 = (\sigma_1\sigma_2\sigma_3\sigma_4)^5,$$

and that the $\widehat{GT}$-action on $\hat{B}_5$ passes to this quotient and to the group $\hat{M}(2,0)$ which is the quotient of $\hat{M}(2,1)$ by the relation $(\sigma_1\sigma_2\sigma_3\sigma_4)^5 = 1$.

According to Grothendieck's conception, these facts should suffice to build up a $\widehat{GT}$-action on all the $\hat{M}(g,n)$, but this remains an open problem. We note here the important discovery by Nakamura of the explicit form of a *Galois* action on the groups $\hat{M}(g,1)$ and $\hat{M}(g,0)$. This naturally gives rise to a conjecture concerning the $\widehat{GT}$-action; see further remarks on this at the end of this article.

§2.2. $\widehat{GT}$-action on groupoids

(1) Braided tensor categories. Let us first of all mention a certain *braided tensor category* C on which $\widehat{GT}$ acts. This groupoid has less structure than the quasi-triangular quasi-Hopf algebras used in [D], and there is also another difference with from the abstract braided tensor category whose objects are *bracketing patterns* on objects $V_1, \ldots, V_n$ used in [D]. Namely, the approach here substitutes numbered trivalent trees for bracketing patterns; the combinatorial difference is slight, and this version has the advantage of reflecting more closely the structure at infinity of the moduli spaces (similarly to the braided-tensor category constructions of Moore and Seiberg [MS]). The category C (suitably completed) possesses the basic elements of structure necessary to ensure a $\widehat{GT}$-action.

Unfortunately, the full definition of the groupoid in question is too long to reproduce here, even assuming the definition of a braided tensor category. It can be found in detail in [PS, chapter II]. We content ourselves with a rather brief description. C is a braided tensor category whose objects are the trivalent trees equipped with a cyclic order on the three edges coming out of each trivalent vertex, and a numbering which consists of a positive integer associated to each tail, except for exactly one distinguished tail numbered 0. The tensor product on such trees is given by attaching the two 0-tails of two trees together and adding a new tail and a cyclic order at the new vertex. The associativity morphisms change the position of an inner edge from horizontal to vertical with respect to the four branches coming out of it (i.e. an H configuration is changed to an I configuration). The commutativity morphisms consist in rigidly switching two branches of the tree which meet at a single vertex. All morphisms of the category are generated (in the sense of composing and taking tensor products) by these morphisms of associativity and commutativity. The local group $\mathrm{Hom}(T,T)$ at a tree T with n tails is exactly the subgroup of the Artin braid group B_n which is the preimage of the subgroup of the permutation group S_n which leaves the tree fixed when acting as permutations on the indices labeling the tails. Thus the local group is larger than K_n only for trees whose indices are not all distinct. In order to define a $\widehat{GT}$-action on such a braided tensor category, it is necessary to take its *profinite completion*, which consists in adding new morphisms, in the sense that each local group is replaced by its profinite completion. Then (very roughly speaking) the action of an element $(\lambda, f) \in \widehat{GT}$ on the profinite category fixes the objects and sends an associativity morphism from T_1 to T_2 to itself precomposed by the image of f in the local group $\widehat{\mathrm{Hom}}(T_1,T_1)$ under a certain natural homomorphism $\hat{F}_2 \to \widehat{\mathrm{Hom}}(T_1,T_1)$, and the image of a commutativity morphism $c(T_1,T_2)$ from T_1 to T_2 is given by $\big(c(T_2,T_1)c(T_1,T_2)\big)^m c(T_1,T_2)$. Note that $c(T_2,T_1)$ is not the inverse of $c(T_1,T_2)$, since the rigid switch of two branches of T_1 performed twice gives a non-trivial morphism from T_1 to itself (visually, one can imagine that in switching, the left-hand branch passes above the right-hand one, so that doing it two times makes a full twist on the third edge connected to the common vertex of the two branches; this is non-trivial). Since all elements of the profinite local groups $\widehat{\mathrm{Hom}}(T,T)$ are generated by the associativity and commutativity moves, this action defines a $\widehat{GT}$-action on the local groups, which turns out to correspond to the $\widehat{GT}$-action on Artin braid groups given above, up to inner automorphisms as one changes from one object to another.

(2) The fundamental groupoids based near infinity of the genus zero moduli spaces. In order to define these groupoids, known as *Teichmüller groupoids*, it is necessary only to define the set of simply connected regions which will serve as base points (a complete and detailed description of them can be found in [PS, I.2].) They are obtained as follows: the neighborhood of each point of *maximal degeneration* in the stable compactification of the genus zero moduli space $\mathcal{M}_{0,n}$ is isomorphic to the $(n-3)$-fold product of a pointed disk with itself; the real locus of this neighborhood naturally falls into 2^{n-3} simply connected pieces. (Simply think of the moduli space $\mathcal{M}_{0,4}$, isomorphic to $\mathbb{P}^1 - \{0,1,\infty\}$: the stable compactification is $\mathbb{P}^1$, the points of maximal degeneration are 0, 1 and ∞, their neighborhoods are pointed disks and the real part of each disk consists of the two segments of the real line on each side of the missing point. This original set of tangential base points was defined by Deligne in [De].) We denote this set of *tangential base points* or *base points near infinity* by $\mathcal{B}_n$, and write $T(0,n) = \pi_1(\mathcal{M}_{0,n};\mathcal{B}_n)$ for the Teichmüller groupoid. Let us give a brief indication of why $\widehat{GT}$ acts on the profinite completion of $T(0,n)$ for $n \geq 4$ (cf. [PS, II] for full proofs). The tangential base points can be visualized as degenerating Riemann spheres with n punctures; they are in bijection with the combinatorial set of trivalent n-tailed trees with tails numbered 1 to n (in any order). Let $\hat{\mathcal{C}}_n$ denote the subgroupoid of the profinite completion $\hat{\mathcal{C}}$ of $\mathcal{C}$ whose objects are the set of trivalent trees with tails numbered from 0 to $n-1$ and whose morphisms are all morphisms of $\hat{\mathcal{C}}$ between these objects. Renumbering the n-th tail to 0 for each tree in $\hat{\mathcal{C}}_n$, we can define a surjective groupoid homomorphism from $\hat{\mathcal{C}}_n$ to $\hat{T}(0,n)$; the associativity and commutativity morphisms have simple geometric interpretations as natural paths on the moduli space $\mathcal{M}_{0,n}$ between tangential base points. The kernel of this homomorphism is not very important; the local groups of $\hat{\mathcal{C}}_n$ are isomorphic to $\hat{K}_n$ and those of $\hat{T}(0,n)$ are the mapping class groups $\hat{K}(0,n)$, and the relations involved in the surjection $\hat{K}_n \to \hat{K}(0,n)$ for all the local groups completely describes the kernel. Once we check that the action of $\widehat{GT}$ on $\hat{\mathcal{C}}$ preserves the subgroupoid $\hat{\mathcal{C}}_n$, which is not difficult, this remark shows that the $\widehat{GT}$-action passes to $\hat{T}(0,n)$.

(3) Higher genus groupoids. As yet, not much is known about a possible $\widehat{GT}$-action on higher genus groupoids. Even the complete structure of the Teichmüller groupoids $T(g,n)$ has not been determined. However, generalizing work by Moore and Seiberg, Lyubashenko recently described an abstract groupoid somewhat in the same vein as $\mathcal{C}$ above, whose objects instead of being trees are trivalent graphs with loops, and which contains besides the associativity and commutativity morphisms, further morphisms related especially to the presence of the loops. The situation is ripe to attempt to determine if there is a $\widehat{GT}$-action on Lyubashenko's groupoid, or conversely if its group of automorphisms can be explicitly determined by analogous simple relations, and the precise relationship of this groupoid with the higher genus Teichmüller groupoids.

(4) The universal Ptolemy-Teichmüller groupoid. The $\widehat{GT}$-action on the profinite completion of a suitable extension of Thompson's group is proved in [Lochak-Schneps] in the companion volume to this one; by the construction of the groupoid, it is closely related to the $\widehat{GT}$-action on the genus zero Teichmüller groupoids. However, the consequences of the various intriguing relations between the Ptolemy-Teichmüller groupoid and higher genus situations are not yet understood.

§2.3. Towers

A *tower* of groups or groupoids is simply a collection of groups or groupoids linked by homomorphisms. An automorphism of a tower is a tuple $(\phi_G)_G$ of automorphisms of all the groups (or groupoids) G in the tower, respecting all the linking homomorphisms. The advantage of towers over isolated groups or groupoids in our situation is that instead of obtaining statements of the form "$\widehat{GT}$ is an automorphism group of such-and-such a group", one obtains stronger statements of the form "$\widehat{GT}$ is *the full* group of automorphisms (of a certain type) of such-and-such a tower".

(1) Artin braid group tower. The simplest example is the small tower of Artin braid groups used in the following theorem proved in [LS]. Let $\hat{A}_3$ be the subgroup of $\hat{B}_3$ generated by σ_1^2 and σ_2; this is exactly the preimage of the subgroup $\{1, (23)\}$ under the natural surjection $\hat{B}_3 \to S_3$. Let κ :

$\hat{A}_3 \hookrightarrow \hat{B}_4$ be defined by $\kappa(\sigma_1^2) = \sigma_2\sigma_1^2\sigma_2$, $\kappa(\sigma_2) = \sigma_3$. Let $\iota : \hat{B}_3 \hookrightarrow \hat{B}_4$ be given by $\iota(\sigma_i) = \sigma_i$ for $i = 1, 2$.

Theorem. $\widehat{GT}_0$ *is the subgroup of* $\mathrm{Aut}(\hat{B}_3)$ *defined by properties (i) and (ii) below, and* $\widehat{GT}$ *is the subgroup of* $\widehat{GT}_0$ *defined by the additional properties (iii) and (iv).*

(i) $\phi(\sigma_1) = \sigma_1^\lambda$ *for some* $\lambda \in \mathbb{Z}^*$;

(ii) ϕ *preserves permutations, i.e. if* $\eta : \hat{B}_3 \to S_3$ *is the natural surjection, then* $\eta \circ \phi = \eta$ *on* $\hat{B}_3$;

(iii) *there exists an automorphism* ϕ' *of* $\hat{B}_4$ *such that* $\phi' \circ \iota = \iota \circ \phi$ *as maps from* $\hat{B}_3$ *to* $\hat{B}_4$;

(iv) *there exists an automorphism* ϕ' *of* $\hat{B}_4$ *such that* $\phi' \circ \kappa = \kappa \circ \phi$ *as maps from* $\hat{A}_3 \to \hat{B}_4$. *(Note that by (ii), ϕ preserves $\hat{A}_3$, so that the map $\kappa \circ \phi$ is defined on $\hat{A}_3$.)*

In other words, $\widehat{GT}$ is the group of automorphisms $(\phi, \phi|_{\hat{A}_3}, \phi')$ of the groups $\hat{B}_3$, $\hat{A}_3$ and $\hat{B}_4$, which send σ_1 to a power of itself, preserve permutations and respect the homomorphisms ι and κ. We note that an analogous tower can be constructed with all the profinite braid groups $\hat{A}_n = \langle \sigma_1^2, \sigma_2, \ldots, \sigma_{n-1} \rangle$ and $\hat{B}_n$ and the inclusion homomorphisms as well as those doubling the first string of every braid; the automorphism group of this tower is still $\widehat{GT}$, reflecting Grothendieck's deep principle that everything is decided on the first two levels.

(2) The tower of the $\hat{T}(0, n)$ with point-erasing homomorphisms. There are natural homomorphisms relating the profinite Teichmüller groupoids $\hat{T}(0, n)$; they do not go *up* the tower as in the Artin braid group tower above, but down from $\hat{T}(0, n)$ to $\hat{T}(0, n-1)$, and are induced by the maps r_i from the moduli space $\mathcal{M}_{0,n}$ to $\mathcal{M}_{0,n-1}$ given by erasing the i-th point of every sphere with n marked points. The main theorem of [S] states that $\widehat{GT}$ is the automorphism group of this tower; the methods of proof are quite similar. A Teichmüller tower in all genera has not yet been constructed (it should be remarked that in order to equip it with adequate homomorphisms, it is probably necessary to use Riemann surfaces with holes, i.e. with boundary components, rather than just punctures).

§3. Relations between $\mathrm{Gal}(\overline{\mathbb{Q}}/\mathbb{Q})$ and $\widehat{GT}$.

The main statement is the major result due to Drinfel'd and Ihara that $\mathrm{Gal}(\overline{\mathbb{Q}}/\mathbb{Q})$ is a subgroup of $\widehat{GT}$; this result suggests the conjecture that the two groups are in fact isomorphic. Below, we indicate the definition of the inclusion map given in [I1] and the proof of the three relations given in [LS2]. The next step thus consists in adducing evidence for the conjecture by showing that various properties of $\mathrm{Gal}(\overline{\mathbb{Q}}/\mathbb{Q})$ are also properties of $\widehat{GT}$. Several such *group-theoretic* properties were considered in §1.4; in §3.2 we explain the compatibility of the $\widehat{GT}$ and $\mathrm{Gal}(\overline{\mathbb{Q}}/\mathbb{Q})$ actions on the various fundamental groups and groupoids described in §2.

§3.1. The injection $\mathrm{Gal}(\overline{\mathbb{Q}}/\mathbb{Q}) \hookrightarrow \widehat{GT}$.

Let $\mathcal{M}_{0,4}$ and $\mathcal{M}_{0,5}$ denote the moduli spaces of Riemann spheres with four and five ordered marked points. We need the two following basic facts about $\mathrm{Gal}(\overline{\mathbb{Q}}/\mathbb{Q})$.

Fact 1. Let $X = \mathcal{M}_{0,4}$ or $\mathcal{M}_{0,5}$. Let

$$\Sigma = \{\text{points defined over } \overline{\mathbb{Q}}\} \cup \{\text{tangential base points}\}.$$

For any α, $\beta \in \Sigma$, the set of *pro-paths* from α to β is given by any path from α to β precomposed with all pro-loops based at α, i.e. elements of the profinite completion of $\pi_1(X; \alpha)$. The group $\mathrm{Gal}(\overline{\mathbb{Q}}/\mathbb{Q})$ acts on the set of pro-paths of X with base points in Σ; a pro-path from α to β is sent by $\sigma \in \mathrm{Gal}(\overline{\mathbb{Q}}/\mathbb{Q})$ to a pro-path from $\sigma(\alpha)$ to $\sigma(\beta)$. σ fixes the tangential base points.

Fact 2. The action of $\mathrm{Gal}(\overline{\mathbb{Q}}/\mathbb{Q})$ on pro-paths of X commutes with the action of $\mathrm{Aut}(X)$. Note that $\mathrm{Aut}(\mathcal{M}_{0,4}) \simeq S_3$, generated by $\Theta(z) = 1 - z$ and $\Omega(z) = 1/1 - z$ and $\mathrm{Aut}(\mathcal{M}_{0,5}) \simeq S_5$; these automorphisms correspond to permuting marked points on spheres.

Let $\hat{F}_2$ be identified with the π_1 of $\mathcal{M}_{0,4}$ based at the tangential base point $\vec{01}$ describing a small open interval along the real axis from 0 towards 1. Consider the paths r and p in the following figure; r goes from the tangential base point $\vec{01}$ to $1/2$ and p from $\vec{01}$ to $\vec{10}$.

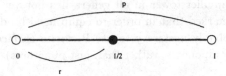

Now, by fact 1, $\sigma \in \mathrm{Gal}(\overline{\mathbb{Q}}/\mathbb{Q})$ acts on p and since it fixes the endpoints of p, p is sent to itself precomposed with a pro-loop based at $\vec{01}$, i.e. $\sigma(p) = pf_\sigma$ for some $f \in \pi_1(\mathcal{M}_{0,4}; \vec{01}) \simeq \hat{F}_2$ (this isomorphism identifies x with a small counterclockwise loop around 0 and y with the $p^{-1}\Theta(x)p$ around 1).

Ihara shows that this f_σ actually lies in the derived subgroup $\hat{F}_2'$, and he associates to σ the couple $(\chi(\sigma), f_\sigma)$ where χ is the cyclotomic character. This defines Ihara's homomorphism $\mathrm{Gal}(\overline{\mathbb{Q}}/\mathbb{Q}) \to \mathbb{Z}^* \times \hat{F}_2'$. Let us show that this homomorphism is injective. There is a natural homomorphism from $\mathbb{Z}^* \times \hat{F}_2'$ to $\mathrm{End}(\hat{F}_2)$ which sends (λ, f) to the endomorphism of $\hat{F}_2$ taking x to x^λ and $y \mapsto f^{-1}y^\lambda f$, so any $\sigma \in \mathrm{Gal}(\overline{\mathbb{Q}}/\mathbb{Q})$ which is in the kernel of Ihara's map would give the trivial automorphism of $\hat{F}_2$. Now, the action of $\mathrm{Gal}(\overline{\mathbb{Q}}/\mathbb{Q})$ on $\hat{F}_2$ is given by $x \mapsto x^{\chi(\sigma)}$ and $y \mapsto f_\sigma^{-1}y^\lambda f_\sigma$, and this is a lifting of the natural outer Galois action on $\hat{F}_2$ coming from the fact that $\hat{F}_2$ is the algebraic fundamental group of a variety defined over $\mathbb{Q}$, namely $\mathcal{M}_{0,4} \simeq \mathbb{P}^1 - \{0, 1, \infty\}$ (cf. §3.2). So the question is, can an element of $\mathrm{Gal}(\overline{\mathbb{Q}}/\mathbb{Q})$ act trivially on $\hat{F}_2$?

The negative answer to this question is a consequence of Belyi's famous theorem stating that a Riemann surface X has a model defined over $\overline{\mathbb{Q}}$ if and only if there exists a covering of Riemann surfaces $\beta : X \to \mathbb{P}^1\mathbb{C}$ ramified over 0, 1 and ∞ only. As a corollary, we see that no element of $\mathrm{Gal}(\overline{\mathbb{Q}}/\mathbb{Q})$ can act trivially on all the subgroups of $\hat{F}_2$. For if $\sigma \in \mathrm{Gal}(\overline{\mathbb{Q}}/\mathbb{Q})$, then there exists $\alpha \in \overline{\mathbb{Q}}$ on which σ acts non-trivially, and letting N be the finite-index subgroup of $\hat{F}_2$ corresponding to a Belyi cover $\beta : X \to \mathbb{P}^1\mathbb{C}$ where X is the elliptic curve of j-invariant equal to α, we see that σ cannot act trivially on N.

Now we sketch part of the proof that the image of $\mathrm{Gal}(\overline{\mathbb{Q}}/\mathbb{Q})$ under this injective homomorphism actually lies in $\widehat{GT}$, i.e. that the couples obtained this way satisfy relations (I), (II) and (III). We restrict ourselves to showing relation (I) and relation (III) in the case $\lambda \equiv \pm 1 \bmod 5$ here and refer to [LS2] for complete details; these two cases certainly give the essence of the proof. Let us do relation (I). The automorphisms θ of $\hat{F}_2$ and Θ of $\mathcal{M}_{0,4}$ introduced earlier are related as follows: for all $z \in \hat{F}_2$, we have $p^{-1}\Theta(z)p = \theta(z)$. Now, by fact 1, we have $\sigma(r) = rg$ for some $g \in \hat{F}_2$, so applying σ to $p = \Theta(r)^{-1}r$ and using fact 2, we have $pf_\sigma = \Theta(rg)^{-1}rg = \Theta(g)^{-1}\Theta(r)^{-1}rg = \Theta(g)^{-1}pg = p\theta(g)^{-1}g$, and therefore $f_\sigma = \theta(g)^{-1}g$. But such an element obviously satisfies relation (I).

Let us show it for relation (III), $\lambda \equiv \pm 1 \bmod 5$. First, we put any sphere with 5 marked points $(x_1, \ldots, x_5)$ into the standard form $(0, y_1, 1, \infty, y_2)$ via an isomorphism (i.e. a unique element of $\mathrm{PSL}_2(\mathbb{C})$). So we give points on $\mathcal{M}_{0,5}$ by pairs (y_1, y_2). Let $\tilde{p}$ denote the following path from the tangential base point (ϵ, μ) on $\mathcal{M}_{0,5}$ to $(1 - \epsilon, \mu)$, where ϵ is a small positive parameter and μ is a negative parameter of very large absolute value: the path $\tilde{p}$ sends ϵ to $1 - \epsilon$ along the real axis and does not move μ. Then $\sigma(\tilde{p}) = \tilde{p}\hat{f}$. Let Z be the point on $\mathcal{M}_{0,5}$ corresponding to the sphere with five marked points $(1, \zeta, \zeta^2, \zeta^3, \zeta^4)$ where $\zeta = \exp(2\pi i/5)$. In standard form, this point

is approximately given by the pair $(.38, -.62)$ and the path v slides ϵ to $.38$
and λ to $-.62$ along the real axis without intersections. Let $P = (12345)^2 \in$
$S_5 \simeq \mathrm{Aut}(\mathcal{M}_{0,5})$. Then P is related to the automorphism ρ of $M(0,5)$
defined in §1.1 by $\tilde{p}^{-1}P\tilde{p} = \rho$, and since P fixed Z, the path $P(v)$ goes from
$(1 - \epsilon, \lambda)$ to Z and $\tilde{p} = P(v)^{-1}v$. If $\lambda \equiv \pm 1 \bmod 5$, then σ fixes the point Z,
so by fact 1, $\sigma(v) = vk$ for some $k \in \hat{\pi}_1(\mathcal{M}_{0,5}; (\epsilon, \mu)) \simeq \hat{K}(0,5)$. Applying
σ to this equality, we obtain

$$\tilde{p}\tilde{f}_\sigma = P(vk)^{-1}vk = P(k)^{-1}P(v)^{-1}vk = P(k)^{-1}\tilde{p}k = \tilde{p}\rho(k)^{-1}k,$$

so $\tilde{f}_\sigma = \rho(k)^{-1}k$. This $\tilde{f}_\sigma$ obviously satisfies relation (III). The proofs of
relation (II) and of relation (III) for $\lambda \equiv \pm 2 \bmod 5$ are analogous, all
showing that the elements f_σ trivially satisfy the relations defining $\widehat{GT}$.

§3.2. Compatibility of the $\mathrm{Gal}(\overline{\mathbb{Q}}/\mathbb{Q})$ and $\widehat{GT}$-actions

For every variety X defined over $\mathbb{Q}$, there is an exact sequence

$$1 \to \hat{\pi}_1(X \otimes \overline{\mathbb{Q}}) \to \hat{\pi}_1(X) \to \mathrm{Gal}(\overline{\mathbb{Q}}/\mathbb{Q}) \to 1,$$

for a suitable choice of base points of the *algebraic* or *geometric* fundamental
group $\hat{\pi}_1(X \otimes \overline{\mathbb{Q}})$ and the *arithmetic* fundamental group $\hat{\pi}_1(X)$ (cf. [Oort] in
this volume for details on this exact sequence). Furthermore, this sequence
is split if X has a rational point, which is certainly the case when X is
one of the moduli spaces $\mathcal{M}_{g,n}$. Thus we have canonical homomorphisms
$\mathrm{Gal}(\overline{\mathbb{Q}}/\mathbb{Q}) \hookrightarrow \mathrm{Out}(\hat{\pi}_1(\mathcal{M}_{g,n}))$ which lift to non-canonical homomorphisms
$\mathrm{Gal}(\overline{\mathbb{Q}}/\mathbb{Q}) \hookrightarrow \mathrm{Aut}(\hat{\pi}_1(\mathcal{M}_{g,n}))$, corresponding to sections of the exact sequence, i.e. to rational points on $\mathcal{M}_{g,n}$. The canonical homomorphisms
into the outer automorphism groups are not known to be injective except
when $g = 0$, although it is plausible. Their liftings into the true automorphism groups are known to be injective for $g = 0$ and also for $g > 0$ and
$n = 0$ or 1 (cf. Nakamura's article in this volume). In the case $g = 0$, it
is natural to ask whether it is possible to choose one of these injections to
make the following diagram commute:

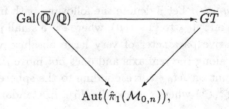

where the injective maps $\widehat{GT} \hookrightarrow \mathrm{Aut}(\hat{\pi}_1(\mathcal{M}_{0,n}))$ and $\mathrm{Gal}(\overline{\mathbb{Q}}/\mathbb{Q}) \hookrightarrow \widehat{GT}$ were explained earlier. The answer is not only that it is possible to do so, but even that there is a natural and elegant geometric explanation for the right choice of section; it corresponds to a "tangential" point on $\mathcal{M}_{0,n}$. This was understood by Ihara and Matsumoto (cf. [IM]). This compatibility of the $\widehat{GT}$ and $\mathrm{Gal}(\overline{\mathbb{Q}}/\mathbb{Q})$-actions extends to all the genus zero Teichmüller groups and groupoids of the $\mathcal{M}_{0,n}$ based at the tangential base points, almost by construction.

What about the higher genus situation? In (3) of §2.1, we mentioned that Ihara and Nakamura computed a $\widehat{GT}$-action on the profinite fundamental groups of all topological surfaces with punctures. Their action is compatible with the $\mathrm{Gal}(\overline{\mathbb{Q}}/\mathbb{Q})$-actions on the exact sequence

$$1 \to \hat{\pi}_1(S_{g,n-1}) \to \hat{\pi}_1(\mathcal{M}_{g,n}) \to \hat{\pi}_1(\mathcal{M}_{g,n-1}) \to 1,$$

arising from every "tangential" Riemann-surface with n-marked points, cf. §6 of Nakamura's article in this volume.

Does the $\mathrm{Gal}(\overline{\mathbb{Q}}/\mathbb{Q})$-action on the $\hat{\pi}_1(\mathcal{M}_{g,n})$ extend to a $\widehat{GT}$-action? Thanks to Nakamura's discovery of the long-awaited and elusive explicit $\mathrm{Gal}(\overline{\mathbb{Q}}/\mathbb{Q})$-action on the higher mapping class groups $\hat{M}(g,1)$ and $\hat{M}(g,0)$, we can at least give an explicit conjecture on the form of such a $\widehat{GT}$-action. Indeed, Nakamura's Galois action is given on the generators of $\hat{M}(g,1)$ and $\hat{M}(g,0)$ by expressions *in which the elements of* $\mathrm{Gal}(\overline{\mathbb{Q}}/\mathbb{Q})$ *act via their components* $(\chi(\sigma), f_\sigma)$, so in order to express the conjectured $\widehat{GT}$-action which should of course be compatible with the subgroup $\mathrm{Gal}(\overline{\mathbb{Q}}/\mathbb{Q})$, we simply replace $\chi(\sigma)$ by λ and f_σ by f in Nakamura's expressions. It remains to show that this action on the generators extends to an automorphism of the mapping class groups, i.e. respects their defining relations, without recourse to the knowledge that an action exists a priori, as for $\mathrm{Gal}(\overline{\mathbb{Q}}/\mathbb{Q})$. This question remains open, as does the possibility that instead of $\widehat{GT}$, we should be considering some subgroup of $\widehat{GT}$ defined by one or more additional relations and still containing $\mathrm{Gal}(\overline{\mathbb{Q}}/\mathbb{Q})$.

Bibliography

This *very restricted* bibliography lists only articles concerned with the profinite version of $\widehat{GT}$. We encourage the reader to consult the beautiful survey article "Galois rigidity of profinite fundamental groups" by H. Nakamura, covering many related aspects of what we loosely term *anabelian algebraic geometry*, including some discussion of $\widehat{GT}$. This article appeared in Japanese in Sugaku (Math. Soc. of Japan), vol. 47 (1995), p. 1-17, and its English translation should appear this year in "Sugaku Expositions", in the AMS translation. Its bibliography contains a list of articles dealing with a wide selection of themes, for instance the pro-ℓ version of $\widehat{GT}$ (a subject rich with results and conjectures unavailable in the profinite case), the study of Galois actions on various fundamental groups, the anabelian conjecture and so on.

The Urtext

[D]   V.G. Drinfel'd, On quasitriangular quasi-Hopf algebras and a group
        closely connected with Gal($\overline{\mathbb{Q}}/\mathbb{Q}$), *Leningrad Math. J.* **2** (1991),
        829-860.

On the injection of Gal($\overline{\mathbb{Q}}/\mathbb{Q}$) *into* $\widehat{GT}$.

[I1]   Y. Ihara, Braids, Galois groups, and some arithmetic functions,
        Proceedings of the ICM, Kyoto, Japan (1990), 99-120.

[I2]   Y. Ihara, On the embedding of Gal($\overline{\mathbb{Q}}/\mathbb{Q}$) into $\widehat{GT}$, in *The Grothen-*
        dieck Theory of Dessins d'Enfants, L. Schneps, ed., Cambridge
        Univ. Press 1994.

[LS2]   P. Lochak, L. Schneps, A cohomological interpretation of the Gro-
        thendieck-Teichmüller group, to appear in *Inv. Math.*.

[N]   H. Nakamura, Galois rigidity of pure sphere braid groups and profi-
        nite calculus, *J. Math. Sci. Univ. Tokyo* **1** (1994), 71-136.

[HS]   D. Harbater and L. Schneps, Approximating Galois orbits of dessins,
        this volume.

On tangential base points, Teichmüller groupoids and Galois and $\widehat{GT}$-*actions on them*

[De]   P.Deligne, Le groupe fondamental de la droite projective moins trois
        points, in *Galois Groups over* $\mathbb{Q}$, MSRI Publ. **16**, Y. Ihara et al.,
        Eds. Springer-Verlag (1989), 79-297.

[IM]   Y. Ihara, M. Matsumoto, On Galois Actions of Profinite Comple-
        tions of Braid Groups, in *Recent Developments in the Inverse Galois*
        Problem, M. Fried et al. Eds., AMS, 1995.

[IN]   Y.Ihara, H.Nakamura, On deformation of maximally degenerate
        stable curves and Oda's problem, RIMS-Preprint

[PS]   *Triangulations, Courbes Arithmétiques et Théorie des Champs*, ed.
        by L. Schneps, to appear in the survey journal *Panoramas et Syn-*
        thèses; see particularly chapters I.2 (tangential base points on the
        genus zero moduli spaces) and II (genus zero Teichmüller groupoids
        and the $\widehat{GT}$-action).

On the $\widehat{GT}$-actions on towers

[LS]   P. Lochak, L. Schneps, The Grothendieck-Teichmüller group as automorphisms of braid groups, in *The Grothendieck Theory of Dessins d'Enfants*, L. Schneps, ed., London Math. Soc. Lecture Note Series **200**, 1994.

[MS]   G. Moore, N. Seiberg, Classical and Quantum Conformal Field Theory, *Commun. Math. Phys.* **123** (1989), 177-254.

[S]   L. Schneps, On the genus zero Teichmüller tower, preprint.

Approximating Galois orbits of dessins

David Harbater* and Leila Schneps**

§0. Introduction

Let $G_{\mathbb{Q}} = \mathrm{Gal}(\overline{\mathbb{Q}}/\mathbb{Q})$. In this paper we study Belyi's action of $G_{\mathbb{Q}}$ on $\pi_1(\mathbb{P}^1 - \{0, 1, \infty\})$ on the level of finite covers . We show that this action can be made effective in terms of the natural outer action on π_1, and that this outer action can itself be approximated so as to obtain information about Galois orbits and fields of moduli of covers and dessins. We begin with some background and motivation, and then describe the structure of the paper.

0.1. Background and motivation. In the theory of branched covers of curves, the action of the arithmetic Galois group on the geometric Galois group is both important and mysterious, and provides a link between number theory and topology. This link arises from the fact that topological covering spaces over a punctured Riemann sphere can be defined as covers of algebraic curves, and that the covers are even arithmetic if the branch locus consists of points defined over $\overline{\mathbb{Q}}$. Namely, as Grothendieck showed [G], such a cover can be defined over $\overline{\mathbb{Q}}$ (and hence over some number field), and the same is true for G-Galois covers. If we fix an algebraic branch locus, then we may first ask how the absolute Galois group $G_{\mathbb{Q}} = \mathrm{Gal}(\overline{\mathbb{Q}}/\mathbb{Q})$ acts on the set of such covers, and secondly ask what the field of moduli of such a cover is. In the G-Galois case, if G has trivial center (and in certain other cases), the cover is defined over its field of moduli, so the latter question is equivalent to asking for the minimum field of definition K. That in turn has applications to the Inverse Galois Problem, via Hilbert's Irreducibility Theorem — for then G is a Galois group over $K(t)$ and hence over K.

These questions can be made more explicit by rephrasing the situation in terms of group theory. Let $P_1, \ldots, P_r \in \mathbb{P}^1_{\overline{\mathbb{Q}}}$ be r distinct points, and let $P_0 \in \mathbb{P}^1 - \{P_1, \ldots, P_r\}$ be a base point. Then we may choose a homotopy basis of counterclockwise loops γ_i at P_0 around P_i such that $\prod \gamma_i \sim 1$. To give a G-Galois cover branched at points $P_1, \ldots, P_r$ is then equivalent to giving its branch cycle description, i.e. an r-tuple of generators $(g_1, \ldots, g_r)$ of G such that $\prod g_i = 1$, determined up to uniform conjugacy (corresponding to the choice of base point on the cover over P_0). The first question above

---

* Department of Mathematics, University of Pennsylvania, Philadelphia, PA 19104

** Faculté des Sciences, Université de Franche-Comté, 25030 Besançon Cedex then becomes the problem of determining how $G_{\mathbb{Q}}$ acts on the set of such equivalence classes of r-tuples (e.g. by giving a formula for the action of $\sigma \in G_{\mathbb{Q}}$ on the set of branch cycle descriptions). Solving this would then answer the second question as well, since the field of moduli of a cover would be the fixed field of the stabilizer of the branch cycle description. We would thus like to have a "formula" for the field of moduli in terms of the branch cycles g_i. (The non-Galois case can be interpreted similarly, in terms of r-tuples of elements in an appropriate S_n, but for simplicity we focus on the Galois case here.)

Using his "branch cycle argument," Fried (cf. [F1]) showed a *branch cycle condition* that gives a weak form of the desired formula for the action of $G_{\mathbb{Q}}$. Namely, if each P_i is defined over $K \subset \overline{\mathbb{Q}}$, and if $\sigma \in G_K$, then the branch cycle description $(g'_1, \ldots, g'_r)$ of $Y^\sigma \to \mathbb{P}^1$ satisfies the relation $g'_i \sim g_i^{\chi(\sigma)}$. (Here χ is the cyclotomic character and $\sim$ denotes conjugacy in G). More generally, if the branch locus is defined over $K \subset \overline{\mathbb{Q}}$ (but the individual P_i's are not necessarily defined over K), then $g'_j \sim g_i^{\chi(\sigma)}$ if $\sigma(P_i) = P_j$.

Later, Belyi [B] considered the case of $r = 3$, where we may take $P_1 = 0$, $P_2 = 1$, $P_3 = \infty$. By considering the special case now known as "rigidity" (and treated independently, about the same time, by Matzat [M] and Thompson [T]), he showed that for certain triples the branch cycle condition $g'_i \sim g_i^{\chi(\sigma)}$ determines the field of moduli — and thus many simple groups can be realized as Galois groups over $\mathbb{Q}^{\mathrm{ab}}$ (or even over $\mathbb{Q}$, in some cases). Also, he showed that for a given $\sigma \in G_{\mathbb{Q}}$, there is a unique $f_\sigma \in \widehat{F}'_2$ (the commutator subgroup) such that σ takes each triple (g_1, g_2, g_3) to an expression of the form $(g_1^{\chi(\sigma)}, f_\sigma^{-1} g_2^{\chi(\sigma)} f_\sigma, \tilde{g}_3^{\chi(\sigma)})$ (for some $\tilde{g}_3 \sim g_3$). This provides a lifting of the natural map $G_{\mathbb{Q}} \to \mathrm{Out}(\widehat{F}_2)$ to a homomorphism $G_{\mathbb{Q}} \to \mathrm{Aut}(\widehat{F}_2)$. Moreover he showed a converse to Grothendieck's theorem, viz. that for every curve Y defined over $\overline{\mathbb{Q}}$ there is a covering map $Y \to \mathbb{P}^1$ that is branched only over $\{0, 1, \infty\}$ — and hence $G_{\mathbb{Q}}$ acts faithfully on the set of étale covers of $\mathbb{P}^1 - \{0, 1, \infty\}$.

In the *Esquisse d'un Programme*, motivated in part by Belyi, Grothendieck suggested studying $G_{\mathbb{Q}}$ as a group of outer automorphisms of $\mathbb{P}^1 - \{0, 1, \infty\}$. More generally, he suggested viewing $G_{\mathbb{Q}}$ as a group of outer automorphisms of the groups $\widehat{K}(g, n) = \pi_1(\mathcal{M}_{g,n})$, where $\mathcal{M}_{g,n}$ is the moduli space of curves of genus g with r ordered marked points, and then trying to understand the outer action on $\widehat{K}(g, n)$ in terms of the one on $\widehat{K}(0, 4) = \widehat{F}_2$ (where $\mathcal{M}_{0,4} = \mathbb{P}^1 - \{0, 1, \infty\}$). In particular, he suggested that the action on the full "Teichmüller tower" of $\widehat{K}(g, n)$'s can be understood in terms of $\widehat{K}(0, 4)$, $\widehat{K}(1, 1)$, $\widehat{K}(0, 5)$, and $\widehat{K}(1, 2)$ (where the actions on the first two would provide the generators of the tower and the second two the relations). In addition, he showed how covers of $\mathbb{P}^1 - \{0, 1, \infty\}$ can be classified not only by equivalence classes of triples, but also by "dessins d'enfants", which encode the same information graphically.

Since the action of an element $\sigma \in G_{\mathbb{Q}}$ is given by $\lambda_\sigma = \chi(\sigma) \in \hat{\mathbb{Z}}^*$ and $f_\sigma \in \hat{F}_2'$ (as in [B]), the above involves characterizing the pairs $(\lambda, f) \in \hat{\mathbb{Z}}^* \times \hat{F}_2'$ that arise from elements of $G_{\mathbb{Q}}$, and characterizing the actions of elements $\sigma \in G_{\mathbb{Q}}$ on the tower of $\hat{K}(g,n)$'s in terms of λ_σ and f_σ. Such characterizations would give an explicit description of $G_{\mathbb{Q}}$ as a subgroup of the outer automorphism group of $\hat{K}(g,n)$, and in particular of $\hat{K}(0,4) = \hat{F}_2'$.

Some progress was made on the first point by the discovery of three necessary conditions for a pair $(\lambda, f) \in \hat{\mathbb{Z}}^* \times \hat{F}_2$ to come from $G_{\mathbb{Q}}$. These properties come from work of Drinfel'd (cf. [D]), where he defined the Grothendieck-Teichmüller group $\widehat{GT}$ as a certain subgroup of $\mathrm{Aut}(\hat{F}_2)$ (constructed in terms of "quasi-triangular quasi-Hopf algebras") that contains the image of $G_{\mathbb{Q}}$ under the Belyi lifting. Later, Ihara reinterpreted the inclusion of $G_{\mathbb{Q}}$ in $\widehat{GT}$ in terms of $\mathcal{M}_{0,4}$ and $\mathcal{M}_{0,5}$, braid groups, and "pro-loops" (cf. [I1], [I2]). The elements of $\widehat{GT}$ are automorphisms corresponding to pairs $(\lambda, f) \in \hat{\mathbb{Z}}^* \times \hat{F}_2'$ satisfying a certain 2-cocycle condition (I), a 3-cocycle condition (II), and a 5-cocycle condition (III) (see the survey on $\widehat{GT}$ in this volume for details). The group $\widehat{GT}$ can be identified with the automorphism group of a tower of profinite Artin braid groups $\hat{B}_n$; cf. [LS].

On the second point, Nakamura recently showed explicitly for $g \geq 1$ and $n = 0$ or 1, how the action of any $\sigma \in G_{\mathbb{Q}}$ on $\hat{K}(g,n)$ can be expressed in terms of the action on $\hat{K}(0,4)$. For $\hat{K}(0,4) = \hat{F}_2 = \langle x, y, z \mid xyz = 1 \rangle$, Belyi's action can be written more explicitly as

$$\sigma : x \mapsto x^\lambda, \quad y \mapsto (y^\lambda)^{f(x,y)}, \quad z \mapsto (z^\lambda)^{f(x,z)x^{-m}}, \qquad (*)$$

where $\lambda = \lambda_\sigma = \chi(\sigma) \in \hat{\mathbb{Z}}^*$, $f = f_\sigma \in \hat{F}_2'$, $m = (\lambda - 1)/2 \in \hat{\mathbb{Z}}$, and $u^v := v^{-1}uv$ for $u, v \in \hat{F}_2$. In [N, Appendix], Nakamura described an analogue of Belyi's lifting that provides an action of $G_{\mathbb{Q}}$ on the group $\hat{K}(0,5)$ (which is generated by elements $x_{i,i+1}$ for i modulo 5) by

$$\sigma : x_{12} \mapsto x_{12}^\lambda, \quad x_{23} \mapsto (x_{23}^\lambda)^{f(x_{12},x_{23})}, \quad x_{34} \mapsto (x_{34}^\lambda)^{f(x_{45},x_{34})},$$

$$x_{45} \mapsto x_{45}^\lambda, \quad x_{51} \mapsto (x_{51}^\lambda)^{f(x_{45},x_{51})f(x_{12},x_{23})}, \qquad (**)$$

where λ, f are as above. In his contribution to this volume, he discovered a similar formula for the action of $G_{\mathbb{Q}}$ on $\hat{K}(g,0)$ and $\hat{K}(g,1)$, with $g \geq 0$.

0.2. Structure of the paper.
In the present paper, we continue the examination of $G_{\mathbb{Q}}$ and $\widehat{GT}$ as groups of outer automorphisms of the fundamental groups $\hat{K}(0,n)$ of the moduli spaces $\mathcal{M}_{0,n}$, and the related problem of finding a "formula" for the action of $\sigma \in G_{\mathbb{Q}}$ on covers — i.e. finding f_σ in terms of σ, at least on finite levels. We break this problem into two parts.

In §1 we treat the first half of this problem, viz. showing that the Belyi lifting $G_{\mathbb{Q}} \to \mathrm{Aut}(\widehat{F}_2)$ can be made "effective" in terms of the natural map $G_{\mathbb{Q}} \to \mathrm{Out}(\widehat{F}_2)$. That is, for every normal subgroup $N \subset \widehat{F}_2$ of finite index, we show that there is a finite index normal subgroup $\tilde{N} \subset \widehat{F}_2$ such that the Belyi lifting modulo N can be computed in terms of the outer action of $G_{\mathbb{Q}}$ modulo $\tilde{N}$. Moreover we show how $\tilde{N}$ can be found explicitly in terms of N, and interpret this in terms of computing the action of $G_{\mathbb{Q}}$ on covers of $\mathbb{P}^1 - \{0, 1, \infty\}$. In §2 we prove analogous results for $\widehat{K}(0,5)$. A basic ingredient is the definition (in 1.1) of a certain group $\mathcal{O}_n^\#$ of symmetric outer automorphisms of $\widehat{K}(0,n)$.

The second half of the problem, i.e. explicitly computing the outer action of a given $\sigma \in G_{\mathbb{Q}}$, remains open. In §3 we obtain partial results in this direction, which compute the $\widehat{GT}$-orbit of a dessin — thus approximating the $G_{\mathbb{Q}}$-orbit — and also yield information about the field of moduli of a dessin (or corresponding cover). This is achieved by using the result (cf. [HS]) that $\widehat{GT} \simeq \mathcal{O}_5^\#$, and by considering the image of $\mathcal{O}_5^\#$ in $\mathrm{Out}(\widehat{K}(0,5)/N)$ for characteristic subgroups $N \subset \widehat{K}(0,5)$ of finite index.

§1. The Belyi lifting and four-point moduli

Let F_2 be the free group on two generators x and y, which is the topological fundamental group of $\mathbb{P}^1 - \{0, 1, \infty\}$ (based at some point), with x and y being counterclockwise loops around 0 and 1 respectively. Let $\widehat{F}_2$ be its profinite completion, which we identify with the algebraic fundamental group of $\mathbb{P}^1 - \{0, 1, \infty\}$ (with the same base point), and hence with $\widehat{K}(0,4)$. Let $\widehat{F}_2'$ denote its commutator subgroup, and consider the *Belyi subgroup* $A \subset \mathrm{Aut}(\widehat{F}_2)$, defined (as in [B]) by

$$A = \big\{ F \in \mathrm{Aut}(\widehat{F}_2) \mid \exists \lambda \in \hat{\mathbb{Z}}^*, f \in \widehat{F}_2' \text{ such that }$$
$$F(x) = x^\lambda, \ F(y) = f^{-1}y^\lambda f, \ F(xy) \sim (xy)^\lambda \big\},$$

where $\sim$ denotes conjugacy in $\widehat{F}_2$. Giving $F \in A$ determines the pair (λ, f) uniquely, so A may also be regarded as a subset of $\hat{\mathbb{Z}}^* \times \widehat{F}_2'$, which is how we consider it henceforth.

Since any $\sigma \in G_{\mathbb{Q}}$ must take the full tower of finite covers (regarded as the "pro-universal cover" of $\mathbb{P}^1 - \{0, 1, \infty\}$) to itself, there is a natural map $\alpha : G_{\mathbb{Q}} \to \mathrm{Out}(\widehat{F}_2)$. Using that the centralizers of x and y are respectively the pro-cyclic subgroups $\langle x \rangle$ and $\langle y \rangle$, Belyi deduced that α may be lifted to a homomorphism $\beta : G_{\mathbb{Q}} \to A \subset \mathrm{Aut}(\widehat{F}_2)$. The map β, known as the *Belyi lifting*, corresponds to a section of the fundamental exact sequence

$$1 \to \widehat{F}_2 \to \pi_1(\mathbb{P}^1_{\mathbb{Q}} - \{0, 1, \infty\}) \to G_{\mathbb{Q}} \to 1$$

obtained via a certain *tangential base point* for π_1 (cf. [IM]).

The goal of this section is to show that the Belyi lifting $\beta : G_{\mathbb{Q}} \to \text{Aut}(\widehat{F}_2)$ is effective in terms of the natural map $\alpha : G_{\mathbb{Q}} \to \text{Out}(\widehat{F}_2)$. We describe our approach to this in 1.2 below; before that, we need to define some important subgroups of the automorphism and outer automorphism groups of the pure mapping class groups $\widehat{K}(0,n) = \pi_1(\mathcal{M}_{0,n})$. (For basic facts about $\widehat{K}(0,n)$, cf. section 1.1 of the survey on $\widehat{GT}$ in this volume.)

1.1. Symmetric automorphisms of pure mapping class groups.

For each n, the symmetric group S_n acts on the moduli space $\mathcal{M}_{0,n}$ by permuting the order of the marked points. For $n = 4$, the automorphism group of $\mathcal{M}_{0,4} = \mathbb{P}^1 - \{0, 1, \infty\}$ is S_3, and the map $\sigma^{(4)} : S_4 \to \text{Aut}(\mathcal{M}_{0,4})$ is surjective with kernel equal to the even involutions in S_4 (which form a Klein four group). On the other hand, for $n > 4$, the map $S_n \to \text{Aut}(\mathcal{M}_{0,n})$ is an isomorphism. For all n, the map $S_n \to \text{Aut}(\mathcal{M}_{0,n})$ induces a homomorphism $\sigma^{(n)} : S_n \to \text{Out}(\widehat{K}(0,n))$, which again is injective for $n > 4$ and has Klein four kernel if $n = 4$. (In fact, by a version of Grothendieck's anabelian conjecture — see the article by Ihara and Nakamura in this volume — the image of this homomorphism is exactly the subgroup of $\text{Out}(\widehat{K}(0,n))$ that commutes with the natural outer action of $G_{\mathbb{Q}}$ on $\widehat{K}(0,n)$.)

For any group G, the outer automorphism group $\text{Out}(G)$ acts on the set of conjugacy classes $[g]$ of elements of G, and we may make the following

Definition. For all $n \geq 4$, let $\mathcal{O}_n^{\#}$ be the subgroup of outer automorphisms $\overline{F} \in \text{Out}\big(\widehat{K}(0,n)\big)$ such that

(i) for each i, j, we have $\overline{F}([x_{ij}]) = [x_{ij}^{\lambda}]$ for some $\lambda \in \widehat{\mathbb{Z}}^{*}$;

(ii) $\overline{F}$ commutes with $\sigma^{(n)}(S_n)$ in $\text{Out}(\widehat{K}(0,n))$.

Note that for $\overline{F} \in \mathcal{O}_n^{\#}$, the value of λ is independent of i, j by the symmetry condition (ii); so we may write $\lambda = \lambda(\overline{F})$. Let $\mathcal{A}_n^{\#}$ denote the inverse image of $\mathcal{O}_n^{\#}$ under the natural map $\text{Aut}(\widehat{K}(0,n)) \to \text{Out}(\widehat{K}(0,n))$, and write $\lambda(F) = \lambda(\overline{F})$ if $F \in \mathcal{A}_n^{\#}$ maps to $\overline{F} \in \mathcal{O}_n^{\#}$. A key fact [HS, 1.2] is that the image of the natural map $G_{\mathbb{Q}} \hookrightarrow \text{Out}(\widehat{K}(0,n))$ is contained in $\mathcal{O}_n^{\#}$.

In this section, we restrict attention to the case $n = 4$, identifying $\widehat{K}(0,4)$ with $\widehat{F}_2$ via $x_{12} = x$ and $x_{23} = y$, and viewing $\mathcal{O}_4^{\#} \subset \text{Out}(\widehat{F}_2)$ and $\mathcal{A}_4^{\#} \subset \text{Aut}(\widehat{F}_2)$. Since $\beta : G_{\mathbb{Q}} \hookrightarrow \text{Aut}(\widehat{F}_2)$ lifts $\alpha : G_{\mathbb{Q}} \hookrightarrow \text{Out}(\widehat{F}_2)$ and since $\alpha(G_{\mathbb{Q}}) \subset \mathcal{O}_4^{\#}$, we have that $\beta(G_{\mathbb{Q}}) \subset \mathcal{A}_4^{\#}$. Our explicit description below of β in terms of α will be based on the following result, which is essentially well-known (cf. [IM]), but is explicitly proved in this form in [HS, 1.2].

Theorem 1. *There is a unique section s of the natural homomorphism $\mathcal{A}_4^{\#} \to \mathcal{O}_4^{\#}$ whose image lies in the Belyi subgroup A of $\text{Aut}(\widehat{F}_2)$. This section satisfies $\beta = s\alpha : G_{\mathbb{Q}} \hookrightarrow \text{Aut}(\widehat{F}_2)$.*

1.2. Explicit computation of the Belyi lifting. Our goal in §1 is to compute β explicitly in terms of α, on the level of finite covers. That is, we will show that for any normal subgroup $N \subset \widehat{F}_2$ of finite index, there is a smaller such subgroup $\tilde{N}$ — which we determine explicitly — such that the reduction of β modulo N is determined by that of α modulo $\tilde{N}$.

Our approach to this is to use Theorem 1, and the section s of $\mathcal{A}_4^{\#} \to \mathcal{O}_4^{\#}$. We show that s is effectively computable, in the sense that $s(\overline{F})$ modulo N is determined by $\overline{F}$ modulo $\tilde{N}$, where $\tilde{N}$ depends only on N (and not on $\overline{F}$). The computation of β in terms of α then follows from the relation $\beta = s\alpha$.

The rest of §1 is thus devoted to finding an $\tilde{N}$ for each N, and describing how to compute $s(\overline{F})$ modulo N in terms of $\overline{F}$ modulo $\tilde{N}$ — and hence β modulo N in terms of α modulo $\tilde{N}$. First, we need to define these reductions of α and β.

Let Γ be the set of normal subgroups N of finite index in $\widehat{F}_2$. For each $N \in \Gamma$, consider the quotient group $G_N = \widehat{F}_2/N$. Then giving the quotient map $\widehat{F}_2 \twoheadrightarrow G_N$ is equivalent to giving a triple (a, b, c) of generators of G_N such that $abc = 1$, corresponding (via Riemann's Existence Theorem) to a pointed G_N-Galois cover $X \to \mathbb{P}_{\mathbb{Q}}^1 - \{0, 1, \infty\}$. The set of equivalence classes of such triples (under uniform conjugacy) can be identified with the set of isomorphism classes of G_N-Galois dessins, each corresponding to a G_N-Galois cover $X \to \mathbb{P}_{\mathbb{Q}}^1 - \{0, 1, \infty\}$.

For $N \in \Gamma$, the action of $G_{\mathbb{Q}}$ on G_N-Galois covers lifts to an action on the set of triples $\{(a, b, c) \mid a, b, c \text{ generate } G_N, \ abc = 1\}$, via the map $\beta : \sigma \mapsto (\lambda_\sigma, f_2) \in A \subset \widehat{\mathbb{Z}}^* \times \widehat{F}_2'$. Since $a, b \in G_N$ determine $c \in G_N$, we can view this as an action on pairs of generators $(a, b) \in G_N$. Note that this action factors through $(\mathbb{Z}/n\mathbb{Z})^* \times G_N'$, where n is the exponent of G_N. Namely, $\sigma \in G_{\mathbb{Q}}$ takes a pair of generators (a, b) of G to $(a^{\lambda_{\sigma,N}}, f_{\sigma,N}^{-1} b^{\lambda_{\sigma,N}} f_{\sigma,N})$, where $\lambda_{\sigma,N}$ is the image of λ_σ in $(\mathbb{Z}/n\mathbb{Z})^*$ and $f_{\sigma,N}$ is the image of f_σ in G_N'. We thus obtain a reduction map $\beta_N : G_{\mathbb{Q}} \to A_N \subset (\mathbb{Z}/n\mathbb{Z})^* \times G_N'$ of β (modulo N), where A_N is the image of A under $\widehat{\mathbb{Z}}^* \times \widehat{F}_2' \to (\mathbb{Z}/n\mathbb{Z})^* \times G_N'$.

Next, we define the reductions α_N of α, for each $N \in \Gamma$. For this, observe that we may identify $\text{Aut}(\widehat{F}_2)$ with the set of pairs of generators (x', y') of $\widehat{F}_2$, via $F \mapsto (F(x), F(y))$. Thus $\text{Out}(\widehat{F}_2)$ becomes identified with the set of equivalence classes of such pairs (with respect to uniform conjugacy), and $\mathcal{O}_4^{\#}$ is identified with a subset of this set of equivalence classes. For each $N \in \Gamma$, let $\mathcal{O}_4^{\#}/N$ be the quotient of this subset under translation by N; this is a set whose elements are equivalence classes of (certain) pairs of generators of G_N. The reduction $\alpha_N : G_{\mathbb{Q}} \to \mathcal{O}_4^{\#}/N$ is then the composition of α with the reduction map $\mathcal{O}_4^{\#} \twoheadrightarrow \mathcal{O}_4^{\#}/N$.

In this section, as noted above, our goal is to find β_N in terms of an appropriate $\alpha_{\tilde{N}}$. This is only part of the larger problem of understanding the

Belyi map completely by computing each β_N directly, thereby bridging the gap from combinatorial group theory to arithmetic in Riemann's Existence Theorem. The other, more difficult part is to compute α_N directly for any given N. A weaker approach to that problem is the subject of §3.

1.3. Construction of $\tilde{N}$. The basic ingredient in the explicit construction of $\tilde{N}$ is a construction of Serre, given in the first of his two letters published in this volume. Namely, let G be a finite group, let $u \in G$, and let r be a positive integer. Following [S, Lemme 0], we consider the abelian group R consisting of maps $f : G \to \mathbb{Z}/r\mathbb{Z}$ satisfying $f(ug) = f(g)$ for all $g \in G$. Let $e \in R$ be the element corresponding to the characteristic function of the subgroup $\langle u \rangle \subset G$. Also, consider the action of G on R given by $f^g(x) := f(xg)$, where $g \in G$, $f \in R$, $x \in G$. With respect to this action we form the semidirect product $R \rtimes G$, and denote this group by $\Sigma_r(G, u)$. For each $g \in G$, let $g^* = eg \in \Sigma_r(G, u)$. The following result slightly generalizes [S, Lemme 0], which considers the case that $r = o = p$, a prime number.

Lemma 2. (cf. [S]) *Let G be a finite group, let $u \in G$, let r be a positive integer, and let $G^* = \Sigma_r(G, u)$.*
(a) Let $o = \mathrm{ord}(u)$ in G. Then $\mathrm{lcm}(r, o) = \mathrm{ord}(u^*)$ in G^*.
(b) If two powers of u^* are conjugate in G^* then they are equal.
(c) Under the quotient map $G^* \to G$, the image of the normalizer $\mathrm{Nor}_{G^*}\langle u^* \rangle$ is $\langle u \rangle$.

Proof. The proofs of (a) and (b) are straightforward computations. By part (b), the normalizer $\mathrm{Nor}_{G^*}\langle u^* \rangle$ is equal to the centralizer $Z_{G^*}(u^*)$. Since (following [S]) the group law on G^* shows that the image of $Z_{G^*}(u^*)$ under $G^* \to G$ is $\langle u \rangle$, part (c) follows.   ◇

The two following lemmas are needed for the construction of $\tilde{N}$.

Lemma 3. *Let G be a finite group with generators a, b. Let $A = \langle a \rangle$ and $B = \langle a \rangle$, and let $\bar{G}$ be the subgroup of $A \times B \times G$ generated by the two elements $\bar{a} := (a, 1, a)$ and $\bar{b} := (1, b, b)$. Then*
(a) The orders of $\bar{a}, \bar{b}$ in $\bar{G}$ are respectively equal to the orders of a, b in G;
(b) The third projection map defines a surjection $\bar{G} \twoheadrightarrow G$ taking $\bar{a}, \bar{b}$ to a, b;
(c) If $i, j \in \mathbb{Z}$ and $\bar{a}^i \bar{b}^j$ lies in the commutator subgroup $\bar{G}'$ of $\bar{G}$, then $\bar{a}^i = \bar{b}^j = 1$.

Proof. Assertions (a) and (b) are immediate. Since the commutator subgroup of $A \times B \times G$ is $1 \times 1 \times G'$, and since $\bar{a}^i \bar{b}^j = (a^i, b^j, a^i b^j)$, the last assertion then follows.   ◇

We may combine these constructions as follows: Let $a = \pi(x), b = \pi(y) \in G$. Using G, a, b, define $\bar{G}, \bar{a}, \bar{b}$ as in Lemma 3 above. Thus we have a surjection $\bar{G} \twoheadrightarrow G$, taking $\bar{a} \mapsto a$ and $\bar{b} \mapsto b$. Choose any positive integer r that is divisible by the exponent of $\bar{G}$. By the construction in Lemma 2, we obtain the group $\Sigma_r(\bar{G}, \bar{b})$ together with elements $\bar{a}^*, \bar{b}^*$. Again applying this construction (but with the order of the two elements reversed), we obtain the group $\Sigma_r(\Sigma_r(\bar{G}, \bar{b}), \bar{a}^*)$, together with elements $\tilde{a} := \bar{a}^{**}$, $\tilde{b} := \bar{b}^{**}$. Let $\tilde{G}$ be the subgroup generated by $\tilde{a}, \tilde{b}$. Thus we have a surjection $\bar{\eta} : \tilde{G} \twoheadrightarrow \bar{G}$ taking $\tilde{a}, \tilde{b}$ to $\bar{a}, \bar{b}$ respectively. Let $\tilde{\pi} : \widehat{F}_2 \twoheadrightarrow \tilde{G}$ be the map taking x to $\tilde{a}$ and y to $\tilde{b}$, so that the composition $\bar{\pi} = \bar{\eta} \circ \tilde{\pi} : \widehat{F}_2 \twoheadrightarrow \bar{G}$ is the map taking x to $\bar{a}$ and y to $\bar{b}$.

Thus for any finite group G and surjection $\pi : \widehat{F}_2 \twoheadrightarrow G$, we obtain in this manner an explicit choice of finite groups $\bar{G}, \tilde{G}$ and a factorization $\widehat{F}_2 \xrightarrow{\tilde{\pi}} \tilde{G} \xrightarrow{\bar{\eta}} \bar{G} \twoheadrightarrow G$ of π. We then have:

Lemma 4. *Let G be a finite group and $\pi : \widehat{F}_2 \twoheadrightarrow G$ a surjection. Consider the groups $\bar{G}, \tilde{G}$ and factorization $\widehat{F}_2 \xrightarrow{\tilde{\pi}} \tilde{G} \xrightarrow{\bar{\eta}} \bar{G} \twoheadrightarrow G$ of π as above. Write $\bar{\pi} = \bar{\eta} \circ \tilde{\pi}$. Then:*

(a) The orders of $\tilde{\pi}(x), \tilde{\pi}(y) \in \tilde{G}$ are divisible by r and the exponent of G.

(b) If two powers of $\tilde{\pi}(x)$ are conjugate in $\tilde{G}$ then they are equal.

(c) $\bar{\eta}(\mathrm{Nor}_{\tilde{G}}\langle \tilde{\pi}(x) \rangle) = \langle \bar{\pi}(x) \rangle$ and $\bar{\eta}(\mathrm{Nor}_{\tilde{G}}\langle \tilde{\pi}(y) \rangle) = \langle \bar{\pi}(y) \rangle$.

(d) If $\bar{\pi}(x)^i \bar{\pi}(y)^j$ lies in the commutator subgroup $\bar{G}'$ of $\bar{G}$ for some $i, j \in \mathbb{Z}$, then $\bar{\pi}(x)^i = \bar{\pi}(y)^j = 1$.

Proof. Condition (a) is clear from Lemma 2(a) and the fact that $\exp(G)$ divides $\exp(\bar{G})$ and hence r. Condition (b) follows from Lemma 2(b), and condition (c) follows from Lemma 2(c) applied to each of the two uses of that construction. Condition (d) is just Lemma 3(c). $\diamondsuit$

Definition of $\tilde{N}$. Given a normal subgroup $N \subset \widehat{F}_2$ of finite index, with $\pi : \widehat{F}_2 \twoheadrightarrow G = \widehat{F}_2/N$ the corresponding quotient map, we have defined $\widehat{F}_2 \xrightarrow{\tilde{\pi}} \tilde{G} \twoheadrightarrow G$ (depending on a choice of positive integer r). Define the subgroup $\tilde{N}$ associated to N by

$$\tilde{N} = \ker\left(\widehat{F}_2 \twoheadrightarrow \tilde{G}\right) \subset N.$$

Let $n, \tilde{n}$ be the exponents of $G, \tilde{G}$ respectively. Write $a = \pi(x)$, $b = \pi(y)$, $\tilde{a} = \tilde{\pi}(x)$, and $\tilde{b} = \tilde{\pi}(y)$. As before we let $\mathcal{O}_4^\#/N$ be the reduction modulo N of the pairs $(x', y') \in \widehat{F}_2^2$ such that $x \mapsto x', y \mapsto y'$ represents an element of $\mathcal{O}_4^\#$; and similarly for $\mathcal{O}_4^\#/\tilde{N}$. For any $\ell \in (\mathbb{Z}/\tilde{n}\mathbb{Z})^*$ and $s|\tilde{n}$ let $\ell_s \in (\mathbb{Z}/s\mathbb{Z})^*$ denote the reduction of ℓ modulo s, and for any $f \in \tilde{G}$ let $f_N \in G$ denote its reduction modulo N (i.e. its image in G).

Lemma 5. *In the above situation, let $(\ell, f), (k, g) \in (\mathbb{Z}/\tilde{n}\mathbb{Z})^* \times \tilde{G}'$, and suppose that the pairs $(\tilde{a}^\ell, f^{-1}\tilde{b}^\ell f), (\tilde{a}^k, g^{-1}\tilde{b}^k g)$ represent the same element of $\mathcal{O}_4^\#/\tilde{N}$. Then*

(a) $\ell_r = k_r$ and hence $\ell_n = k_n$;
(b) $f_N = g_N$.

Proof. Since $\tilde{a}^\ell$, $\tilde{a}^k$ are conjugate in $\tilde{G}$, Lemma 4(b) says that these two elements are equal. Thus Lemma 4(a) yields part (a) of the proposition.

So there is an element $h \in \tilde{G}$ such that conjugation by h takes $(\tilde{a}^\ell, f^{-1}\tilde{b}^\ell f)$ to $(\tilde{a}^\ell, g^{-1}\tilde{b}^k g)$. Thus $h \in Z_{\tilde{G}}(\tilde{a}^\ell) = Z_{\tilde{G}}(\tilde{a})$, using that $\ell \in (\mathbb{Z}/\tilde{n}\mathbb{Z})^*$. Also, $ghf^{-1} \in \mathrm{Nor}_{\tilde{G}}\langle \tilde{b}^\ell \rangle = \mathrm{Nor}_{\tilde{G}}\langle \tilde{b}\rangle$. Let $\tilde{G} \twoheadrightarrow \bar{G}$ be as in the construction above, and write $\bar{N} = \ker(\widehat{F}_2 \twoheadrightarrow \bar{G})$, $\bar{a} = \tilde{a}_{\bar{N}}$, $\bar{b} = \tilde{b}_{\bar{N}}$. By Lemma 4(c), we have

$$h_{\bar{N}} = \bar{a}^i, \qquad g_{\bar{N}}\bar{a}^i f_{\bar{N}}^{-1} = (ghf^{-1})_{\bar{N}} = \bar{b}^j \qquad (*)$$

for some integers i, j. Since $f, g \in \tilde{G}'$, we have that $f_{\bar{N}}, g_{\bar{N}} \in \bar{G}'$. So $\bar{a}^i \bar{b}^{-j} \in \bar{G}'$, and hence $\bar{a}^i = \bar{b}^{-j} = 1$ by Lemma 4(d). Thus $(*)$ yields $g_{\bar{N}} f_{\bar{N}}^{-1} = 1$. So $f_{\bar{N}} = g_{\bar{N}}$ and thus $f_N = g_N$. $\diamond$

1.4. The main result. Recall that the goal of this section is to make the section $s : \mathcal{O}_4^\# \to A \cap \mathcal{A}_4^\#$ of Theorem 1 explicit. This is done in the following theorem: for $N \in \Gamma$ and $\tilde{N}$ as above, it shows how to compute $s(\overline{F})$ modulo N in terms of $\overline{F}$ modulo $\tilde{N}$.

Theorem 6. *Let N be a normal subgroup of $\widehat{F}_2$ of finite index, and let n be the exponent of $G = \widehat{F}_2/N$. Define $\widehat{F}_2 \overset{\tilde{\pi}}{\twoheadrightarrow} \tilde{G} \twoheadrightarrow G$ as in the construction above, and let $\tilde{N} = \ker(\widehat{F}_2 \twoheadrightarrow \tilde{G})$. For $\overline{F} \in \mathcal{O}_4^\#$, write $F = s(\overline{F}) = (\lambda, f) \in A \subset \widehat{\mathbb{Z}}^* \times \widehat{F}_2'$, and let F_N be the image of F in $A_N \subset (\mathbb{Z}/n\mathbb{Z})^* \times G'$.*

(a) F_N depends only on $\overline{F}_{\tilde{N}}$, the image of $\overline{F}$ in $\mathcal{O}_4^\#/\tilde{N}$.

(b) Explicitly, F_N is given in terms of $\overline{F}_{\tilde{N}}$ as follows, where we write $a = \pi(x)$, $b = \pi(y)$, $\tilde{a} = \tilde{\pi}(x)$, $\tilde{b} = \tilde{\pi}(y)$: Let $(\tilde{a}', \tilde{b}')$ be any element of the equivalence class $\overline{F}_{\tilde{N}}$. Choose $\tilde{d}, \tilde{e} \in \tilde{G}$ and $\ell \in \mathbb{Z}$ such that $\tilde{a}' = \tilde{d}^{-1}\tilde{a}^\ell \tilde{d}$ and $\tilde{b}' = \tilde{e}^{-1}\tilde{b}^\ell \tilde{e}$, and choose $i, j \in \mathbb{Z}$ such that $\tilde{e}\tilde{d}^{-1} \equiv \tilde{b}^j \tilde{a}^{-i} \bmod \tilde{G}'$. Then $F_N = (\ell_n, b^{-j}ed^{-1}a^i)$, where ℓ_n is the reduction of ℓ modulo n, and where d, e are the images of $\tilde{d}, \tilde{e}$ in G.

Proof. (a) Assume that $\overline{F}, \overline{E} \in \mathcal{O}_4^\#$ have the same image in $\mathcal{O}_4^\#/\tilde{N}$. Let $F = s(\overline{F}), E = s(\overline{E}) \in A \cap \mathcal{A}_4^\#$, and identify these lifts with the corresponding pairs $(\lambda_F, f_F), (\lambda_E, f_E) \in \widehat{\mathbb{Z}}^* \times \widehat{F}_2'$. By hypothesis, the images of $(x^{\lambda_F}, f_F^{-1}y^{\lambda_F}f_F)$ and $(x^{\lambda_E}, f_E^{-1}y^{\lambda_E}f_E)$ in $\tilde{G} \times \tilde{G}$ represent the same element of $\mathcal{O}_4^\#/\tilde{N}$. So by Lemma 5, λ_F and λ_E have the same image in $(\mathbb{Z}/n\mathbb{Z})^*$, and f_F and f_E have the same image in G'. Thus F, E have the same image in A_N, i.e. $F_N = E_N$.

(b) We show that the construction described can be carried out, and that the asserted equality holds regardless of choices made in the construction.

Let $\tilde{n}$ be the exponent of $\tilde{G}$. Viewing $F \in \mathrm{Aut}(\widehat{F}_2)$, we have that $F(x) = x^\lambda$ and $F(y) = f^{-1}yf$, with $f \in \widehat{F}_2'$. So each element of the equivalence class

David Harbater and Leila Schneps

$\overline{F}_{\tilde{N}}$ is of the form $(\tilde{a}^{\lambda_{\tilde{n}}}, f_{\tilde{N}}^{-1}\tilde{b}^{\lambda_{\tilde{n}}}f_{\tilde{N}})$, where $\lambda_{\tilde{n}} \in (\mathbb{Z}/\tilde{n}\mathbb{Z})^*$ is the reduction of $\lambda \in \hat{\mathbb{Z}}^*$ modulo $\tilde{n}$ and where $f_{\tilde{N}} \in \tilde{G}'$ is the reduction of f modulo $\tilde{N}$. Thus $(\tilde{a}', \tilde{b}')$ is uniformly conjugate to $(\tilde{a}^{\lambda_{\tilde{n}}}, f_{\tilde{N}}^{-1}b^{\lambda_{\tilde{n}}}f_{\tilde{N}})$, and so the desired $\tilde{d}, \tilde{e}, \ell$ exist (though are not necessarily unique). Since $\tilde{a}, \tilde{b}$ generate $\tilde{G}$, it follows that the abelianization $\tilde{G}^{\mathrm{ab}} = \tilde{G}/\tilde{G}'$ is generated by the images of these elements; hence the desired i, j exist (but again are not unique). Thus the pair $(\ell_n, b^{-j}ed^{-1}a^i) \in (\mathbb{Z}/n\mathbb{Z})^* \times G$ can be constructed. Moreover the second entry lies in G', because $\tilde{b}^{-j}\tilde{e}\tilde{d}^{-1}\tilde{a}^i \in \tilde{G}'$. It remains to show that this pair equals $F_N = (\lambda_N, f_N)$, and so is independent of the above choices.

By Lemma 5, it suffices to show that $(\tilde{a}^{\lambda_{\tilde{n}}}, f_{\tilde{N}}^{-1}\tilde{b}^{\lambda_{\tilde{n}}}f_{\tilde{N}})$ and $(\tilde{a}^{\lambda_{\tilde{n}}}, \tilde{g}^{-1}\tilde{b}^\ell\tilde{g})$ represent the same element of $\mathcal{O}_4^\#/\tilde{N}$, where $\tilde{g} = \tilde{b}^{-j}\tilde{e}\tilde{d}^{-1}\tilde{a}^i$. The first of these pairs represents the element $\overline{F}_{\tilde{N}} \in \mathcal{O}_4^\#/\tilde{N}$, as does $(\tilde{a}', \tilde{b}') = (\tilde{d}^{-1}\tilde{a}^\ell\tilde{d}, \tilde{e}^{-1}\tilde{b}^\ell\tilde{e})$. But the latter pair is uniformly conjugate, via $\tilde{d}^{-1}\tilde{a}^i$, to $(\tilde{a}^\ell, \tilde{a}^{-i}\tilde{d}\tilde{e}^{-1}\tilde{b}^\ell\tilde{e}\tilde{d}^{-1}\tilde{a}^i) = (\tilde{a}^\ell, \tilde{g}^{-1}\tilde{b}^\ell\tilde{g})$. So the two given pairs represent the same element of $\mathcal{O}_4^\#/\tilde{N}$ (regardless of choices made above). ◇

As a consequence, we obtain the desired result that the Belyi lifting $\beta : G_\mathbb{Q} \to \mathrm{Aut}(\widehat{F}_2)$ is determined, and can be constructed, at finite levels via $\alpha : G_\mathbb{Q} \to \mathrm{Out}(\widehat{F}_2)$:

Corollary. *Given $N \in \Gamma$ of exponent n, let $\tilde{N} \in \Gamma$ be as in the above construction, and let n be the exponent of $G_N = \widehat{F}_2/N$. Then for every $\sigma \in G_\mathbb{Q}$, the image $\beta_N(\sigma) \in A_N \subset (\mathbb{Z}/n\mathbb{Z})^* \times G'_N$ is determined by $\alpha_{\tilde{N}}(\sigma) \in \mathcal{O}_4^\#/\tilde{N}$, and is computed by applying the procedure of Theorem 6(b) to that element.*

Proof. This is an immediate consequence of Theorem 6, using the result from Theorem 1 that $\beta = s\alpha$. ◇

Theorem 6 can be reinterpreted as saying that the section s is uniformly continuous. Namely, as above the Belyi group $A \subset \mathrm{Aut}(\widehat{K}(0,4))$ can be regarded as a subset of $\hat{\mathbb{Z}}^* \times \widehat{F}'_2$ (though the group law on A — given by composition — is not the one induced by the usual group law on $\hat{\mathbb{Z}}^* \times \widehat{F}'_2$). Given $N \in \Gamma$, let n be the exponent of $G = \widehat{F}_2/N$, and consider the map $A \twoheadrightarrow A_N$ induced by restricting the map $\hat{\mathbb{Z}}^* \times \widehat{F}'_2 \twoheadrightarrow (\mathbb{Z}/n\mathbb{Z})^* \times G'$ to A. Consider the weakest topology on A such that these maps $A \twoheadrightarrow A_N$ (for $N \in \Gamma$) are continuous with respect to the discrete topology on A_N. In fact, this is a uniform structure on A, with respect to the partial ordering of normal subgroups under inclusion. Similarly, we obtain a topology, and uniform structure, on $\mathcal{O}_4^\#$, via its maps to the sets $\mathcal{O}_4^\#/N$. Theorem 6(a) can then be restated as:

Corollary. *The section $s : \mathcal{O}_4^\# \to A$ is uniformly continuous.*

Here the uniformity of the continuity corresponds to the fact that $s(\overline{F})$ modulo N is determined by $\overline{F}$ modulo $\tilde{N}$, where $\tilde{N}$ is independent of $\overline{F}$ (and depends only on N).

§2. Generalization of the Belyi lifting to five-point moduli

This section is an analogue of §1 for the moduli space $\mathcal{M}_{0,5}$ and its fundamental group $\widehat{K}(0,5)$ (the profinite pure mapping class group), which replace $\mathcal{M}_{0,4} = \mathbb{P}^1 - \{0, 1, \infty\}$ and $\widehat{K}(0,4) = \widehat{F}_2$ considered before. To begin with, we have a natural map $\mu : G_{\mathbb{Q}} \to \text{Out}(\widehat{K}(0,5))$, which will play the role of the natural map $\alpha : G_{\mathbb{Q}} \to \text{Out}(\widehat{K}(0,4))$ used in §1. Furthermore, Nakamura [N, Theorem A20] generalized the Belyi lifting β to a lifting $\nu : G_{\mathbb{Q}} \to \text{Aut}(\widehat{K}(0,5))$ of μ. The image of the lifting ν is contained in a certain group $A_5 \subset \text{Aut}(\widehat{K}(0,5))$ which generalizes the Belyi group A considered in §1. Specifically,

$$
\begin{aligned}
A_5 := \{ F \in \text{Aut}(\widehat{K}(0,5)) \mid {}&\exists \lambda \in \hat{\mathbb{Z}}^*, f \in \widehat{F}_2' : F(x_{12}) = x_{12}^{\lambda}, \\
&F(x_{23}) = f(x_{12}, x_{23})^{-1} x_{23}^{\lambda} f(x_{12}, x_{23}), \\
&F(x_{34}) = f(x_{45}, x_{34})^{-1} x_{34}^{\lambda} f(x_{45}, x_{34}), \\
&F(x_{45}) = x_{45}^{\lambda}, \ F(x_{51}) \sim (x_{51})^{\lambda} \}.
\end{aligned}
$$

(This is a slightly different expression than in [N, Appendix], due to minor differences in choices made in the set-ups.) As in the case of $\mathcal{M}_{0,4}$, an element of A_5 determines the pair $(\lambda, f) \in \hat{\mathbb{Z}}^* \times \widehat{F}_2'$. Furthermore the liftings β and ν satisfy a certain agreeable compatibility relation: they associate the same pair (λ, f) to a given element of $G_{\mathbb{Q}}$. The groups $\mathcal{O}_5^{\#}$ and $\mathcal{A}_5^{\#}$ were defined in 1.1, and the images of μ and ν lie in $\mathcal{O}_5^{\#}$ and $\mathcal{A}_5^{\#}$ respectively.

The structure of this section is exactly parallel to that of §1. In Theorem 7 of 2.1, we give a result from [HS], analogous to Theorem 1 of 1.1, asserting the existence of a section s_5 of the homomorphism $\mathcal{A}_5^{\#} \to \mathcal{O}_5^{\#}$ such that $\nu = s_5 \mu$. In 2.2 we use Theorem 7 to outline the precise nature of the "explicit determination" of ν in terms of μ which is the goal of this section, and in 2.3 we give a generalization of Serre's construction to show that ν is effective in terms of μ, leading to the main result (Theorem 10 and its corollaries) given in 2.4.

2.1. Symmetric automorphisms of $\widehat{K}(0,5)$.

Let $\mathcal{O}_5^{\#}$ and $\mathcal{A}_5^{\#}$ be the subgroups of $\text{Out}(\widehat{K}(0,5))$ and $\text{Aut}(\widehat{K}(0,5))$ respectively defined in 1.1. The following result generalizes to the case $n = 5$ the statement of Theorem 1 which concerned the case $n = 4$. As above, $\mu : G_{\mathbb{Q}} \to \text{Out}(\widehat{K}(0,5))$ is the natural map, and $\nu : G_{\mathbb{Q}} \to \text{Aut}(\widehat{K}(0,5))$ is Nakamura's lifting of μ.

Theorem 7. ([HS], 2.2) *There is a unique section s_5 of the natural homomorphism $\mathcal{A}_5^{\#} \to \mathcal{O}_5^{\#}$ whose image lies in the generalized Belyi subgroup A_5 of $\mathrm{Aut}\big(\widehat{K}(0,5)\big)$. This section satisfies $\nu = s_5\mu : G_{\mathbb{Q}} \hookrightarrow \mathrm{Aut}\big(\widehat{K}(0,5)\big)$.*

We will also need the following result from [HS, 2.2].

Lemma 8. *For each $F \in \mathcal{A}_5^{\#}$, let $\overline{F}$ be the image of F in $\mathcal{O}_5^{\#}$ and let $\xi(\overline{F})$ be the equivalence class of $(F(x_{12}), F(x_{23})) \in \widehat{K}(0,5) \times \widehat{K}(0,5)$ with respect to the equivalence relation $\sim$ of uniform conjugacy. Then $\xi : \mathcal{O}_5^{\#} \to \widehat{K}(0,5) \times \widehat{K}(0,5)/\sim$ is a well-defined injection.*

2.2. Explicit computation of the generalized Belyi lifting.

Here we provide an analogue of our remarks in 1.2, with $\widehat{K}(0,5)$ playing the role of $\widehat{F}_2 = \widehat{K}(0,4)$. The goal of §2 is to show that Nakamura's lifting $\nu : G_{\mathbb{Q}} \to \mathrm{Aut}(\widehat{K}(0,5))$ is effective in terms of the natural map $\mu : G_{\mathbb{Q}} \to \mathrm{Out}(\widehat{K}(0,5))$. That is, we show that for any normal subgroup $M \subset \widehat{K}(0,5)$ of finite index, there exists a smaller such subgroup $\tilde{M}$ — which we give explicitly in terms of M — such that the reduction of ν modulo M is determined by that of μ modulo $\tilde{M}$. More generally, consider the map $s_5 : \mathcal{O}_4^{\#} \to \mathcal{A}_5^{\#}$ as in Theorem 7. Thus s_5 is the unique section of $\mathcal{A}_5^{\#} \to \mathcal{O}_5^{\#}$ whose image lies in A_5, and $\nu = s_5 \circ \mu$. We show that the map s_5 is effective (thus implying the effectivity of ν in terms of μ), and interpret this (in the second corollary to Theorem 10) as showing that s_5 is uniformly continuous.

Thus, parallel to §1, the goal of this section is to find an explicit $\tilde{M}$ for each M having the above property, and to describe how to compute $s_5(\overline{F})$ modulo M in terms of $\overline{F}$ modulo $\tilde{M}$ — and thus ν modulo M in terms of μ modulo $\tilde{M}$. As before, we first need to define the reductions of μ and ν.

So let Γ_5 be the set of normal subgroups M of finite index in $\widehat{K}(0,5)$, and for $M \in \Gamma_5$ consider the quotient group $G_M = \widehat{K}(0,5)/M$. For each $\sigma \in G_{\mathbb{Q}}$, the map $\nu : G_{\mathbb{Q}} \to A_5 \cap \mathcal{A}_5^{\#} \subset \mathrm{Aut}(\widehat{K}(0,5))$ assigns an automorphism of $\widehat{K}(0,5)$ that is characterized by a unique pair $(\lambda_\sigma, f_\sigma) \in \widehat{\mathbb{Z}}^* \times \widehat{F}_2'$. (Here, as before, we regard $\widehat{F}_2$ as a subgroup of $\widehat{K}(0,5)$ via the inclusion $\iota : x \mapsto x_{12}, y \mapsto x_{23}$.) As in §1, we identify elements $F \in A_5$ with the associated pairs (λ, f). For $M \in \Gamma_5$ with exponent m, let $A_{5,M}$ be the image of A_5 under the map $\widehat{\mathbb{Z}}^* \times \widehat{F}_2' \twoheadrightarrow (\mathbb{Z}/m\mathbb{Z})^* \times H_M' \subset (\mathbb{Z}/m\mathbb{Z})^* \times G_M'$, where $H_M \subset G_M$ is the subgroup generated by the images of x_{12}, x_{23}. We then obtain the reduction $\nu_M : G_{\mathbb{Q}} \to A_{5,M}$ by composing ν with this map.

Continuing as in 1.2, we define the reduction μ_M of μ for $M \in \Gamma_5$: For each $\overline{F} \in \mathcal{O}_5^{\#}$ and any lift $F \in \mathcal{A}_5^{\#}$ of $\overline{F}$, consider the pair $(F(x_{12}), F(x_{23})) \in \widehat{K}(0,5) \times \widehat{K}(0,5)$. This is well defined in terms of $\overline{F}$ up to uniform conjugacy in $\widehat{K}(0,5)$. Conversely, this equivalence class of pairs uniquely determines $\overline{F}$ by Lemma 8. So we may identify $\mathcal{O}_5^{\#}$ with the set of equivalence classes of pairs arising in this manner. For each $M \in \Gamma_5$, let $\mathcal{O}_5^{\#}/M$ be the quotient of this set under translation by M; this is a set whose elements are equivalence classes of (certain) pairs of elements of G_M. The reduction $\mu_M : G_{\mathbb{Q}} \to \mathcal{O}_5^{\#}/M$ is then the composition of μ with the reduction map $\mathcal{O}_5^{\#} \twoheadrightarrow \mathcal{O}_5^{\#}/M$.

In 2.3 below we construct the desired $\tilde{M}$ in terms of M, and we use this to compute ν_M in terms of $\mu_{\tilde{M}}$ in 2.4. The related problem of understanding μ modulo M directly is considered in §3 below, in the weaker form of finding information about the images of μ and ν modulo M.

2.3. Construction of $\tilde{M}$. We draw here on the construction used in 1.3, which relied on an idea of Serre. There, for each normal subgroup N of finite index in $\widehat{F}_2 = \widehat{K}(0,4)$, we let $G = G_N$ be the quotient $\widehat{F}_2/N$. We then associated a certain subgroup $\bar{N} \subset N$ that is also normal in $\widehat{F}_2$ of finite index, and the corresponding quotient group $\bar{G} = \widehat{F}_2/\bar{N}$, satisfying the properties of Lemma 2. Choosing a positive integer r that is divisible by the exponent of $\bar{G}$ (and thus in particular by the exponent of G), we then applied Serre's construction twice and obtained a certain group $\tilde{G}$, corresponding to a normal subgroup $\tilde{N}$ of finite index in $\widehat{F}_2$ contained in N.

Using this, we consider the following analogous construction for normal subgroups M of $\widehat{K}(0,5)$ of finite index. First, recall that for each positive $n \geq 5$ and each i modulo n, there is a surjection $p_i : \widehat{K}(0,n) \twoheadrightarrow \widehat{K}(0,n-1)$ that suppresses the generators x_{ij} of $\widehat{K}(0,n)$ for all j; in particular we may consider p_4 with $n = 5$. In the other direction, there is an inclusion $\iota : \widehat{F}_2 \hookrightarrow \widehat{K}(0,5)$ given by $x \mapsto x_{12}$, $y \mapsto x_{23}$. Thus ι is a section of $p_4 : \widehat{K}(0,5) \twoheadrightarrow \widehat{K}(0,4) = \widehat{F}_2$. Next, given $M \in \Gamma_5$, let $G = G_M = \widehat{K}(0,5)/M$ and let $\pi : \widehat{K}(0,5) \twoheadrightarrow G$ be the quotient map. Let $H \subset G$ be the image of $\phi = \pi \circ \iota : \widehat{F}_2 \to G$; i.e. H is generated by the two elements $a = \pi(x_{12}) = \phi(x)$ and $b = \pi(x_{23}) = \phi(y)$. Let N be the kernel of $\widehat{F}_2 \twoheadrightarrow H$, so that $H = \widehat{F}_2/N$. The above construction first yields a certain finite group $\bar{H}$. Then, taking a positive integer r that is divisible by the exponents of $\bar{H}$ and of G, we obtain a finite group $\tilde{H}$, together with a factorization $\widehat{F}_2 \xrightarrow{\tilde{\phi}} \tilde{H} \xrightarrow{\tilde{\theta}} \bar{H} \twoheadrightarrow H$ of ϕ. Let $\tilde{N} = \ker \tilde{\phi}$ and $\bar{N} = \ker \bar{\phi}$, where $\bar{\phi} = \bar{\theta} \circ \tilde{\phi}$. Thus $\tilde{H} = \widehat{F}_2/\tilde{N}$ and $\bar{H} = \widehat{F}_2/\bar{N}$. Finally, let $\tilde{M} = M \cap p_4^{-1}(\tilde{N})$ and $\bar{M} = M \cap p_4^{-1}(\bar{N})$, and let $\tilde{G} = \widehat{K}(0,5)/\tilde{M}$ and $\bar{G} = \widehat{K}(0,5)/\bar{M}$.

Observe that the inclusion $\iota : \widehat{F}_2 \hookrightarrow \widehat{K}(0,5)$ compatibly lifts the inclusions $\tilde{\epsilon} : \tilde{H} \hookrightarrow \tilde{G}$ and $\bar{\epsilon} : \bar{H} \hookrightarrow \bar{G}$ that correspond to the inclusions $\tilde{M} \subset p_4^{-1}(\tilde{N})$ and $\bar{M} \subset p_4^{-1}(\bar{N})$ of normal subgroups of $\widehat{K}(0,5)$. Similarly, ι lifts the natural inclusion $\epsilon : H \hookrightarrow G$. Also, using the definitions of $\tilde{G}, \bar{G}$, it is easy to check that $p_4 : \widehat{K}(0,5) \twoheadrightarrow \widehat{K}(0,4)$ descends compatibly to surjections $\tilde{q} : \tilde{G} \twoheadrightarrow \tilde{H}$, $\bar{q} : \bar{G} \twoheadrightarrow \bar{H}$ such that $\tilde{\epsilon}, \bar{\epsilon}$ are sections of $\tilde{q}, \bar{q}$ respectively.

Thus given a normal subgroup $M \subset \widehat{K}(0,5)$ of finite index, we have a quotient map $\pi : \widehat{K}(0,5) \twoheadrightarrow G = \widehat{K}(0,5)/M$, a factorization $\widehat{K}(0,5) \overset{\tilde{\pi}}{\twoheadrightarrow} \tilde{G} \twoheadrightarrow G$ of π (depending on a choice of positive integer r), and a finite index subgroup $\tilde{M} = \ker \tilde{\pi} \subset M$. We also have corresponding quotients $\tilde{H} = \widehat{F}_2/\tilde{N}$ and $H = \widehat{F}_2/N$. With notation as above, these yield an explicit commutative diagram

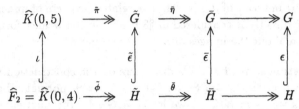

and we identify the lower groups with their images in the upper groups. Under this identification, $\tilde{\pi}(x_{12}), \tilde{\pi}(x_{23})$ lie in $\tilde{H}$, since $x_{12}, x_{23} \in \widehat{K}(0,5)$ are the images of $x, y \in \widehat{F}_2$. Let $m, \tilde{m}$ be the exponents of $G, \tilde{G}$ respectively. Write $a = \pi(x_{12})$, $b = \pi(x_{23})$, $\tilde{a} = \tilde{\pi}(x_{12})$, and $\tilde{b} = \tilde{\pi}(x_{23})$. As above we have the sets $\mathcal{O}_5^{\#}/M$ and $\mathcal{O}_5^{\#}/\tilde{M}$, which consist of certain equivalence classes of pairs in G^2 and $\tilde{G}^2$ respectively, under uniform conjugacy. For any $\ell \in (\mathbb{Z}/\tilde{m}\mathbb{Z})^*$ and $s|\tilde{m}$, let $\ell_s \in (\mathbb{Z}/s\mathbb{Z})^*$ denote the reduction of ℓ modulo s, and for any $f \in \tilde{G}$ let $f_M \in G$ denote its reduction modulo M (i.e. its image in G). We then have the following analogue of Lemma 5 of 1.3:

Lemma 9. *In the above situation, let $(\ell, f), (k, g) \in (\mathbb{Z}/\tilde{m}\mathbb{Z})^* \times \tilde{H}'$, and suppose that $(\tilde{a}^{\ell}, f^{-1}\tilde{b}^{\ell}f)$ and $(\tilde{a}^k, g^{-1}\tilde{b}^kg)$ represent the same element of $\mathcal{O}_5^{\#}/\tilde{M}$. Then the pairs also represent the same element of $\mathcal{O}_4^{\#}/\tilde{N}$. Moreover*
(a) $\ell_r = k_r$ and hence $\ell_m = k_m$;
(b) $f_M = g_M$.

Proof. The elements $\tilde{a}, \tilde{b}, f, g$ lie in the image of $\tilde{\pi}\iota = \tilde{\epsilon}\tilde{\phi}$ and hence in the image of $\tilde{\epsilon}$. By hypothesis, there is an element $h \in \tilde{G}$ such that conjugation by h takes $(\tilde{a}^{\ell}, f^{-1}\tilde{b}^{\ell}f)$ to $(\tilde{a}^{\ell}, g^{-1}\tilde{b}^kg)$ in $\tilde{G}^2$. Applying $\tilde{q} : \tilde{G} \twoheadrightarrow \tilde{H}$, and using that $\tilde{\epsilon}$ is a section of $\tilde{q}$, we deduce that conjugation by $\tilde{q}(h) \in \tilde{H} \subset \tilde{G}$ takes $(\tilde{a}^{\ell}, f^{-1}\tilde{b}^{\ell}f)$ to $(\tilde{a}^{\ell}, g^{-1}\tilde{b}^kg)$ in $\tilde{H}^2$. Since these two pairs are uniformly conjugate with respect to $\tilde{H}$ (and not just with respect to $\tilde{G}$), it follows that they represent the same element of $\mathcal{O}_4^{\#}/\tilde{N}$ (and not just of $\mathcal{O}_5^{\#}/\tilde{M}$).

Thus the situation here is a special case of that of Lemma 5, so Lemma 5(a) implies that $\ell_r = k_r$. Thus $\ell_m = k_m$, since $m = \exp(G)$ divides r by assumption. This yields assertion (a) of the proposition. Also, Lemma 5(b) says that $f_N = g_N$. By the commutativity of the above diagram, $f_M = \epsilon(f_N) \in G$ and similarly for g; this yields assertion (b). $\diamondsuit$

2.4. The main result. Using this, we obtain the following analogue of Theorem 6, which makes explicit the section $s_5 : \mathcal{O}_5^{\#} \to A_5 \cap \mathcal{A}_5^{\#} \subset \mathcal{A}_5^{\#}$

of Theorem 7. Namely, for $M \in \Gamma_5$ and $\tilde{M}$ as above, it computes $s_5(\overline{F})$ modulo M in terms of $\overline{F}$ modulo $\tilde{M}$.

Theorem 10. *Let M be a normal subgroup of $\widehat{K}(0,5)$ of finite index, and let m be the exponent of $G = \widehat{K}(0,5)/M$. Define $\widehat{K}(0,5) \overset{\tilde{\pi}}{\twoheadrightarrow} \tilde{G} \twoheadrightarrow G$ as in the construction above, and let $\tilde{M} = \ker(\widehat{K}(0,5) \twoheadrightarrow \tilde{G})$. For $\overline{F} \in \mathcal{O}_5^\#$, write $F = s_5(\overline{F}) = (\lambda, f) \in A_5 \subset \hat{\mathbb{Z}}^* \times \hat{F}_2'$, and let F_M be the image of F in $A_{5,M} \subset (\mathbb{Z}/m\mathbb{Z})^* \times G'$.*

(a) Then F_M depends only on $\overline{F}_{\tilde{M}}$, the image of $\overline{F}$ in $\mathcal{O}_5^\#/\tilde{M}$.

(b) Explicitly, F_M is given in terms of $\overline{F}_{\tilde{M}}$ as follows, where we write $a = \pi(x_{12})$, $b = \pi(x_{23})$, $\tilde{a} = \tilde{\pi}(x_{12})$, $\tilde{b} = \tilde{\pi}(x_{23})$: Let $(\tilde{a}', \tilde{b}')$ be any element of the equivalence class $\overline{F}_{\tilde{M}}$. Choose $\tilde{d}, \tilde{e} \in \tilde{H} = \langle \tilde{a}, \tilde{b} \rangle$ and $\ell \in \mathbb{Z}$ such that $\tilde{a}' = \tilde{d}^{-1}\tilde{a}^\ell\tilde{d}$ and $\tilde{b}' = \tilde{e}^{-1}\tilde{b}^\ell\tilde{e}$, and choose $i, j \in \mathbb{Z}$ such that $\tilde{e}\tilde{d}^{-1} \equiv \tilde{b}^j\tilde{a}^{-i} \bmod \tilde{H}'$. Then $F_M = (\ell_m, b^{-j}ed^{-1}a^i)$, where ℓ_m is the reduction of ℓ modulo m, and where d, e are the images of $\tilde{d}, \tilde{e}$ in G.

Proof. Theorem 10 follows from Lemma 9 exactly as Theorem 6 followed from Lemma 5. ◇

Thus parallel to §1, we obtain the desired consequence that Nakamura's lifting $\nu : G_\mathbb{Q} \to \mathrm{Aut}(\widehat{K}(0,5))$ is determined, and can be constructed, at finite levels via $\mu : G_\mathbb{Q} \to \mathrm{Out}(\widehat{K}(0,5))$:

Corollary. *Given $M \in \Gamma_5$ of exponent m, let $\tilde{M} \in \Gamma_5$ be as in the above construction, and let m be the exponent of $G_M = \widehat{K}(0,5)/M$. Then for every $\sigma \in G_\mathbb{Q}$, the image $\nu_M(\sigma) \in A_{5,M} \subset (\mathbb{Z}/m\mathbb{Z})^* \times G'_M$ is determined by $\mu_{\tilde{M}}(\sigma) \in \mathcal{O}_5^\#/\tilde{M}$, and is computed by applying the procedure of Theorem 10(b) to that element.*

Proof. This is an immediate consequence of Theorem 10, using the result from Theorem 7 that $\nu = s_5 \circ \mu$. ◇

As in §1, we may put a topology and uniform structure on A_5 via the normal subgroups $M \in \Gamma_5$, and similarly on $\mathcal{O}_5^\#$ via the maps to the sets $\mathcal{O}_5^\#/M$. Parallel to the second corollary to Theorem 6, we then may restate Theorem 10 as:

Corollary. *The section $s_5 : \mathcal{O}_5^\# \to A_5$ is uniformly continuous.*

§3. The $\widehat{GT}_0$ and $\widehat{GT}$-orbits of dessins

As discussed in the introduction, we would like to compute the action of $G_\mathbb{Q}$ on Galois dessins, thus relating the arithmetic of covers to their combinatorics. In the previous sections, we described how to compute canonical actions of $G_\mathbb{Q}$ on the fundamental groups of $\mathcal{M}_{0,4} = \mathbb{P}^1 - \{0,1,\infty\}$ and

$\mathcal{M}_{0,5}$ in terms of the natural outer actions. To complete the description we would like to be able to describe the outer action of an element of $G_{\mathbb{Q}}$ on these fundamental groups $\widehat{K}(0,4)$ and $\widehat{K}(0,5)$. Unfortunately this seems beyond reach at present, and here we content ourselves with somewhat less — namely finding the $\widehat{GT}_0$- and $\widehat{GT}$-orbits of a dessin, which approximate the $G_{\mathbb{Q}}$-orbit and yield information about the field of moduli of the corresponding cover. We do so by using the relationship of $\widehat{GT}_0$ and $\widehat{GT}$ to the groups $\mathcal{O}_n^{\#}$ considered above.

The $\widehat{GT}_0$- and $\widehat{GT}$-orbits of a dessin will be found by an iterative process, computing somewhat larger orbits under computable finite groups, which converge to the true $\widehat{GT}_0$- and $\widehat{GT}$-orbits after finitely many steps (and thus, in the latter case, to the $G_{\mathbb{Q}}$-orbit, if $\widehat{GT}$ turns out to be the same as $G_{\mathbb{Q}}$). The key idea is that we can explicitly find subgroups of the outer automorphism groups of finite quotients of $\widehat{K}(0,n)$ that are defined by finite-level properties analogous to the defining properties of $\mathcal{O}_n^{\#}$. We explain the procedure in 3.4 and 3.5, after giving an overview in 3.1 and some necessary definitions and results in 3.2 and 3.3. Then, in 3.6, we use these orbits to obtain information about the field of moduli of a Galois dessin.

3.1. Approximation of the Galois orbit of a dessin. Given a dessin D, it is unknown how to find its $G_{\mathbb{Q}}$-orbit effectively, since this involves understanding the connection between the combinatorics of a dessin (or of a branch cycle description) and the arithmetic of a cover. But some $G_{\mathbb{Q}}$-orbits can be separated, using known invariants such as the geometric Galois group and valency list of the dessin (as well as less obvious invariants, e.g. via obstructed components of modular towers [F2]). In the special case of rigidity, it is possible to understand the $G_{\mathbb{Q}}$-orbit and the field of moduli. In general, though, the known invariants give only *approximative* orbits which contain the Galois orbit but may be quite crude.

However, even approximations are useful for the inverse Galois problem. For instance the order of an approximative orbit is greater or equal to the order of the true Galois orbit, and so gives an upper bound on the degree of the field of moduli of the dessin. Moreover, if the approximative orbit turns out to consist only of D, then this is also true of the Galois orbit, so D has field of moduli $\mathbb{Q}$ — and its geometric Galois group G is thus a Galois group over $\mathbb{Q}$, under mild hypotheses on G [CH, 2.8(c)]. This viewpoint provides our motivation for finding approximative orbits of dessins. In keeping with the approach of the preceding sections, we use the fact that $G_{\mathbb{Q}}$ is contained in the larger groups $\mathcal{O}_n^{\#} \subset \mathrm{Out}\big(\widehat{K}(0,n)\big)$ [HS, 1.2]. By finding the orbits under these larger groups, we thus obtain approximate $G_{\mathbb{Q}}$-orbits. This approach takes advantage of the fact that the action of $\mathcal{O}_n^{\#}$ on dessins is known, and extends that of the subgroup $G_{\mathbb{Q}}$ (though what is not known explicitly is the precise injection $G_{\mathbb{Q}} \hookrightarrow \mathcal{O}_n^\#$). In addition, this approach reflects the conjecture that the Grothendieck-Teichmüller group $\widehat{GT}$ is isomorphic to $G_{\mathbb{Q}}$, together with the following theorem:

Theorem 11. [HS, 1.3, 2.3] $\widehat{GT}_0 \simeq \mathcal{O}_4^\#$ and $\widehat{GT} \simeq \mathcal{O}_5^\#$.

Thus if the conjecture on $\widehat{GT}$ is correct, then by finding the $\mathcal{O}_5^\#$-orbit of a dessin we are actually finding the $G_{\mathbb{Q}}$-orbit. (For definitions and background on $\widehat{GT}$ and the larger group $\widehat{GT}_0$, see the survey on $\widehat{GT}$ in this volume.)

3.2. Automorphism groups on finite levels. Before computing approximations, let us define some important groups of outer automorphisms of the pure profinite mapping class groups $\widehat{K}(0,n)$.

Let $\mathcal{O}_n = \mathrm{Out}(\widehat{K}(0,n))$ be the group of outer automorphisms. As in 1.1, there is a natural outer S_n-action on $\widehat{K}(0,n)$, i.e. a homomorphism $\sigma^{(n)} : S_n \to \mathcal{O}_n$. In terms of this, we defined the groups $\mathcal{O}_n^\#$ in 1.1.

If N is any characteristic subgroup of finite index of $\widehat{K}(0,n)$, then automorphisms of $\widehat{K}(0,n)$ preserve N and thus pass to the quotient $\widehat{K}(0,n)/N$. The same is true for outer automorphisms, giving rise to a natural homomorphism

$$\psi_N^n : \mathcal{O}_n \to \mathrm{Out}(\widehat{K}(0,n)/N).$$

Similarly, for characteristic subgroups $M \subset N$ of finite index in $\widehat{K}(0,n)$, we have a homomorphism

$$\psi_{M,N}^n : \mathrm{Out}(\widehat{K}(0,n)/M) \to \mathrm{Out}(\widehat{K}(0,n)/N).$$

We then define the mod N reductions of the groups $\mathcal{O}_n$ and $O_n^\#$:

- $\mathcal{O}_{n,N} = \psi_N^n(\mathcal{O}_n)$, $\mathcal{O}_{n,N}^\# = \psi_N^n(\mathcal{O}_n^\#)$.

Let $\sigma_N^{(n)}$ denote the composition of homomorphisms $\psi_N^n \circ \sigma^{(n)}$ for $n \geq 4$. As sketched above, we will define the *approximations* $\mathcal{O}_n(N)$ and $\mathcal{O}_n^\#(N)$ of the finite groups $\mathcal{O}_{n,N}$ and $\mathcal{O}_{n,N}^\#$ as groups of outer automorphisms of $K(0,n)/N$ having properties analogous to those of $\mathcal{O}_n$ and $\mathcal{O}_n^\#$. (We indicate "approximations" by putting N in parentheses, whereas the actual images are indicated by an N in the subscript.) Namely, writing $(x_{ij})_N$ for the image of x_{ij} in $K(0,n)/N$ and e for the exponent of $K(0,n)/N$, we define

- $\mathcal{O}_n(N) = \mathrm{Out}(K(0,n)/N)$,
- $\mathcal{O}_n^\#(N) = \{\overline{F} \in \mathcal{O}_n(N) \mid$ (i) $\exists \lambda \in (\mathbb{Z}/e\mathbb{Z})^* : (\forall i,j)\, \overline{F}([(x_{ij})_N]) = [(x_{ij})_N^\lambda];$
  (ii) $\overline{F}$ commutes with $\sigma_N^{(n)}(S_n)$ in $\mathcal{O}_n(N)\}$.

It is immediate from the definitions that $\mathcal{O}_{n,N} \subset \mathcal{O}_n(N)$; it is also easily seen that $\mathcal{O}_{n,N}^\# \subset \mathcal{O}_n^\#(N)$ since elements of $\mathcal{O}_{n,N}^\#$ inherit properties (i) and (ii) from the analogous properties in $\mathcal{O}_n^\#$. The main point is that the approximative groups $\mathcal{O}_n(N)$ and $\mathcal{O}_n^\#(N)$ can be *computed* (unlike $\mathcal{O}_{n,N}$ and $\mathcal{O}_{n,N}^\#$). Moreover these two groups serve as good approximations since (as we will show in 3.3) the two above containments become equalities in the limit, with respect to the inverse systems given by the maps $\psi_{M,N}^n$.

3.3. Inverse systems and inverse limits. In this section we prove some results on inverse limits of the automorphism groups defined in 3.2.

Lemma 12. *Let G be any finitely generated group, and let $\widehat{G}$ denote the profinite completion of G, i.e. the inverse limit of the quotients of G by all the normal subgroups of finite index. Then the inverse limit of the quotients of G by all the characteristic subgroups of finite index is also isomorphic to $\widehat{G}$, i.e. the characteristic subgroups of G form a cofinal system.*

Proof. It suffices to show that every normal subgroup N of finite index in G contains a characteristic subgroup $\check{N}$ of finite index. In particular we can let $\check{N}$ be the intersection of all the normal subgroups N^* of G such that $G/N^* \simeq G/N$. There are only finitely many of these; indeed, they are the kernels of the homomorphisms of G into a finite group, and since G is finitely generated there are only finitely many such homomorphisms. $\diamond$

Corollary. *For all $n \geq 4$, we have*
$$\varprojlim_N \widehat{K}(0,n)/N = \widehat{K}(0,n)$$
as N runs over the characteristic subgroups of finite index of $\widehat{K}(0,n)$.

Proof. By the lemma, we know that since the discrete pure mapping class group $K(0,n)$ is finitely generated, it is the inverse limit of the finite quotients $K(0,n)/N$ for characteristic subgroups N of $K(0,n)$. But the characteristic subgroups of finite index of $\widehat{K}(0,n)$ are in bijection with those of $K(0,n)$, so the finite quotients of $K(0,n)$ and of $\widehat{K}(0,n)$ by corresponding subgroups are in one-to-one correspondence. $\diamond$

If N is any characteristic subgroup of finite index in $\widehat{K}(0,n)$, then along with the homomorphism ψ_N^n of 3.2, there are natural homomorphisms
$$\Psi_N^n : \widehat{K}(0,n) \to \widehat{K}(0,n)/N$$
$$\tilde{\psi}_N^n : \mathrm{Aut}\big(\widehat{K}(0,n)\big) \to \mathrm{Aut}\big(\widehat{K}(0,n)/N\big).$$

Similarly, if $M \subset N$ are characteristic subgroups of finite index of $\widehat{K}(0,n)$, then along with $\psi_{M,N}^n$ of 3.2 we have natural homomorphisms:
$$\Psi_{M,N}^n : \widehat{K}(0,n)/M \to \widehat{K}(0,n)/N$$
$$\tilde{\psi}_{M,N}^n : \mathrm{Aut}\big(\widehat{K}(0,n)/M\big) \to \mathrm{Aut}\big(\widehat{K}(0,n)/N\big).$$

The diagram

$$\mathrm{Aut}\big(\widehat{K}(0,n)/M\big) \xrightarrow{\tilde{\psi}^n_{M,N}} \mathrm{Aut}\big(\widehat{K}(0,n)/N\big)$$

$$(*)$$

$$\mathrm{Out}\big(\widehat{K}(0,n)/M\big) \xrightarrow{\psi^n_{M,N}} \mathrm{Out}\big(\widehat{K}(0,n)/N\big)$$

commutes, and for all $x \in \widehat{K}(0,n)/M$ and $\Phi \in \mathrm{Aut}\big(\widehat{K}(0,n)/M\big)$, we have

$$\tilde{\psi}^n_{M,N}(\Phi)(\Psi^n_{M,N}(x)) = \Psi^n_{M,N}(\Phi(x)) \qquad (**)$$

Theorem 13. *(a) For all $n \geq 4$, the groups $\mathcal{O}_{n,N}$ form an inverse system as N runs over the characteristic subgroups of finite index of $\widehat{K}(0,n)$, and so do the groups $\mathcal{O}_n(N)$. Moreover, we have*

$$\varprojlim_N \mathcal{O}_{n,N} = \varprojlim_N \mathcal{O}_n(N) = \mathcal{O}_n.$$

(b) For all $n \geq 4$, the groups $\mathcal{O}^{\#}_{n,N}$ form an inverse system as N runs over the characteristic subgroups of $K(0,n)$, and so do the groups $\mathcal{O}^{\#}_n(N)$. Moreover, we have

$$\varprojlim_N \mathcal{O}^{\#}_{n,N} = \varprojlim_N \mathcal{O}^{\#}_n(N) = \mathcal{O}^{\#}_n.$$

Proof. (a) The groups $\mathcal{O}_n(N)$ form an inverse system under the maps $\psi^n_{M,N}$ since these maps are well-defined. Moreover, whenever $M \subset N$, we have $\psi^n_N = \psi^n_{M,N} \circ \psi^n_M$, so $\mathcal{O}_{n,N} = \psi^n_N(\mathcal{O}_n) = (\psi^n_{M,N} \circ \psi^n_M)(\mathcal{O}_n) = \psi^n_{M,N}(\mathcal{O}_{n,M})$ by the definition of the $\psi^n_{M,n}$; so the $\mathcal{O}_{n,N}$ also form an inverse system under the maps $\psi^n_{M,N}$. Next we define homomorphisms from $\mathcal{O}_n$ to $\varprojlim \mathcal{O}_{n,N}$ and to $\varprojlim \mathcal{O}_n(N)$; then we will define homomorphisms in the other direction and show that they are mutual inverses, so isomorphisms. Let $\phi \in \mathcal{O}_n$. For N a finite-index characteristic subgroup, $\phi_N = \psi^n_N(\phi)$ is in $\mathcal{O}_{n,N} \subset \mathcal{O}_n(N)$; so $\phi \mapsto (\phi_N)_N$ is a group homomorphism from $\mathcal{O}_n$ to each of the two inverse limits. Conversely, consider an element $(\phi_N)_N$ in either inverse limit, where the ϕ_N's are compatible elements of $\mathcal{O}_{n,N}$ (resp. $\mathcal{O}_n(N)$). Then $(\phi_N)_N$ gives an outer automorphism ϕ of $\widehat{K} = \widehat{K}(0,n)$, defined by $\phi((\alpha_N)_N) = (\phi_N(\alpha_N))_N$; here we use the identification in the corollary to Lemma 12. (This ϕ really is an outer automorphism since it is trivial on the identity, respects relations in $\widehat{K}$ and is invertible — all consequences of the analogous behavior on the finite levels.) This map $(\phi_N)_N \mapsto \phi$ is inverse to the previous map $\phi \mapsto (\phi_N)_N$, so both are isomorphisms. (The difference between the inverse systems of $\mathcal{O}_{n,N}$ and of $\mathcal{O}_n(N)$ is that the $\psi^n_{M,N}$ restrict to surjections on the $\psi^n_{M,N}$, but not necessarily for the $\mathcal{O}_n(N)$.)

(b) The $\mathcal{O}^\#_{n,N}$ form an inverse system under the restriction to the subgroups $\mathcal{O}^\#_{n,N} \subset \mathcal{O}_{n,N}$ of the homomorphisms $\psi^n_{M,N}$ for the same reason as the $\mathcal{O}_{n,N}$ in (a). To show that $\left(\{\mathcal{O}^\#_n(N)\}, \{\psi^n_{M,N}\}\right)$ also forms an inverse system, we check that $\psi^n_{M,N}\left(\mathcal{O}^\#_n(M)\right) \subset \mathcal{O}^\#_n(N)$. Let $\phi \in \mathcal{O}^\#_n(M)$, so ϕ has the properties (i) and (ii) in the definition of the groups $\mathcal{O}^\#_n(M)$; we need to check that $\psi^n_{M,N}(\phi)$ has the corresponding two properties.

Let $\phi \in \mathcal{O}^\#_n(M)$. Let us show that $\psi^n_{M,N}(\phi)$ has property (i), viz. that it sends the conjugacy class of $(x_{ij})_N$ to that of $(x_{ij})^\lambda_N$ for each pair i, j. Since ϕ has property (i) (for M), there is a lifting Φ of ϕ in $\mathrm{Aut}\left(K(0,n)/M\right)$ such that $\Phi(x_{ij}) = \alpha_{ij}^{-1} x_{ij}^\lambda \alpha_{ij}$ for some $\alpha_{ij} \in \widehat{K}(0,n)/M$. By (**) with this Φ,

$$\tilde{\psi}^n_{M,N}(\Phi)\left((x_{ij})_N\right) = \tilde{\psi}^n_{M,N}(\Phi)\left(\Psi^n_{M,N}((x_{ij})_M)\right) = \Psi^n_{M,N}\left(\Phi((x_{ij})_M)\right)$$
$$= \Psi^n_{M,N}\left(\alpha_{ij}^{-1}(x_{ij})^\lambda_M \alpha_{ij}\right) = \Psi^n_{M,N}(\alpha_{ij})^{-1}(x_{ij})^\lambda_N \Psi^n_{M,N}(\alpha_{ij}).$$

But $\tilde{\psi}^n_{M,N}(\Phi)$ is a lifting of $\psi^n_{M,N}(\phi)$ to $\mathrm{Aut}\left(\widehat{K}(0,n)/N\right)$ by the diagram (*), and the existence of a lifting sending each $(x_{ij})_N$ to a conjugate of $(x_{ij})^\lambda_N$ shows that $\psi^n_{M,N}(\phi)$ has property (i). Property (ii), i.e. that $\psi^n_{M,N}$ commutes with $\sigma^{(n)}_M(S_n)$, follows immediately since $\psi^n_{M,N} \circ \sigma^{(n)}_M = \sigma^{(n)}_N$.

Let us show that the inverse limits of the two systems $\mathcal{O}^\#_{n,N}$ and $\mathcal{O}^\#_n(N)$ are both isomorphic to $\mathcal{O}^\#_n$. For any $\phi \in \mathcal{O}^\#_n$, setting $\phi_N = \psi^n_N(\phi)$, the element $(\phi_N)_N$ belongs to both $\varprojlim \mathcal{O}^\#_{n,N}$ and $\varprojlim \mathcal{O}^\#_n(N)$. Conversely, let $(\phi_N)_N$ lie in $\varprojlim \mathcal{O}^\#_{n,N}$ (resp. $\varprojlim \mathcal{O}^\#_n(N)$). We showed in the proof of (a) that $(\phi_N)_N$ determines an element ϕ of $\mathcal{O}_n$, so we only need to show that ϕ has properties (i) and (ii) of the definition of $\mathcal{O}^\#_n$. But these two properties hold if they hold on every finite level, which proves the result. (Again here, as in (a), the inverse system of the $\mathcal{O}^\#_{n,N}$ consists of surjective homomorphisms whereas this is not necessarily true for the $\mathcal{O}^\#_n(N)$.)   ◊

Corollary. *The groups* $\widehat{GT}_0$ *and* $\widehat{GT}$ *are profinite groups.*

Proof. We know by Theorem 11 that $\widehat{GT}_0 = \mathcal{O}^\#_4$ and $\widehat{GT} = \mathcal{O}^\#_5$. By the preceding theorem $\mathcal{O}^\#_4$ is the inverse limit of the finite groups $\mathcal{O}^\#_4(N)$ or $\mathcal{O}^\#_{4,N}$ as N runs through the characteristic subgroups of $\widehat{K}(0,4)$, and $\mathcal{O}^\#_5$ is the inverse limit of the groups $\mathcal{O}^\#_5(M)$ or $\mathcal{O}^\#_{5,M}$ as M runs through the characteristic subgroups of $\widehat{K}(0,5)$.   ◊

3.4. Approximation of the $\widehat{GT}_0$-orbit of a dessin.

We explain here our approach to the approximation and computation of the $\widehat{GT}_0$- and $\widehat{GT}$-orbits of a dessin D; for simplicity we assume that D is Galois. The somewhat easier case of $\widehat{GT}_0$ and $\widehat{K}(0,4)$ is actually carried out here, while that of $\widehat{GT}$ and $\widehat{K}(0,5)$ is treated in 3.5. As before we identify $\widehat{F}_2 = \langle x, y, z \,|\, xyz = 1 \rangle$ with $\widehat{K}(0,4)$ via the isomorphism $x \mapsto x_{12}$, $y \mapsto x_{23}$.

For any finite group G with two generators, the G-Galois dessins D are in bijection with normal subgroups $N \subset \widehat{K}(0,4)$ together with isomorphisms $\widehat{K}(0,4)/N \xrightarrow{\sim} G$ (determined up to an inner automorphism). They are also in bijection with uniform conjugacy classes of triples (a, b, c) of generators of G such that $abc = 1$; viz. the branch cycle description of the corresponding G-Galois cover, where $a = x_N$, $b = y_N$, and $c = z_N$. Since $abc = 1$, we may classify G-Galois dessins simply by the pair (x_N, y_N). Now there is a natural action of $\mathcal{O}_4^\#$ on the set of G-Galois dessins, which by Theorem 1 extends the action of $G_{\mathbb{Q}}$ on this set. Also, given D, N, G as above, if $\check{N} \subset \widehat{K}(0,4)$ is any finite index characteristic subgroup contained in N, the action of $\mathcal{O}_4^\#$ on the orbit of D factors through the finite group $\mathcal{O}_{4,\check{N}}^\#$ — and thus the $\widehat{GT}_0$-orbit of D is the same as the $\mathcal{O}_{4,\check{N}}^\#$-orbit (and is equal to the set $\mathcal{O}_4^\#/N$ of §1). Unfortunately, it is unclear how to find exactly which elements of the finite group $\mathrm{Out}(G)$ lie in $\mathcal{O}_{4,\check{N}}^\#$, since the definition of $\mathcal{O}_{4,\check{N}}^\#$ involves the infinite group $\widehat{GT}_0 \simeq \mathcal{O}_4^\#$.

So instead, we take an indirect approach, using the fact that we can find the elements of the (possibly) larger finite group $\mathcal{O}_4^\#(\check{N}) \subset \mathrm{Out}(G)$, along with the fact that the $\mathcal{O}_{4,\check{N}}^\#$-action extends to an $\mathcal{O}_4^\#(\check{N})$-action. Namely, for $\overline{F} \in \mathcal{O}_4^\#(\check{N})$, we define $D_{\overline{F}}$ to be the G-Galois dessin corresponding to the equivalence class of the pair $((F(x_{\check{N}}))_N, (F(y_{\check{N}}))_N)$ in G^2, where $F \in \mathrm{Aut}(\widehat{K}(0,4)/\check{N})$ is a lifting of $\overline{F}$. (The equivalence class of the pair is independent of the choice of lifting F.) We call the orbit of D under $\mathcal{O}_4^\#(\check{N})$ the $\check{N}$-approximative $\widehat{GT}_0$-orbit of D.

We can thus explicitly compute the $\check{N}$-approximative $\widehat{GT}_0$-orbit of any G-Galois dessin D, where $G = \widehat{K}(0,4)/N$:

(1) Let $\check{N}$ be any characteristic subgroup of finite index in $\widehat{K}(0,4)$ that is contained in N — e.g. the $\check{N}$ in the proof of Lemma 12.

(2) Compute the finite group $\mathcal{O}_4^\#(\check{N})$ from the definition.

(3) For each element $\overline{F}$ of $\mathcal{O}_4^\#(\check{N})$, compute $\left((F(x_{\check{N}}))_N, (F(y_{\check{N}}))_N\right) \in G^2$.

By taking smaller and smaller choices of $\check{N}$, we can compute the actual $\widehat{GT}_0$-orbit of D in finitely many steps. Namely, by Lemma 12, there is a cofinal sequence of finite index characteristic subgroups $N_0 \subset N_1 \subset \cdots$ of $\widehat{K}(0,4)$ inside any given normal subgroup N of finite index. In this situation we have the following corollary of Theorem 13:

Corollary. *Let D be a G-Galois dessin, corresponding to a normal subgroup $N \subset \widehat{K}(0,4)$ of finite index. Let $\{N_i\}$ be a cofinal sequence of finite index characteristic subgroups of $\widehat{K}(0,4)$ contained in N. Then for all $i \gg 0$, the N_i-approximative $\widehat{GT}_0$-orbit of D is equal to the $\widehat{GT}_0$-orbit of D.*

Proof. By Theorem 13(b) together with Theorem 11, the inverse limit of the groups $\mathcal{O}_4^{\#}(N_i)$ is isomorphic to $\mathcal{O}_4^{\#} \simeq \widehat{GT}_0$. Thus the descending intersection of the finite groups $\psi_{N_i,N_0}^4(\mathcal{O}^{\#}(N_i))$ is equal to $\psi_{N_0}^4(\mathcal{O}_4^{\#}) = \mathcal{O}_{4,N_0}^{\#} \subset \mathcal{O}_4^{\#}(N_0)$. By finiteness, there exists an integer $I \geq 0$ such that $\psi_{N_i,N_0}^4(\mathcal{O}^{\#}(N_i)) = \mathcal{O}_{4,N_0}^{\#} \subset \mathcal{O}_4^{\#}(N_0)$ for all $i \geq I$. Since the actions of $\mathcal{O}_4^{\#}(N_i)$ and of $\mathcal{O}_4^{\#}$ factor through $\mathcal{O}_4^{\#}(N_0)$, the result follows. $\diamond$

Thus this procedure computes the $\widehat{GT}_0$-orbit of a Galois dessin D, after finitely many steps. Since $G_{\mathbb{Q}} \subset \widehat{GT} \subset \widehat{GT}_0$, the $\widehat{GT}_0$-orbit of D contains the $G_{\mathbb{Q}}$-orbit — but an even better approximation to the $G_{\mathbb{Q}}$-orbit would be obtained by computing the $\widehat{GT}$-orbit. That analogous construction is the subject of 3.5.

3.5. Approximation of the $\widehat{GT}$-orbit of a dessin.

By using an analogue of the procedure of 3.4 with $\widehat{K}(0,5)$ replacing $\widehat{K}(0,4)$, we can find the $\widehat{GT}$-orbit of a dessin D — thus separating more $G_{\mathbb{Q}}$-orbits and obtaining finer information about the field of moduli of the corresponding cover of $\mathbb{P}^1 - \{0, 1, \infty\}$. As in 3.4 we rely on Theorem 11, to identify $\widehat{GT}$ with $\widehat{K}(0,5)$. As before, we identify $\widehat{K}(0,4)$ with a subgroup of $\widehat{K}(0,5)$ via the injection $\iota : x_{12} \mapsto x_{12}$, $x_{23} \mapsto x_{23}$. Recall that ι is a section of the surjection $p_4 : \widehat{K}(0,5) \twoheadrightarrow \widehat{K}(0,4) \simeq \langle x_{12}, x_{23} \rangle$, and yields a decomposition $\widehat{K}(0,5) \simeq \langle x_{14}, x_{24}, x_{34} \rangle \rtimes \langle x_{12}, x_{23} \rangle$.

So let D be a G-Galois dessin and let N be the corresponding normal subgroup in $\widehat{K}(0,4) \simeq \widehat{F}_2$. Thus $G \simeq \widehat{K}(0,4)/N \simeq \widehat{K}(0,5)/M$, where $M = p_4^{-1}(N) \simeq \langle x_{14}, x_{24}, x_{34} \rangle \rtimes N$. By Theorem 7, there is a natural action of $\mathcal{O}_5^{\#}$ on the set of G-Galois dessins. So the $\mathcal{O}_5^{\#}$-orbit of D contains the $G_{\mathbb{Q}}$-orbit, and is in turn contained in the $\mathcal{O}_4^{\#}$-orbit (by the compatibility of the $\mathcal{O}_4^{\#}$- and $\mathcal{O}_5^{\#}$-actions [HS, 2.2]). If $\check{M} \subset \mathcal{O}_5^{\#}$ is any finite index characteristic subgroup contained in M, the action of $\mathcal{O}_5^{\#}$ factors through the finite group $\mathcal{O}_{5,\check{M}}^{\#}$ — and so the $\widehat{GT}$-orbit of D is the same as the $\mathcal{O}_{5,\check{M}}^{\#}$-orbit (and is equal to the set $\mathcal{O}_5^{\#}/M$ of §2). As in 3.4, though, it is unclear how to find exactly which elements of the finite group $\mathrm{Out}(G)$ lie in $\mathcal{O}_{5,\check{M}}^{\#}$.

Thus, as with $\widehat{GT}_0$, we take an indirect approach to finding the $\widehat{GT}$-orbit, by using that we can find the elements of the (possibly) larger finite group $\mathcal{O}_5^{\#}(\check{M}) \subset \mathrm{Out}(G)$, along with the fact that the $\mathcal{O}_{5,\check{M}}^{\#}$-action extends to an $\mathcal{O}_5^{\#}(\check{M})$-action. Namely, for $\overline{F} \in \mathcal{O}_5^{\#}(\check{M})$, if $F \in \mathrm{Aut}(\widehat{K}(0,5)/\check{M})$ is any lifting of $\overline{F}$, then we define $D_{\overline{F}}$ to be the G-Galois dessin corresponding to

the uniform conjugacy class of the pair $\left(\left(F((x_{12})_{\check{M}}) \right)_M, \left(F((x_{23})_{\check{M}}) \right)_M \right) \in$
$(\widehat{K}(0,5)/M)^2 \simeq (\widehat{K}(0,4)/N)^2 \simeq G^2$. This equivalence class is well-defined,
since conjugacy in $\widehat{K}(0,5)/\check{M}$ maps to conjugacy in $G = \widehat{K}(0,5)/M$. We
call the orbit of D under $\mathcal{O}_5^{\#}(\check{M})$ the $\check{M}$-*approximative* $\widehat{GT}$-*orbit of* D.

As in 3.4, we can thus explicitly compute the $\check{M}$-approximative $\widehat{GT}$-orbit
of any G-Galois dessin D, where $G = \widehat{K}(0,4)/N$:

($1'$) Let $\check{M}$ be any characteristic subgroup of finite index in $\widehat{K}(0,5)$ that is
contained in $M = p_4^{-1}(N)$ — e.g. choose $\check{M}$ as in the proof of Lemma 12.

($2'$) Compute the finite group $\mathcal{O}_5^{\#}(\check{M})$ from the definition.

($3'$) Compute $\left(\left(F((x_{12})_{\check{M}}) \right)_M, \left(F((x_{23})_{\check{M}}) \right)_M \right) \in G^2$ for each $\overline{F} \in \mathcal{O}_5^{\#}(\check{M})$.

Again as in 3.4, by taking smaller and smaller choices of $\check{M}$, we can compute the actual $\widehat{GT}$-orbit of D in finitely many steps. Namely, again by
Lemma 12, there is a cofinal sequence of finite index characteristic sub-
groups $M_0 \subset M_1 \subset \cdots$ of $\widehat{K}(0,5)$ inside $M = p_4^{-1}(N)$ for any given normal
subgroup $N \subset \widehat{K}(0,4)$ of finite index. We then obtain the following $\widehat{GT}$-
version of the corollary in 3.4, whose proof is essentially the same as before:

Corollary. *Let D be a G-Galois dessin, corresponding to a normal subgroup
$N \subset \widehat{K}(0,4)$ of finite index. Let $\{M_i\}$ be a cofinal sequence of finite index
characteristic subgroups of $\widehat{K}(0,5)$ contained in $M = p_4^{-1}(N)$. Then for all
$i \gg 0$, the M_i-approximative $\widehat{GT}$-orbit of D is equal to the $\widehat{GT}$-orbit of D.*

This procedure thus computes the $\widehat{GT}$-orbit of a Galois dessin D after
finitely many steps, and so approximates (and conjecturally equals) the $G_\mathbb{Q}$-
orbit of D. This complements the construction of §2 (just as that of 3.4
complements the construction of §1). Unlike the constructions in sections 1
and 2, though, the procedure here (and in 3.4) is not effective, because we
do not know how to determine *a priori* how large i should be, or even when
we have reached the goal! To be better able to exploit this construction, we
would thus like to be able to solve the following

Problem. Given a normal subgroup $N \subset \widehat{F}_2$ of finite index, corresponding
to a G-Galois dessin D (where $G = \widehat{F}_2/N$), find a characteristic subgroup
$M \subset \widehat{K}(0,5)$ of finite index such that the M-approximative $\widehat{GT}$-orbit of D
is equal to the $\widehat{GT}$-orbit of D.

Still, successive steps of the above refining procedure either improve the
approximation or leave it unchanged, and thus (even without solving this
problem) they provide increasingly better information about the field of
moduli of D — which can be used as in 3.6 below.

3.6. Stabilizers and fields of moduli. The above approach may be used to obtain explicit information about the field of moduli K of a G-Galois dessin D (i.e. of the corresponding G-Galois cover $X \to \mathbb{P}^1 - \{0, 1, \infty\}$). For example, the degree $[K : \mathbb{Q}]$ is equal to the order of the $G_{\mathbb{Q}}$-orbit of D (or of X), since that order is the index of the stabilizer $G_{\mathbb{Q}}^D$ of D in $G_{\mathbb{Q}}$, and hence is the degree of the fixed field of $G_{\mathbb{Q}}^D$ — i.e. the degree of the field of moduli. Since the $\widehat{GT}$-orbit contains the $G_{\mathbb{Q}}$-orbit, and since any $\check{M}$-approximative $\widehat{GT}$-orbit of D contains the $\widehat{GT}$-orbit, we obtain

Proposition 14. *Let K be the field of moduli of a G-Galois dessin D, where $G = \widehat{K}(0,4)/N$. Then $[K : \mathbb{Q}]$ is bounded above by the order of the $\check{M}$-approximative $\widehat{GT}$-orbit of D, for any finite index characteristic subgroup $\check{M} \subset M = p_4^{-1}(N)$ of $\widehat{K}(0,5)$.*

The approximative $\widehat{GT}$-actions, i.e. the actions of the $\mathcal{O}_5^\#(\check{M})$'s on dessins, also provide computable information about the Galois group of the field of moduli K of a G-Galois dessin D. Below, we write $\psi_N^{5\#} : \mathcal{O}_5^\# \twoheadrightarrow \mathcal{O}_{5,\check{M}}^\#$ for the restriction of the map $\psi_N^5 : \mathcal{O}_5 \twoheadrightarrow \mathcal{O}_{5,\check{M}}$ (cf. 3.2 and Theorem 13). Also, if a group Γ acts on a set of G-Galois dessins, we denote the stabilizer of D in Γ by Γ^D, and the intersection of the conjugates of $\Gamma^D \subset \Gamma$ by $C^D(\Gamma)$.

Proposition 15. *(a) With notation as above, let $\tilde{K}$ be the Galois closure of K. Then $\mathrm{Gal}(\tilde{K}/\mathbb{Q})$ is a subquotient of each $\mathcal{O}_5^\#(\check{M})/C^D(\mathcal{O}_5^\#(\check{M}))$.*

(b) If $\mathcal{O}_5^\#(\check{M})^D$ is normal in $\mathcal{O}_5^\#(\check{M})$, then $K/\mathbb{Q}$ is Galois, and $\mathrm{Gal}(K/\mathbb{Q})$ is a subgroup of $\mathcal{O}_5^\#(\check{M})/\mathcal{O}_5^\#(\check{M})^D$.

Proof. Since $C^D(\mathcal{O}_5^\#(\check{M}))$ is normal in $\mathcal{O}_5^\#(\check{M})$ and contained in $\mathcal{O}_5^\#(\check{M})^D$, we have that $C_{\check{M}}^D := C^D(\mathcal{O}_5^\#(\check{M})) \cap \mathcal{O}_{5,\check{M}}^\#$ is normal in $\mathcal{O}_{5,\check{M}}^\#$ and contained in $(\mathcal{O}_{5,\check{M}}^\#)^D$. So $C_{\check{M}}^D \subset C^D(\mathcal{O}_{5,\check{M}}^\#)$, and $\mathcal{O}_{5,\check{M}}^\#/C_{\check{M}}^D \subset \mathcal{O}_5^\#(\check{M})/C^D(\mathcal{O}_5^\#(\check{M}))$. Since the action of $\widehat{GT} \simeq \mathcal{O}_5^\#$ on the orbit of D factors through $\mathcal{O}_{5,\check{M}}^\#$ via $\psi_N^{5\#}$ (cf. 3.4), we have that $(\psi_N^{5\#})^{-1}(C_{\check{M}}^D) \subset C^D(\mathcal{O}_5^\#)$ and $\mathcal{O}_5^\#/(\psi_N^{5\#})^{-1}(C_{\check{M}}^D) = \mathcal{O}_{5,\check{M}}^\#/C_{\check{M}}^D$. Since the action of $G_{\mathbb{Q}}$ on dessins factors through $\mu : G_{\mathbb{Q}} \hookrightarrow \mathcal{O}_5^\#$, $(\psi_N^{5\#}\mu)^{-1}(C_{\check{M}}^D) \subset C^D(G_{\mathbb{Q}})$ and $G_{\mathbb{Q}}/(\psi_N^{5\#}\mu)^{-1}(C_{\check{M}}^D) \subset \mathcal{O}_5^\#/(\psi_N^{5\#})^{-1}(C_{\check{M}}^D)$. Combining the above equalities and inclusions, we have that $\mathrm{Gal}(\tilde{K}/\mathbb{Q}) = G_{\mathbb{Q}}/C^D(G_{\mathbb{Q}})$ is a quotient of $G_{\mathbb{Q}}/(\psi_N^{5\#}\mu)^{-1}(C_{\check{M}}^D)$, which in turn is a subgroup of $\mathcal{O}_5^\#(\check{M})/C^D(\mathcal{O}_5^\#(\check{M}))$. This proves (a).

For (b), observe that $G_{\mathbb{Q}}^D$ is the inverse image of $\mathcal{O}_5^\#(\check{M})^D$ under the composition $G_{\mathbb{Q}} \hookrightarrow \widehat{GT} \xrightarrow{\sim} \mathcal{O}_5^\# \twoheadrightarrow \mathcal{O}_{5,\check{M}}^\# \hookrightarrow \mathcal{O}_5^\#(\check{M})$, since the action of $G_{\mathbb{Q}}$ on the orbit of D factors through these maps. Since $\mathcal{O}_5^\#(\check{M})^D$ is normal in $\mathcal{O}_5^\#(\check{M})$, it follows that $G_{\mathbb{Q}}^D$ is normal in $G_{\mathbb{Q}}$, so that K is Galois over $\mathbb{Q}$.

It also follows that $\mathcal{O}_5^{\#}(\check{M})^D = C^D(\mathcal{O}_5^{\#}(\check{M}))$. Thus $(\psi_N^{5\#}\mu)^{-1}(C_{\check{M}}^D)$, which is the inverse image of $C^D(\mathcal{O}_5^{\#}(\check{M}))$ under the above composition, is equal to $G_{\mathbb{Q}}^D$. So $\mathrm{Gal}(K/\mathbb{Q})$ equals $(\psi_N^{5\#}\mu)^{-1}(C_{\check{M}}^D)$, which was observed to be a subgroup of $\mathcal{O}_5^{\#}(\check{M})/C^D(\mathcal{O}_5^{\#}(\check{M})) = \mathcal{O}_5^{\#}(\check{M})/\mathcal{O}_5^{\#}(\check{M})^D$. $\quad\diamond$

It is also possible to use the actions of the $\mathcal{O}_5^{\#}(\check{M})$'s in order to find cyclotomic number fields that contain the field of moduli of a given G-Galois dessin (and thus, when G has trivial center, over which the corresponding G-Galois cover is defined [CH, 2.8(c)]). For $m \in \mathbb{Z}$, let $\mathcal{O}_n^{\#}(\check{M})_m \subset \mathcal{O}_n^{\#}(\check{M})$ be the subgroup of elements for which we may take $\lambda \equiv 1 \pmod{m}$ in (i) of the definition of $\mathcal{O}_n^{\#}(\check{M})$ (in 3.2).

Proposition 16. *(a) If some $\mathcal{O}_n^{\#}(\check{M})_0$ stabilizes a G-Galois dessin D, then the corresponding G-Galois cover is defined over $\mathbb{Q}^{ab}$.*

(b) If $m > 0$ and some $\mathcal{O}_n^{\#}(\check{M})_m$ stabilizes a G-Galois dessin D, then the field of moduli of the corresponding G-Galois cover is contained in $\mathbb{Q}(\zeta_m)$.

Proof. Under $G_{\mathbb{Q}} \to \mathcal{O}_5^{\#}(\check{M})$, $G_{\mathbb{Q}^{ab}}$ maps to $\mathcal{O}_n^{\#}(\check{M})_0$ and $G_{\mathbb{Q}(\zeta_m)}$ maps to $\mathcal{O}_n^{\#}(\check{M})_m$ for $m > 0$. So $\mathbb{Q}^{ab}$ and $\mathbb{Q}(\zeta_m)$ contain the respective fields of moduli. This proves (b), and (a) then follows by [CH, 2.8(a)]. $\quad\diamond$

In particular, taking $m = 1$, we conclude that if some $\mathcal{O}_5^{\#}(\check{M})$ stabilizes a G-Galois dessin D, then the field of moduli of D is equal to $\mathbb{Q}$.

Note that one may obtain stronger conclusions from the above results not only by shrinking $\check{M}$, but also by replacing $\mathcal{O}_5^{\#}(\check{M})$ by a smaller subgroup $\mathcal{O}_5^{\circ}(\check{M})$ that is known to contain $\mathcal{O}_{5,\check{M}}^{\#}$. Because of Theorem 7, the definition of A_5, and the fact that the automorphism $\theta \in \mathrm{Aut}(\widehat{K}(0,5))$ defined by $\theta(x_{i,i+1}) = x_{5-i,6-i}$ descends to an automorphism θ of $\widehat{K}(0,5)/\check{M}$, we may in particular take $\mathcal{O}_5^{\circ}(\check{M})$ to consist of the elements $\overline{F} \in \mathcal{O}_{5,\check{M}}^{\#} \subset \mathrm{Out}(\widehat{K}(0,5)/\check{M})$ having a lifting $F \in \mathrm{Aut}(\widehat{K}(0,5)/\check{M})$ for which there is an $f \in \langle (x_{12})_{\check{M}}, (x_{23})_{\check{M}} \rangle$ with $F((x_{12})_{\check{M}}) = (x_{12})_{\check{M}}^{\lambda}$; $F((x_{23})_{\check{M}}) = f^{-1}(x_{23})_{\check{M}}^{\lambda}f$; $F((x_{34})_{\check{M}}) = \theta(f)^{-1}(x_{34})_{\check{M}}^{\lambda}\theta(f)$; and $F((x_{45})_{\check{M}}) = (x_{45})_{\check{M}}^{\lambda}$. (One could shrink $\mathcal{O}_5^{\circ}(\check{M})$ further by also using $F((x_{51})_{\check{M}})$, via the expression (**) in 0.1 and the analogous result for $\mathcal{O}_5^{\#}$ in [HS, 2.2].)

The above raises several questions: For a given characteristic subgroup $\check{M} \subset M = p_4^{-1}(N)$, to what extent does $\mathcal{O}_5^{\circ}(\check{M})$ yield better information about the field of moduli and the Galois orbit than just using $\mathcal{O}_5^{\#}(\check{M})$? As $\check{M}$ shrinks, how fast does the information given by $\mathcal{O}_5^{\#}(\check{M})$ improve? To what extent does the information given by $\mathcal{O}_5^{\#}(\check{M})$ go beyond that given by $\mathcal{O}_4^{\#}(\check{N})$, where $\check{N} = p_4(\check{M}) \subset \widehat{K}(0,4)$? And to what extent do computations using the above give information beyond that obtainable by rigidity?

References

[B]   G.V. Belyi, On Galois extensions of a maximal cyclotomic field, *Math. USSR Izvestija* (translations) **14** (1980), No. 2, 247-256.

[CH]   K. Coombes and D. Harbater, Hurwitz families and arithmetic Galois groups, *Duke Math. J.* **52** (1985), 821-839.

[D]   V.G. Drinfel'd, On quasitriangular quasi-Hopf algebras and a group closely connected with $\mathrm{Gal}(\overline{\mathbb{Q}}/\mathbb{Q})$, *Leningrad Math. J.* Vol. 2 (1991), No. 4, 829-860.

[F1]   M. Fried, Fields of definition of function fields and Hurwitz families — groups as Galois groups, *Comm. Alg.* **5** (1977), 17-82.

[F2]   M. Fried, Introduction to Modular Towers: Generalizing dihedral group–modular curve connections, in *Recent Developments in the Inverse Galois Problem*, M. Fried et al., Eds., AMS Contemp. Math. Series, vol. 186, 1995, 111-171.

[G]   A. Grothendieck, Revêtements étales et groupe fondamental (SGA 1). Lecture Notes in Math. **224**, Springer-Verlag, Berlin-Heidelberg-New York, 1971.

[GTD]   *The Grothendieck Theory of Dessins d'Enfants*, L. Schneps, ed., London Math. Soc. Lecture Notes **200**, Cambridge University Press, 1994.

[HS]   D. Harbater and L. Schneps, Fundamental groups of moduli and the Grothendieck-Teichmüller group, preprint.

[I1]   Y. Ihara, Braids, Galois groups, and some arithmetic functions, Proceedings of the ICM, Kyoto, Japan, 1990, 99-120.

[I2]   Y. Ihara, On the embedding of $\mathrm{Gal}(\overline{\mathbb{Q}}/\mathbb{Q})$ into $\widehat{GT}$, in [GTD].

[IM]   Y. Ihara and M. Matsumoto, On Galois Actions of Profinite Completions of Braid Groups, in *Recent Developments in the Inverse Galois Problem*, M. Fried et al., Eds., AMS Contemp. Math. Series, vol. 186, 1995, 173-200.

[LS]   P. Lochak and L. Schneps, The Grothendieck-Teichmüller group as automorphisms of braid groups, in [GTD].

[M]   B.H. Matzat, Zur Konstruction von Zahl- und Funktionkorpern mit vorgegebener Galoisgruppe, *J. reine angew. Math.* **349** (1984), 179-220.

[N]   H. Nakamura, Galois rigidity of pure sphere braid groups and profinite calculus, *J. Math. Sci. Univ. Tokyo* **1** (1994), 71-136.

[S]   J-P. Serre, Deux lettres sur la cohomologie non abélienne, this volume.

[T]   J. Thompson, Some finite groups which appear as Gal L/K, where $K \subset \mathbb{Q}(\mu_n)$, *J. Algebra* **89** (1984), 437-499.

Tame and Stratified Objects

Bernard Teissier

§0. Introduction.

In §5 of the *Esquisse*, Grothendieck describes his approach to the foundations of what he calls "tame topology", motivated by his wish to develop,
with a view to a study of compactifications of moduli spaces by stratified
dévissage, a sufficiently general theory of stratified spaces and maps. He
envisions there an axiomatic approach to a hierarchy of (tame) geometric
categories, from semi-algebraic sets defined over $\overline{\mathbb{Q}}$ to the subanalytic sets
studied by Lojasiewicz, Gabrielov, Hironaka. In §6 he goes on to describe
a hierarchy of "elementary functions" or, as he says, "differentiable categories", and explicitly states that such categories should be associated to a
tame geometric category.

The ideas in these two paragraphs have high contact with the theory
of Whitney and Thom stratifications of spaces and maps and their use in
algebraization, and with recent work in real algebraic geometry. In these
notes I shall endeavour to make some of it explicit, and point out some
references. There are many relevant references which I did not include for
lack of knowledge, and I apologize to their authors.

§1. Tame sets

The tameness of algebraic, semi-algebraic and analytic sets, is quite old,
going back at least to the triangulation of algebraic varieties at the beginning
of the century; one should also mention Lelong's theorem on integration on
singular analytic sets. However, it is perhaps in Lojasiewicz's 1965 paper
([L]) on semi-analytic sets, i.e. subsets of $\mathbb{R}^n$ defined locally near every point
of $\mathbb{R}^n$ by a finite number of equalities and inequalities of real analytic functions, that a rather systematic study of a class of subsets of Euclidean space
from the viewpoint of tameness (local finiteness of the number of connected
components, semi-analyticity of the boundary and interior, triangulability,
Lojasiewicz exponents, etc.) first appears. Tameness is not easily defined by
a single property; indeed one would have thought perhaps that the triangulability of a semi-analytic subset of $\mathbb{R}^n$, i.e. the fact that it is homeomorphic
to a simplicial subcomplex of a linear simplicial subdivision of $\mathbb{R}^n$, is the
strongest tameness property. It has turned out, however, that the homeomorphism can be chosen to be subanalytic (see below), and this is much
stronger; it implies for example that the homeomorphism is locally Hölder.

This suggests the idea that tameness in geometry has several avatars; one may think of topological tameness, and then it means triangulability, and "looking locally like a semi-algebraic set". One may think in a more metric way and then it means the triangulability is Hölder and we have Lojasiewicz inequalities, and so on.

The theory of semi-analytic sets suffers from the basic defect that the image of a semi-analytic set by a proper real analytic map is not in general semi-analytic. This contrasts of course with the theorem of Tarski-Seidenberg for semi-algebraic sets, and led Gabrielov to prove in [G1] that the complement of the image of a semi-analytic set by a proper analytic map is also the image of a semi-analytic set by a proper map (theorem of the complement), which was the key result needed in Lojasiewicz's approach to extend the basic finiteness theorems to these images. Thom needed these finiteness properties for his theory of stratified maps, and in [Th] he used under the name PSA (projections de semi-analytiques) the proper images of semi-analytic sets. A little later, with different motivations, Hironaka developed from scratch ([Hi1]) the theory of what he called subanalytic sets, locally Boolean combinations of images of analytic sets by proper analytic maps*. One of his main tools was a theorem of resolution of singularities which enabled him to show that subanalytic sets could be described as those subsets of $\mathbb{R}^n$ which were locally near every point of $\mathbb{R}^n$ in the Boolean algebra generated by images of spheres by real analytic maps.

By the mid-seventies the theory was well established, and its usefulness gradually became more and more apparent. Very soon, the need to extend the finiteness properties of subanalytic sets to sets defined by equalities and inequalities between the elements of a class of functions wider than real-analytic functions (and the images of these sets) appeared to geometers in connection with the search for tameness results for sets appearing as orbits of analytic dynamical systems, especially in connection with Hilbert's 16th problem (see [F]). Here one may mention the work of Françoise and Pugh on Hilbert's 16th problem and the pioneering work of Khovanskii [Kh] and its development by the Dijon school. The facts that the class of sets under consideration should be closed under passing to the complement and under taking images by proper maps whose graph is in the class are key facts for proving tameness by induction on the dimension.

I am much less familiar with the motivations of logicians which led to their study of systems of subsets of $\mathbb{R}^n$ defined axiomatically, closer in spirit apparently to the preoccupations of the *Esquisse*, but in the last few years the two streams have been rapidly converging and a rather general theory corresponding well to a part of Grothendieck's wishes has been worked out

---

* In the *Esquisse* Grothendieck calls subanalytic sets "semi-analytic".

by the efforts notably of Coste, Gabrielov, van den Dries and Miller, Shiota, and is about to be published (see [vdDM], [Sh]) while several of the geometric finiteness problems which motivated it have been solved ([Wi]).

Let me give the flavour by quoting the definition from [vdDM]:
An analytic-geometric category is the datum for each real-analytic manifold M of a collection $\mathcal{C}(M)$ of subsets of M satisfying for any two real-analytic manifolds M, N the following axioms:

AG 1) $\mathcal{C}(M)$ is a Boolean algebra of subsets of M, and $M \in \mathcal{C}(M)$.

AG 2) If $A \in \mathcal{C}(M)$, then $A \times \mathbb{R} \in \mathcal{C}(M \times \mathbb{R})$.

AG 3) If $f: M \to N$ is a proper analytic map, then $A \in \mathcal{C}(M)$ implies $f(A) \in \mathcal{C}(N)$.

AG 4) If A is a subset of M and $(U_i)_{i \in I}$ is an open covering of M then $A \in \mathcal{C}(M)$ if and only if $A \cap U_i \in \mathcal{C}(U_i)$ for all $i \in I$.

AG 5) Every bounded set in $A \in \mathcal{C}(\mathbb{R})$ has finite boundary.

One can make a hierarchy of such categories, simply by saying that $\mathcal{C} \leq \mathcal{C}'$ if for all M one has $\mathcal{C}(M) \subset \mathcal{C}'(M)$. The maps in the category are maps $(M, A) \to (N, B)$, continuous and such that $A \in \mathcal{C}(M), B \in \mathcal{C}(N)$ and the graph of $f|A: A \to B$ is in $\mathcal{C}(M \times N)$.

Subanalytic sets in real analytic manifolds form such an analytic-geometric category, denoted by $\mathcal{C}_{an}$, and in fact any analytic-geometric category contains $\mathcal{C}_{an}$.

We may call "tame" the elements of an analytic-geometric category; in [vdDM] it is shown that they are Whitney-stratifiable, triangulable, and have the tameness properties which we associate to subanalytic sets. The challenge is of course to find useful analytic-geometric categories.

Here the basic idea, due to Khovanskii [Kh], is to consider "Pfaff manifolds", essentially obtained by glueing up Pfaffian subsets of $\mathbb{R}^n$, described as Boolean combinations of separating solutions of algebraic or analytic dynamical systems. For example a limit cycle of a polynomial vector field in the plane is a Pfaffian subset, the graph of a solution of a differential equation

$$dY = \sum_{i=1}^{n} P_i(x_1, \ldots, x_n, Y) dx_i, \quad P \in \mathbb{R}[x_1, \ldots, x_n, Y]$$

is a Pfaffian subset in $\mathbb{R}^{n+1}$, etc. However, it is necessary to understand the affine theory well before attacking the analytic-geometric category. In [Kh], this is done with many geometric lemmas. In [vdDM] the link is made explicitly between analytic-geometric categories and what one may call the "affine" theory of tame sets, which is a more direct generalization of the theory of semi-algebraic sets. The problem is that while a semi-algebraic set in $\mathbb{R}^n$ remains semi-algebraic at infinity (i.e. in $\mathbb{P}^n$), this is of course no longer the case as soon as one allows analytic functions. The logical framework is then the following (for all this, see [vdDM]):

A Structure on the real field $\mathbb{R}$ is the datum of a Boolean algebra $\mathcal{S}_n$ of subsets of $\mathbb{R}^n$ for each integer $n \geq 0$, in such a way that:

S 1) For each n, $\mathbb{R}^n \in \mathcal{S}_n$.

S 2) $\mathcal{S}_n$ contains the diagonals $\{(x_1, \ldots, x_n)/x_i = x_j\}$ for $1 \leq i < j \leq n$.

S 3) If $A \in \mathcal{S}_n$, then $A \times \mathbb{R}$ and $\mathbb{R} \times A$ are in $\mathcal{S}_{n+1}$.

S 4) If $A \in \mathcal{S}_{n+1}$ then $\pi(A) \in \mathcal{S}_n$ where $\pi: \mathbb{R}^{n+1} \to \mathbb{R}^n$ is the projection on the first n coordinates.

S 5) $\mathcal{S}_3$ contains the graphs of addition and multiplication.

One says that a subset $A \subset \mathbb{R}^n$ is in $\mathcal{S}$ if it belongs to $\mathcal{S}_n$, and a map $f: A \to \mathbb{R}^m$ is in $\mathcal{S}$ if its graph is in $\mathcal{S}_{n+m}$.

Given functions $(f_j: \mathbb{R}^{n(j)} \to \mathbb{R})_{j \in J}$ one denotes by $\mathcal{S}(\mathbb{R}, (f_j)_{j \in J}$ the smallest structure containing the graphs of the functions $(f_j)_{j \in J}$. Remark that by the Tarski-Seidenberg theorem, $\mathcal{S}(\mathbb{R})$ is the theory of semi-algebraic sets defined over $\overline{\mathbb{Q}}$. More generally one can obtain semi-algebraic sets defined over an algebraically closed subfield k of $\mathbb{R}$ by taking for f_j the constant functions $\mathbb{R}^0 = \{0\} \to \mathbb{R}$ having images in k; compare with the *Esquisse*, p. 41. If one takes for the set $(f_j)_{j \in J}$ all restricted analytic functions, i.e. functions $\mathbb{R}^n \to \mathbb{R}$ for all n such that f vanishes identically outside the unit n-cube and is analytic on this n-cube, the elements of the corresponding structure are the subsets of $\mathbb{R}^n$ which are subanalytic not only in $\mathbb{R}^n$, but also when viewed in $\mathbb{P}^n$ via the natural embedding; compare with the *Esquisse*, p. 37.

A Structure is called o-minimal if $\mathcal{S}_1$ consists only of unions of intervals, including singletons.

Now to an analytic-geometric category one can associate an o-minimal structure by taking

$$\mathcal{S}(\mathbb{R}^n) = \{X \subset \mathbb{R}^n : X \in \mathcal{C}(\mathbb{P}^n)\}$$

where X is viewed as a subset of $\mathbb{P}^n$ via the natural embedding.

Conversely, given an o-minimal structure $\mathcal{S}$, we obtain an analytic-geometric category by deciding that given a real-analytic manifold M, a subset $A \subset M$ is in $\mathcal{C}(M)$ if for any point $x \in M$ there are a neighbourhood U of x in M and an analytic isomorphism $f: U \to V$ to an open set of $\mathbb{R}^m$ such that $f(U \cap A) \in \mathcal{S}_m$.

In particular, this allows one to define an analytic-geometric category from algebras of "tame" functions, and conversely, according to the wish of the *Esquisse*, p. 42.

Again, the most important fact is to show that certain analytic-geometric categories associated to classes of functions have all the nice finiteness properties: locally a finite number of connected components, each belonging to the category, stratifiability (see below), triangulability (see below), Lojasiewicz-type inequalities, tame variation of the fibres of a map $f: A \to B$, etc. For this, the fact that the complement of the image by a proper map of a set of the category is also in the category (theorem of the complement, see above) is often the key.

In the model-theoretic langage, it is called "model completeness" and amounts to the reduction of formulas to formulas with only existential quantifiers, while the fact that the proper image of an element of the category is in the category amounts to elimination of quantifiers.

Many of the finiteness results concerning Pfaffian sets should follow from Wilkie's proof ([Wi]) of the model-completeness of the structure $(\mathbb{R}, \mathcal{G})$ where $\mathcal{G}$ is the set of restricted Pfaffian chains, defined as follows (see [Wi]):

For given $m, p \geq 1$ let $U \subset \mathbb{R}^m$ be an open set containing the closed unit cube; and let $G_1, \ldots, G_p$ be analytic functions satisfying a system of partial differential equations

$$\frac{\partial G_i}{\partial x_j} = p_{i,j}\big(x, G_1(x), \ldots G_p(x)\big) \text{ for } x \in U,$$

where $p_{i,j} \in k[x_1, \ldots, x_{m+p}]$, where k is a subfield of $\mathbb{R}$. Such a sequence of functions is called a Pfaffian chain, and we consider the functions $f_1, \ldots, f_p$ where f_i is defined as the function equal to 0 outside the unit cube of $\mathbb{R}^m$ and to G_i in that cube.

Wilkie also proved that the structure model completeness of the structure defined by adding exponential functions, which implies that if one defines exponential polynomials on $\mathbb{R}^n$ as polynomials in $x_1, \ldots, x_n, e^{x_1}, \ldots, e^{x_n}$ with real coefficients and a semi-EA set as defined by Boolean combination of sets of the form $p(x) = 0$, $q(x) > 0$ for exponential polynomials p, q, a semi-EA map as a map having a semi-EA graph, then the complement of the image of a semi-EA set by a semi-EA map is also the image of a semi-EA set by a semi-EA map. Such sets may be called sub-EA sets. For Pfaffian sets as well as sub EA sets one deduces that the closure, boundary, interior are also Pfaffian and sub-EA respectively.

Shiota adopted in [Sh] a slightly different viewpoint by defining $\mathcal{X}$-sets as a collection of subsets of Euclidean spaces satisfying the following axioms:
A 1) Any algebraic subset in Euclidean space is an $\mathcal{X}$-set.

A 2) If X_1, X_2 are $\mathcal{X}$-sets in $\mathbb{R}^n$, so are $X_1 \cap X_2$, $X_1 \setminus X_2$, and $X_1 \times X_2$ is an $\mathcal{X}$-set in $\mathbb{R}^{2n}$.

A 3) If X is an $\mathcal{X}$-set in $\mathbb{R}$ every point of X has a neighbourhood in X which is a finite union of points and intervals.

Shiota notes that o-minimal structures are made of $\mathcal{X}$-sets, but the converse is not true. He proves all the basic tameness results for $\mathcal{X}$-sets, and notably the main conjecture on triangulations (see below).

From now on I will call tame the $\mathcal{C}$ sets of [vdDM] and, for subsets of $\mathbb{R}^n$, the elements of an o-minimal structure or the $\mathcal{X}$-sets of Shiota.

§2. Stratifications

The weakest definition of a stratified set is the following: A topological space X is stratified if it can be represented as a locally finite union of non-singular spaces, called strata, such that the closure of each stratum is a union of strata. Here the definition of non-singular space depends on the category in which we work. It may be a differentiable manifold, or we may be in a situation where X is a subset of k^n, where $k = \mathbb{R}$ or $\mathbb{C}$ and each stratum is smooth and semi-algebraic, semi-analytic or subanalytic in $\mathbb{R}^n$, or a locally closed complex submanifold in $\mathbb{C}^n$ with analytic closure in the complex case.

However with such a definition there is very little one can do in terms of "dévissage" since there is practically no information on the way strata attach to one another; in particular there are no relations between tubular neighbourhoods of the various strata. One must add "incidence conditions" which imply some constancy in the way a "big stratum" X_α attaches to a "small stratum" $X_\beta \subset \overline{X_\alpha}$. The first example of an incidence condition, due to Whitney in 1964, is still the most widely used. Given a pair of strata $X_\beta \subset \overline{X_\alpha}$ as above, and a point $y \in X_\beta$, we may assume that a neighbourhood U of y in X is embedded in k^n, and then we say that (X_α, X_β) satisfies Whitney's condition at y if for any sequence of pairs of points $(x_i, y_i) \in X_\alpha \times X_\beta$ converging to (y, y), for any subsequence such that the limit position $\ell \in \mathbb{P}^{n-1}$ of directions of secant lines $k(x_i - y_i)$ and the limit position $T \in G(n, d_\alpha)$ of the tangent spaces T_{X_α, x_i} both exist, we have

$$\ell \subset T.$$

Here $d_\alpha = \dim X_\alpha$ and $G(n, d_\alpha)$ denotes the Grassmannian of d_α-dimensional subspaces of k^n.

The motivation for this condition is perhaps best described by considering the case where X_β is a point y ; there of course the condition should be satisfied, and indeed it is in most geometric cases. But this means that the limit directions of secant lines $k(x_i - y)$ are contained in the limit of the tangent spaces T_{X_α, x_i}; in the case where $\overline{X_\alpha}$ is a cone with vertex y, this is a well-known consequence of the Euler relation for homogeneous functions, and the fact that Whitney's condition is satisfied means that it is satisfied asymptotically for sequences of points tending to y. It means that $\overline{X_\alpha}$ is in a neighbourhood of y like a cone with vertex y. When X_β is no longer assumed to be a point, looking in a small enough neighbourhood of y we may assume that X_β is a linear subspace of k^n and we see that Whitney's condition is the spreading along X_β of that conicity condition, and indeed it implies that near y, the space $\overline{X_\alpha}$ is like a cone with "vertex" X_β. The usefulness of the Whitney condition is summarized in the following theorem.

Theorem. (Thom-Mather) *Let $X = \bigcup_{\alpha \in A} X_\alpha$ be a stratified set, let $x \in X$ and let X_β be the stratum containing x. There exist a neighbourhood U of x in X, a retraction $r: U \to X_\beta \cap U$, a closed stratified set $X_x \subset U$ and a homeomorphism*

$$\phi: U \to X_x \times (X_\beta \cap U)$$

such that $\mathrm{pr}_2 \circ \phi = r$, mapping each $X_\alpha \cap U$ to the product with $(X_\beta \cap U)$. In particular the topology of the germ at $y \in X_\beta$ of the stratified set X is locally constant on X_β.

A stratification $X = \bigcup X_\alpha$ such that whenever $X_\beta \subset \overline{X_\alpha}$ the pair satisfies the Whitney condition at every point of X_β is called a *Whitney stratification*. The other basic theorem concerning Whitney stratifications is that they exist for tame spaces; in particular for complex analytic, semi-algebraic, semi-algebraic over $\overline{\mathbb{Q}}$, semi-analytic, subanalytic sets. The key is a formal construction which shows that given a condition on triples (X, Y, y) where $y \in Y \subset \overline{X}$ and X, Y belong to a given geometric category (semi-algebraic, subanalytic, etc.) which satisfies the following properties:

a)   it is of a local nature on X;

b)   it is hereditary in the sense that if (X, Y, y) satisfies the condition, and $Y' \subset Y$ is a subspace with $y \in Y'$, then (X, Y', y) also satisfies the condition;

c)   given (X, Y) with $Y \subset \overline{X}$, the set of points $y \in Y$ where the condition is not satisfied is contained in a closed subspace (in the same category) which is of dimension $< \dim Y$;

then given X there exists a stratification of X such that any pair of strata (X, Y) with $Y \subset \overline{X}$ satisfies the incidence condition at every point of Y.

In recent work ([vdDM], [Sh]) the existence of Whitney stratifications for tame sets is proved.

The reason to insist on Whitney stratifications is that from the origin of the proof of Thom-Mather's topological triviality theorem, the construction of a coherent system of tubular neighbourhoods of the various strata ("control data") has been essential (see [Th1], [Ma1], [Ma2]), and this fits with the *Esquisse*, pp. 26-28 and 33-34. However these depend on a metric structure and I do not know for the moment of any intrinsic definition, although the fact that there is a completely algebraic characterization of Whitney conditions in the complex analytic case ([Te1], Chap. 5, Th. 1.2) by the equimultiplicity of polar varieties suggests that at least in this case a definition of tubular neighbourhood of a stratum as some kind of privileged neighbourhood à la Douady for suitable sheaves should be possible and lead to a more intrinsic description. It also fits with the fact that although the converse of the Thom-Mather topological triviality theorem is not true, one can describe a more precise topological triviality theorem which follows from the Whitney conditions, and which does have a converse (see [LT]).

Finally, as a consequence of the algebraic characterization one can prove the existence of a *minimal* Whitney stratification for any complex-analytic space, such that the strata of any other Whitney stratification are contained in its strata. If X is an algebraic variety defined over an algebraically closed subfield k of $\mathbb{C}$, the strata are defined over k. According to Mebkhout, this is the result to which Grothendieck alludes on p. 38 of the *Esquisse*.

However, in the differentiable category, a Whitney-stratified set may exhibit tangential wildness: M. Kwieciński and D. Trotman ([KT]) have shown the following: Given any compact connected continuum K in $\mathbb{P}^2(\mathbb{R})$ there exists a real analytic surface S in $\mathbb{R}^3 \setminus \{0\}$ having 0 in its closure and such that $S \cup \{0\}$ is a Whitney stratified set, and such that the set of limit positions at 0 of tangent spaces $T_{S,s}$ as $s \to 0$ is equal to K. Note that S is not subanalytic in $\mathbb{R}^3$, and is not tame in any reasonable sense. This example is quoted to show that the existence of Whitney stratifications does not imply tameness since it does not prevent wild wrinkling of the strata.

Concerning the stratification of tame maps, this was the theme of a lot of work by Thom ([Th1]), Mather ([Ma1], [Ma2]) and others, in the differentiable or analytic case, and in [Sh] one finds a proof that given a proper tame map $f: X \to Y$ the existence of tame Whitney stratifications $X = \bigcup X_i$, $Y = \bigcup Y_j$ such that the restriction of f to each X_i is a (tame) submersion onto a $Y_{j(i)}$ (see p. 37 of the *Esquisse*). However, as was noted by Thom long ago, this is far from sufficient to allow topological dévissage of the map, since it does not imply continuity of the behaviour of the tangent spaces of the fibers; think of the natural stratification of the blowing-up of the origin in affine space. For example one cannot define in general vanishing cycles at a point of a fiber for a proper tame map, Whitney stratified as above.

For this reason, Thom introduced a "condition de non éclatement" for maps of stratified sets which, from a technical point of view, allows one to compare "control data", i.e. tubular neighbourhoods of strata on X and on Y. This allowed him to state and prove a topological triviality theorem for maps (which I believe should be a basic ingredient for the dévissage of maps evoked in the *Esquisse*), his "second isotopy theorem" (see [Th1], [Ma1], [Ma2]). The other basic application belongs in fact to the next topic, triangulability.

§3. Triangulations of sets and of maps

Triangulability is perhaps the first tameness property to have been considered. A triangulation of a subset $X \subset \mathbb{R}^n$ is a simplicial linear decomposition of $\mathbb{R}^n$, a simplicial subcomplex Δ of this decomposition, and a homeomorphism $\pi: X \to \Delta$. Historically the proofs of triangulability for semi-algebraic or semi-analytic sets use the following fact: Given a tame subset $X \subset \mathbb{R}^n$, to prove the existence of a triangulation of $\mathbb{R}^n$ compatible with X, it is sufficient to prove the existence of a triangulation compatible with the boundary of X, which is of smaller dimension. Then one proves a good projection lemma asserting that if $\dim X < n$, a sufficiently general linear projection $p: \mathbb{R}^n \to \mathbb{R}^{n-1}$ induces on X a map with discrete fibers, and finally one triangulates this map (see below, and [Ha]).

In any case the tame triangulability of tame objects in $\mathbb{R}^n$ (*Esquisse*, p. 32) is established ([vdDM], [Sh]).

A map $f: X \to \mathbb{R}^m$, where $X \subset \mathbb{R}^m \times \mathbb{R}^p$, is *triangulated* if there exist linear simplicial decompositions of $\mathbb{R}^m \times \mathbb{R}^p$ and $\mathbb{R}^m$, simplicial subcomplexes $\Delta \subset \mathbb{R}^m \times \mathbb{R}^p$ and $T \subset \mathbb{R}^m$, and homeomorphisms $\pi: \Delta \to X$ and $\tau: T \to Y$, such that $\tau^{-1} \circ f \circ \pi: \Delta \to T$ is a simplicial map. Note that the blowing up of a point in the plane is *not* triangulable.

Thom conjectured that any proper analytic map Whitney stratified "sans éclatement" is triangulable.

Hardt showed that given a proper real-algebraic map $f: X \to \mathbb{R}^m$, one could cover $\mathbb{R}^m$ by semi-algebraic regions in such a way that the restriction of f to the counterimage of each region is triangulable with a semi-algebraic homeomorphism. This result does not control what happens to the triangulations near the boundaries, but it played an important historical role, and suffices for a good many finiteness results for maps. Another important result in that direction is the semi-algebraic version of Thom's first isotopy lemma ([CS]).

It can be shown ([Te2]) that given a proper subanalytic map $f: X \to Y$, for every compact subset K of Y there exists a finite collection of maps $g_i: Z_i \to Y$, each composed of a finite number of local blowing ups, having the property that each Z_i contains a relatively compact subset K_i in such a way that the union of the $g_i(K_i)$ contains a neighbourhood of K and the maps $f_i = f \times_Y g_i : X \times_Y Z_i \to Z_i$ are all triangulable.

Shiota gave in [Sh] a spectacular confirmation of the view of the *Esquisse* by proving the Hauptvermutung (essential uniqueness of triangulations) for all $\mathcal{X}$-sets (see above). The statement he proves is that two $\mathcal{X}$-homeomorphic compact polyhedra in $\mathbb{R}^n$ are PL homeomorphic.

§4. Are they algebraic?

One of the deepest problems evoked by Grothendieck in the *Esquisse* is what he calls "comparison theorems" (pp. 31, 33) according to which all tame categories should be isotopically equivalent (i.e. after replacing mappings in the category by their isotopy classes). It is not clear why the inclusion of one tame category in another should be essentially surjective up to isotopy. At least in the local case, this fits well with some problems raised by Thom ([Th2], see also [Te3]). In order to see the relationship, one must realize the following: given a Whitney stratified complex algebraic or analytic set $X = \bigcup X_i$, say embedded in an open set of $\mathbb{C}^n$, and a point $x \in X_i$, for all non-singular germs of subspaces H of $(\mathbb{C}^n, x)$ which are *transversal* to X_i at x, the topological type as a stratified set of the germ at x of the intersection $H \cap X$ is the same. This means that they are all homeomorphic, and moreover the intersection $H \cap X_j$ are Whitney strata of $H \cap X$ and the homeomorphisms send strata to strata and are analytic on each stratum.

There is a similar statement in the real case, taking into account the fact that there may be several connected components of the space of transversal H's. Then Thom asks in loc. cit. questions which amount to this:

Given a complex (or real-) analytic germ $(X, 0) \subset (\mathbb{C}^n, 0)$, endow it with its minimal Whitney stratification (resp. the one induced by the minimal Whitney stratification of its complexification) $X = \bigcup X_i$. Then there exists an integer m, an *algebraic* germ $(\mathcal{Z}, 0) \subset (\mathbb{C}^{n+m}, 0)$ and an analytic germ of map $s : (\mathbb{C}^n, 0) \to (\mathbb{C}^m, 0)$ such that its graph $(H, 0)$ is transversal to the stratum of the minimal Whitney stratification of $(\mathcal{Z}, 0)$ and $(X, 0)$ is analytically isomorphic to $(\mathcal{Z} \cap H, 0)$

It means that our germ is isomorphic to an analytic section of an algebraic space transversal to its Whitney stratification; it will then suffice to approximate s by an algebraic section to produce an algebraic space germ which is isotopic to $(X, 0)$. There is a similar statement for maps. The only statement of this nature has been proved by Mostowski ([Mo]) in a different context.

For more global statements, from a different viewpoint compatible with the preoccupations of the *Esquisse*, p. 40, one may consult ([CRS]).

References

[CS] M. Coste, M. Shiota, *Thom's first isotopy lemma: a semi-algebraic version with uniform bounds, Real analytic and algebraic geometry,* Trento 1992, De Gruyter 1995.

[CRS] M. Coste, J. Ruiz, M. Shiota, *Approximation in compact Nash manifolds,* Amer. J. Math., **117**, 4, (1995), 905-927.

[vdDM]L. van den Dries and C. Miller, Geometric categories and o-minimal structures, to appear in *Duke Math. J..*

[F] J-P. Françoise and C. Pugh, Keeping track of limit cycles, *Journ. Diff. Equ.* **65**, (1986) 139-157.

[G1] A.M. Gabrielov, Projections of semi-analytic sets, *Funct. Analiz i Pril.* **2**, 4 (1968), 282-291.

[G2] A.M. Gabrielov, Complements of subanalytic sets and existential formulas for analytic functions, *Inv. Math.* **125** 1 (1996), 1-12.

[Ha] R.M. Hardt, Semi-algebraic local triviality in semi-algebraic mappings, *Amer.J. Math.* **102** (1980), 291-302.

[Hi1] H. Hironaka, Introduction to real-analytic sets and real-analytic maps, *Quaderni dei gruppi di ricerca matematica del CNR,* Inst. Leonida Tonelli, Pisa (1973).

[Hi2] H. Hironaka, Triangulation of real algebraic sets, *Proc. A.M.S. Symp. in Pure Math.,* No. 29, A.M.S. (1975), 165-185.

[Kh] A.G. Khovanskii, *Fewnomials,* Translations of Mathematical Monographs, No. 88, A.M.S, Providence 1991.

[KT] M. Kwieciński and D. Trotman, Scribbling continua in $\mathbb{R}^n$ and constructing singularities with prescribed Nash fibre and tangent cone, *Topology Appl.* **64** (1995), 177-189.

[LT] D.T. Lê et B. Teissier, Cycles évanescents, sections planes et conditions de Whitney II, in *Proceedings of A.M.S Symposia in Pure Mathematics,* Vol 40 , A.M.S.,1983, part 2, pp 65-103.

[L] S. Lojasiewicz, *Ensembles semi-analytiques,* IHES, 1965.

[Ma1] J. Mather, *Notes on Topological stability,* Harvard University, 1970.

[Ma2] J. Mather, Stratifications and mappings, in *Dynamical systems,* Academic Press, 1973.

[MR] R. Moussu, C. Roche, Théorie de Hovanskii et problème de Dulac, *Inv. Math.* **105**, 2 (1991), 431-441.

[M] T. Mostowski, Topological equivalence between algebraic and analytic sets, *Bull. Pol. Acad. Sc. Math.* **32**, 7-8 (1984), 393-400.

[S]   M. Shiota, *Geometry of subanalytic and semi-algebraic sets*, a book
       to appear.

[Te1] B. Teissier, Variétés polaires II, in *Algebraic Geometry, Proceedings,*
       La Rábida 1981, Springer LNM 961 (1982), 314-491.

[Te2] B. Teissier, Sur la triangulation des morphismes sous-analytiques,
       Publ. Math. IHES, **70** (1989), 169-198.

[Te3] B. Teissier, Travaux de Thom sur les singularités, *Publ. Math.*
       IHES, **68**, (1989), 19-25.

[Th1] R. Thom, Ensembles et morphismes stratifiés, *Bull. A.M.S.* **75**
       (1969), 240-284.

[Th2] R. Thom, Local topological properties of differentiable mappings,
       Coll. on Diff. Analysis, Bombay 1964, 191-202.

[Wh]   H. Whitney, Tangents to an analytic variety, *Ann. Math.* **81** (1964)
       496-549.

[Wi]   A.J. Wilkie, Model completeness results for extensions of the or-
       dered field of real numbers by restricted Pfaffian functions and the
       exponential function, *J. AMS* **9**, 4 (1996), 1051-1094.

SKETCH OF A PROGRAMME

by Alexandre Grothendieck

N.B. The asterisks (*) refer to the footnotes on the same page, the super-scripts numbered from (^{1}) to (^{7}) refer to the notes (added later) collected at the end of this report.

SKETCH OF A PROGRAMME

by Alexandre Grothendieck

1. As the present situation makes the prospect of teaching at the research level at the University seem more and more illusory, I have resolved to apply for admission to the CNRS, in order to devote my energy to the development of projects and perspectives for which it is becoming clear that no student (nor even, it seems, any mathematical colleague) will be found to develop them in my stead.

In the role of the document "Titles and Articles", one can find after this text the complete reproduction of a sketch, by themes, of what I considered to be my principal mathematical contributions at the time of writing that report, in 1972. It also contains a list of articles published at that date. I ceased all publication of scientific articles in 1970. In the following lines, I propose to give a view of at least some of the principal themes of my mathematical reflections since then. These reflections materialised over the years in the form of two voluminous boxes of handwritten notes, doubtless difficult to decipher for anyone but myself, and which, after several successive stages of settling, are perhaps waiting for their moment to be written up together at least in a temporary fashion, for the benefit of the mathematical community. The term "written up" is somewhat incorrect here, since in fact it is much more a question of developing the ideas and the multiple visions begun during these last twelve years, to make them more precise and deeper, with all the unexpected rebounds which constantly accompany this kind of work – a work of discovery, thus, and not of compilation of piously accumulated notes. And in writing the "Mathematical Reflections", begun since February 1983, I do intend throughout its pages to clearly reveal the process of thought, which feels and discovers, often blindly in the shadows, with sudden flashes of light when some tenacious false or simply inadequate image is finally shown for what it is, and things which seemed all crooked fall into place, with that mutual harmony which is their own.

In any case, the following sketch of some themes of reflection from the last ten or twelve years will also serve as a sketch of my programme of work for the coming years, which I intend to devote to the development of these themes, or at least some of them. It is intended on the one hand for my colleagues of the National Committee whose job it is to decide the fate of my application, and on the other hand for some other colleagues, former students, friends, in the possibility that some of the ideas sketched here might interest one of them.

2. The demands of university teaching, addressed to students (including those said to be "advanced") with a modest (and frequently less than modest) mathematical baggage, led me to a Draconian renewal of the themes of reflection I proposed to my students, and gradually to myself as well. It seemed important to me to start from an intuitive baggage common to everyone, independent of any technical language used to express it, and anterior to any such language – it turned out that the geometric and topological intuition of shapes, particularly two-dimensional shapes, formed such a common ground. This consists of themes which can be grouped under the general name of "topology of surfaces" or "geometry of surfaces", it being understood in this last expression that the main emphasis is on the topological properties of the surfaces, or the combinatorial aspects which form the most down-to-earth technical expression of them, and not on the differential, conformal, Riemannian, holomorphic aspects, and (from there) on to "complex algebraic curves". Once this last step is taken, however, algebraic geometry (my former love!) suddenly bursts forth once again, and this via the objects which we can consider as the basic building blocks for all other algebraic varieties. Whereas in my research before 1970, my attention was systematically directed towards objects of maximal generality, in order to uncover a general language adequate for the world of algebraic geometry, and I never restricted myself to algebraic curves except when strictly necessary (notably in etale cohomology), preferring to develop "pass-key" techniques and statements valid in all dimensions and in every place (I mean, over all base schemes, or even base ringed topoi...), here I was brought back, via objects so simple that a child learns them while playing, to the beginnings and origins of algebraic geometry, familiar to Riemann and his followers!

Since around 1975, it is thus the geometry of (real) surfaces, and starting in 1977 the links between questions of geometry of surfaces and the algebraic geometry of algebraic curves defined over fields such as $\mathbb{C}$, $\mathbb{R}$ or extensions of $\mathbb{Q}$ of finite type, which were my principal source of inspiration and my constant guiding thread. It is with surprise and wonderment that over the years I discovered (or rather, doubtless, rediscovered) the prodigious, truly inexhaustible richness, the unsuspected depth of this theme, apparently so anodine. I believe I feel a central sensitive point there, a privileged point of convergence of the principal currents of mathematical ideas, and also of the principal structures and visions of things which they express, from the most specific (such as the rings $\mathbb{Z}$, $\mathbb{Q}$, $\overline{\mathbb{Q}}$, $\mathbb{R}$, $\mathbb{C}$ or the group $\mathrm{Sl}(2)$ over one of these rings, or general reductive algebraic groups) to the most "abstract", such as the algebraic "multiplicities", complex analytic or real analytic. (These are naturally introduced when systematically studying "moduli varieties" for the geometric objects considered, if we want to go farther than the notoriously insufficient point of view of "coarse moduli" which comes down to most unfortunately killing the automorphism groups of these objects.) Among these modular multiplicities, it is those of Mumford-Deligne for "stable" algebraic curves of genus g with ν marked points, which I denote by $\widehat{M}_{g,\nu}$ (compactification of the "open" multiplicity $M_{g,\nu}$ corresponding to non-singular curves), which for the last two or three years have exercised a particular fascination over me, perhaps even stronger than any other mathematical object to this day. Indeed, it is more the system of all the multiplicities $M_{g,\nu}$ for variable g, ν, linked together by a certain number of fundamental operations (such as the operations of "plugging holes", i.e. "erasing" marked points, and of "glueing", and the inverse operations), which are the reflection in absolute algebraic geometry in characteristic zero (for the moment) of geometric operations familiar from the point of view of topological or conformal "surgery" of surfaces. Doubtless the principal reason of this fascination is that this very rich geometric structure on the system of "open" modular multiplicities $M_{g,\nu}$ is reflected in an analogous structure on the corresponding fundamental groupoids, the "Teichmüller groupoids" $\widehat{T}_{g,\nu}$, and that these operations on the level of the $\widehat{T}_{g,\nu}$ are sufficiently intrinsic for the Galois group $\mathbb{\Gamma}$ of $\overline{\mathbb{Q}}/\mathbb{Q}$ to act on this whole "tower" of Teichmüller groupoids, respecting all these structures. Even more extraordinary, this action is faithful – indeed, it is already faithful on the first non-trivial "level" of this tower, namely $\widehat{T}_{0,4}$ – which also means, essentially, that the outer action of $\mathbb{\Gamma}$ on the fundamental group $\hat{\pi}_{0,3}$ of the standard projective line $\mathbb{P}^1$ over $\mathbb{Q}$ with the three points 0, 1 and ∞ removed, is already faithful. Thus the Galois group $\mathbb{\Gamma}$ can be realised as an automorphism group of a very concrete profinite group, and moreover respects certain essential structures of this group. It follows that an element of $\mathbb{\Gamma}$ can be "parametrised" (in various equivalent ways) by a suitable element of this profinite group $\hat{\pi}_{0,3}$ (a free profinite group on two generators), or by a system of such elements, these elements being subject to certain simple necessary (but doubtless not sufficient) conditions for this or these elements to really correspond to an element of $\mathbb{\Gamma}$. One of the most fascinating tasks here is precisely to discover necessary and sufficient conditions on an exterior automorphism of $\hat{\pi}_{0,3}$, i.e. on the corresponding parameter(s), for it to come from an element of $\mathbb{\Gamma}$ – which would give a "purely algebraic" description, in terms of profinite groups and with no reference to the Galois theory of number fields, to the Galois group $\mathbb{\Gamma} = \mathrm{Gal}(\overline{\mathbb{Q}}/\mathbb{Q})$.

Perhaps even a conjectural characterisation of $\mathbb{\Gamma}$ as a subgroup of $\mathrm{Autext}\,\hat{\pi}_{0,3}$ is for the moment out of reach [1]; I do not yet have any conjecture to propose. On the other hand another task is immediately accessible, which is to describe the action of Γ on all of the Teichmüller tower, in terms of its action on the "first level" $\hat{\pi}_{0,3}$, i.e. to express an automorphism of this tower, in terms of the "parameter" in $\hat{\pi}_{0,3}$ which picks out the element γ running through Γ. This is linked to a representation of the Teichmüller tower (considered as a groupoid equipped with an operation of "glueing") by generators and relations, which will in particular give a presentation by generators and relations in the usual sense of each of the $\widehat{T}_{g,\nu}$ (as a profinite groupoid). Here, even for $g = 0$ (so when the corresponding Teichmüller groups are "well-known" braid groups), the generators and relations known to date which I have heard of appear to me to be unusable as they stand, because they do not present the characteristics of invariance and of symmetry indispensable for the action of Γ to be directly legible on the presentation. This is particularly linked to the fact that people still obstinately persist, when calculating with fundamental groups, in fixing a single base point, instead of cleverly choosing a whole packet of points which is invariant under the symmetries of the situation, which thus get lost on the way. In certain situations (such as descent theorems for fundamental groups à la van Kampen) it is much more elegant, even indispensable for understanding something, to work with fundamental groupoids with respect to a suitable packet of base points, and it is certainly so for the Teichmüller tower. It would seem (incredible, but true!) that even the geometry of the first level of the Teichmüller tower (corresponding thus to "moduli" either for projective lines with four marked points, or to elliptic curves (!)) has never been explicitly described, for example the relation between the genus 0 case and the geometry of the octahedron, and that of the tetrahedron. A fortiori the modular multiplicities $M_{0,5}$ (for projective lines with five marked points) and $M_{1,2}$ (for curves of genus 1 with two marked points), which actually are practically isomorphic, appear to be virgin territory – braid groups will not enlighten us on their score! I have begun to look at $M_{0,5}$ at stray moments; it is a real jewel, with a very rich geometry closely related to the geometry of the icosahedron.

$\frac{5}{6}$

The a priori interest of a complete knowledge of the two first levels of the tower (i.e., the cases where the modular dimension $N = 3g - 3 + \nu$ is ≤ 2) is to be found in the principle that <u>the entire tower can be reconstituted from these two first levels</u>, in the sense that via the fundamental operation of "glueing", level 1 gives a complete system of generators, and level 2 a complete system of relations. There is a striking analogy, and I am certain it is not merely formal, between this principle and the analogous principle of Demazure for the structure of reductive algebraic groups, if we replace the term "level" or "modular dimension" with "semi-simple rank of the reductive group". The link becomes even more striking, if we recall that the Teichmüller group $T_{1,1}$ (in the discrete, transcendental context now, and not in the profinite algebraic context, where we find the profinite completions of the former) is no other than $Sl(2,\mathbb{Z})$, i.e. the group of integral points of the simple group scheme of "absolute" rank 1 $Sl(2)_{\mathbb{Z}}$. Thus, the fundamental building block for the Teichmüller tower is essentially the same as for the "tower" of reductive groups of all ranks – a group of which, moreover, we may say that it is doubtless present in all the essential disciplines of mathematics.

This principle of construction of the Teichmüller tower is not proved at this time – but I have no doubt that it is valid. It would be a consequence (via a theory of dévissage of stratified structures – here the $\widehat{M}_{g,\nu}$ – which remains to be written, cf. par. 5) of an extremely plausible property of the open modular multiplicities $M_{g,\nu}$ in the complex analytic context, namely that for modular dimension $N \geq 3$, the fundamental group of $M_{g,\nu}$ (i.e. the usual Teichmüller group $T_{g,\nu}$) is isomorphic to the "fundamental group at infinity", i.e. that of a "tubular neighbourhood of infinity". This is a very familiar thing (essentially due to Lefschetz) for a non-singular affine variety of dimension $N \geq 3$. True, the modular multiplicities are not affine (except for small values of g), but it would suffice if such an $M_{g,\nu}$ of dimension N (or rather, a suitable finite covering) were a union of $N-2$ affine open sets, making $M_{g,\nu}$ "not too near a compact variety".

Having no doubt about this principle of construction of the Teichmüller tower, I prefer to leave to the experts, better equipped than I am, the task of proving the necessary (if it so happens that any are interested), to rather study, with all the care it deserves, the structure which ensues for the Teichmüller tower by generators and relations, this time in the discrete, not the profinite framework – which essentially comes down to a complete understanding of the four modular multiplicities $M_{0,4}$, $M_{1,1}$, $M_{0,5}$, $M_{1,2}$ and their fundamental groupoids based at suitably chosen "base points". These offer themselves quite naturally, as the complex algebraic curves of the type (g,n) under consideration, having automorphism group (necessarily finite) larger than in the generic case (*). Including the holomorphic

---

(*) It is also necessary to add the "base points" coming from operations of glueing of "blocks" of the same type in smaller modular dimension. On the other hand, in modular dimension 2 (the cases of $M_{0,5}$ and $M_{1,2}$), it is advisable to exclude the points of certain one-parameter families of curves admitting an exceptional automorphism of order 2. These families actually constitute remarkable rational curves on the multiplicities considered, which appear to me to be an important ingredient in the structure of these multiplicities.

sphere with three marked points (coming from $M_{0,3}$, i.e. from level 0), we find twelve fundamental "building blocks" (6 of genus 0, 6 of genus 1) in a "game of Lego-Teichmüller" (large box), where the points marked on the surfaces considered are replaced by "holes" with boundary, so as to have surfaces with boundary, functioning as building blocks which can be assembled by gentle rubbing as in the ordinary game of Lego dear to our children (or grandchildren...). By assembling them we find an entirely visual way to construct every type of surface (it is essentially these constructions which will be the "base points" for our famous tower), and also to visualise the elementary "paths" by operations as concrete as "twists", or automorphisms of blocks in the game, and to write the fundamental relations between composed paths. According to the size (and the price!) of the construction box used, we can even find numerous different descriptions of the Teichmüller tower by generators and relations. The smallest box is reduced to identical blocks, of type $(0,3)$ – these are the Thurston "pants", and the game of Lego-Teichmüller which I am trying to describe, springing from motivations and reflections of absolute algebraic geometry over the field $\mathbb{Q}$, is very close to the game of "hyperbolic geodesic surgery" of Thurston, whose existence I learned of last year from Yves Ladegaillerie. In a microseminar with Carlos Contou-Carrère and Yves Ladegaillerie, we began a reflection one of whose objects is to confront the two points of view, which are mutually complementary.

I add that each of the twelve building blocks of the "large box" is equipped with a canonical cellular decomposition, stable under all symmetries, having as its only vertices the "marked points" (or centres of the holes), and as edges certain geodesic paths (for the canonical Riemannian structure on the sphere or the torus considered) between certain pairs of vertices (namely those which lie on the same "real locus", for a suitable real structure of the complex algebraic curve considered). Consequently, all the surfaces obtained in this game by assembling are equipped with canonical cellular

structures, which in their turn (cf. §3 below) enable us to consider these surfaces as associated to complex algebraic curves (and even over $\overline{\mathbb{Q}}$) which are canonically determined. There is here a typical game of intertwining of the combinatorial and the complex algebraic (or rather, the algebraic over $\overline{\mathbb{Q}}$).

The "small box" with identical blocks, which has the charm of economy, will doubtless give rise to a relatively complex description for the relations (complex, but not at all inextricable!). The large box will give rise to more numerous relations (because there are many more base points and remarkable paths between them), but with a more transparent structure. I foresee that in modular dimension 2, just as in the more or less familiar case of modular dimension 1 (in particular with the description of Sl(2, $\mathbb{Z}$) by ($\rho, \sigma \mid \rho^3 = \sigma^2, \rho^4 = \sigma^6 = 1$)), we will find a generation by the automorphism groups of the three types of relevant blocks, with simple relations which I have not clarified as I write these lines. Perhaps we will even find a principle of this type for all the $T_{g,\nu}$, as well as a cellular decomposition of $\widehat{M}_{g,\nu}$ generalising those which present themselves spontaneously for $\widehat{M}_{0,4}$ and $\widehat{M}_{1,1}$, and which I already perceive for modular dimension 2, using the hypersurfaces corresponding to the various <u>real structures</u> on the complex structures considered, to effect the desired cellular decomposition.

3. Instead of following (as I meant to) a rigorous thematic order, I let myself be carried away by my predilection for a particularly rich and burning theme, to which I intend to devote myself prioritarily for some time, starting at the beginning of the academic year 84/85. Thus I will take the thematic description up again where I left it, at the very beginning of the preceding paragraph.

My interest in topological surfaces began to appear in 1974, when I proposed to Yves Ladegaillerie the theme of the isotopic study of embeddings of a topological 1-complex into a compact surface. Over the two following years, this study led him to a remarkable isotopy theorem, giving a complete algebraic description of the isotopy classes of embeddings of such 1-complexes, or compact surfaces with boundary, in a compact oriented surface, in terms of certain very simple combinatorial invariants, and the fundamental groups of the protagonists. This theorem, which should be easily generalisable to embeddings of any compact space (triangulable to simplify) in a compact oriented surface, gives as easy corollaries several deep classical results in the theory of surfaces, and in particular Baer's isotopy theorem. It will finally be published, separately from the rest (and ten years later, seeing the difficulty of the times...), in Topology. In the work of Ladegaillerie there is also a purely algebraic description, in terms of fundamental groups, of the "isotopic" category of compact surfaces X, equipped with a topological 1-complex K embedded in X. This description, which had the misfortune to run counter to "today's taste" and because of this appears to be unpublishable, nevertheless served (and still serves) as a precious guide in my later reflections, particularly in the context of absolute algebraic geometry in characteristic zero.

The case where (X, K) is a 2-dimensional "map", i.e. where the connected components of $X\backslash K$ are open 2-cells (and where moreover K is equipped with a finite set S of "vertices", such that the connected components of $K\backslash S$ are open 1-cells) progressively attracted my attention over the following years. The isotopic category of these maps admits a particularly simple algebraic description, via the set of "markers" (or "flags", or "biarcs") associated to the map, which is naturally equipped with the structure of a set with a group of operators, under the group

$$\underline{C}_2 =< \sigma_0, \sigma_1, \sigma_2 \mid \sigma_0^2 = \sigma_1^2 = \sigma_2^2 = (\sigma_0\sigma_2)^2 = 1 >,$$

which I call the (non-oriented) cartographic group of dimension 2. It admits as a subgroup of index 2 the oriented cartographic group, generated by the products of an even number of generators, which can also be described by

$$\underline{C}_2^+ =< \rho_s, \rho_f, \sigma \mid \rho_s\rho_f = \sigma, \ \sigma^2 = 1 >,$$

(with

$$\rho_s = \sigma_2\sigma_1, \quad \rho_f = \sigma_1\sigma_0, \quad \sigma = \sigma_0\sigma_2 = \sigma_2\sigma_0,$$

$\frac{10}{11}$ operations of elementary rotation of a flag around a vertex, a face and an edge respectively). There is a perfect dictionary between the topological situation of compact maps, resp. oriented compact maps, on the one hand, and finite sets with group of operators $\underline{C}_2$ resp. $\underline{C}_2^+$ on the other, a dictionary whose existence was actually more or less known, but never stated with the necessary precision, nor developed at all. This foundational work was done with the care it deserved in an excellent DEA thesis, written jointly by Jean Malgoire and Christine Voisin in 1976.

This reflection suddenly takes on a new dimension, with the simple remark that the group $\underline{C}_2^+$ can be interpreted as a quotient of the fundamental group of an oriented sphere with three points, numbered 0, 1 and 2, removed; the operations ρ_s, σ, ρ_f are interpreted as loops around these points, satisfying the familiar relation

$$\ell_0\ell_1\ell_2 = 1,$$

while the additional relation $\sigma^2 = 1$, i.e. $\ell_1^2 = 1$ means that we are interested in the quotient of the fundamental group corresponding to an imposed ramification index of 2 over the point 1, which thus classifies the coverings of the sphere ramified at most over the points 0, 1 and 2 with ramification equal to 1 or 2 at the points over 1. Thus, the compact oriented maps form an isotopic category equivalent to that of these coverings, subject to the additional condition of being finite coverings. Now taking the Riemann sphere, or the projective complex line, as reference sphere, rigidified by the three points 0, 1 and ∞ (this last thus replacing 2), and recalling that every finite ramified covering of a complex algebraic curve itself inherits the structure of a complex algebraic curve, we arrive at this fact, which eight years later still appears to me as extraordinary: every "finite" oriented map is canonically realised on a complex algebraic

$\frac{11}{12}$ curve! Even better, as the complex projective line is defined over the absolute base field $\mathbb{Q}$, as are the admitted points of ramification, the algebraic curves we obtain are defined not only over $\mathbb{C}$, but over the algebraic closure $\overline{\mathbb{Q}}$ of $\mathbb{Q}$ in $\mathbb{C}$. As for the map we started with, it can be found on the algebraic curve, as the inverse image of the real segment $[0, 1]$ (where 0 is considered as a vertex, and 1 as the middle of a "folded edge" of centre 1), which itself is the "universal oriented 2-map" on the Riemann sphere (*). The points of the algebraic curve X over 0, 1 and ∞ are neither more nor less than the vertices, the "centres" of the edges and those of the faces of the map (X, K), and the orders of the vertices and the faces are exactly the multiplicities of the zeros and the poles of the rational function (defined over $\mathbb{Q}$) on X, which expresses its structural projection to $\mathbb{P}^1_{\mathbb{C}}$.

This discovery, which is technically so simple, made a very strong impression on me, and it represents a decisive turning point in the course of my reflections, a shift in particular of my centre of interest in mathematics, which suddenly found itself strongly focused. I do not believe that a mathematical fact has ever struck me quite so strongly as this one, nor had a comparable psychological impact ([2]). This is surely because of the very familiar, non-technical nature of the objects considered, of which any child's drawing scrawled on a bit of paper (at least if the drawing is made without lifting the pencil) gives a perfectly explicit example. To such a dessin, we find associated subtle arithmetic invariants, which are completely turned topsy-turvy as soon as we add one more stroke. Since these are spherical
$\frac{12}{13}$ maps, giving rise to curves of genus 0 (which thus do not lead to "moduli"), we can say that the curve in question is "pinned down" if we fix three of its points, for instance three vertices of the map, or more generally three centres of facets (vertices, edges or faces) – and then the structural map $f : X \rightarrow \mathbb{P}^1_{\mathbb{C}}$ can be interpreted as a well-determined rational function

$$f(z) = P(z)/Q(z) \in \mathbb{C}(z),$$

quotient of two well-determined relatively prime polynomials, with Q unitary, satisfying algebraic conditions which in particular reflect the fact that f is unramified outside of 0, 1 and ∞, and which imply that the coefficients

---

(*) There is an analogous description of finite non-oriented maps, possibly with boundary, in terms of <u>real</u> algebraic curves, more precisely of coverings of $\mathbb{P}^1_{\mathbb{R}}$ ramified only over 0, 1, ∞, the surface with boundary associated to such a covering being $X(\mathbb{C})/\tau$, where τ is complex conjugation. The "universal" non-oriented map is here the disk, or upper hemisphere of the Riemann sphere, equipped as before with the embedded 1-complex $K = [0, 1]$.

of these polynomials are <u>algebraic numbers</u>; thus their zeros are algebraic numbers, which represent respectively the vertices and the centres of the faces of the map under consideration.

Returning to the general case, since finite maps can be interpreted as coverings over $\overline{\mathbb{Q}}$ of an algebraic curve defined over the prime field $\mathbb{Q}$ itself, it follows that the Galois group Γ of $\overline{\mathbb{Q}}$ over $\mathbb{Q}$ acts on the category of these maps in a natural way. For instance, the operation of an automorphism $\gamma \in \Gamma$ on a spherical map given by the rational function above is obtained by applying γ to the coefficients of the polynomials P, Q. Here, then, is that mysterious group Γ intervening as a transforming agent on topologico-combinatorial forms of the most elementary possible nature, leading us to ask questions like: are such and such oriented maps "conjugate" or: exactly which are the conjugates of a given oriented map? (Visibly, there is only a finite number of these).

I considered some concrete cases (for coverings of low degree) by various methods, J. Malgoire considered some others – I doubt that there is a uniform method for solving the problem by computer. My reflection quickly took a more conceptual path, attempting to apprehend the nature of this action of Γ. One sees immediately that roughly speaking, this action is expressed by a certain "outer" action of Γ on the profinite compactification of the oriented cartographic group $\underline{C}_2^+$, and this action in its turn is deduced by passage to the quotient of the canonical outer action of Γ on the profinite fundamental group $\hat{\pi}_{0,3}$ of $(U_{0,3})_{\overline{\mathbb{Q}}}$, where $U_{0,3}$ denotes the typical curve of genus 0 over the prime field $\mathbb{Q}$, with three points removed. This is how my attention was drawn to what I have since termed "<u>anabelian algebraic geometry</u>", whose starting point was exactly a study (limited for the moment to characteristic zero) of the action of "absolute" Galois groups (particularly the groups $\mathrm{Gal}(\overline{K}/K)$, where K is an extension of finite type of the prime field) on (profinite) geometric fundamental groups of algebraic varieties (defined over K), and more particularly (breaking with a well-established tradition) fundamental groups which are very far from abelian groups (and which for this reason I call "<u>anabelian</u>"). Among these groups, and very close to the group $\hat{\pi}_{0,3}$, there is the profinite compactification of the modular group $\mathrm{Sl}(2, \mathbb{Z})$, whose quotient by its centre ± 1 contains the former as congruence subgroup mod 2, and can also be interpreted as an oriented "cartographic" group, namely the one classifying <u>triangulated</u> oriented maps (i.e. those whose faces are all triangles or monogons).

Every finite oriented map gives rise to a projective non-singular algebraic curve defined over $\overline{\mathbb{Q}}$, and one immediately asks the question: which are the algebraic curves over $\overline{\mathbb{Q}}$ obtained in this way – do we obtain them all,

13
14

who knows? In more erudite terms, could it be true that every projective
non-singular algebraic curve defined over a number field occurs as a possible "modular curve" parametrising elliptic curves equipped with a suitable
rigidification? Such a supposition seemed so crazy that I was almost em-
barrassed to submit it to the competent people in the domain. Deligne
when I consulted him found it crazy indeed, but didn't have any counterex-
ample up his sleeve. Less than a year later, at the International Congress
in Helsinki, the Soviet mathematician Bielyi announced exactly that result,
with a proof of disconcerting simplicity which fit into two little pages of a
letter of Deligne – never, without a doubt, was such a deep and disconcerting
result proved in so few lines!

In the form in which Bielyi states it, his result essentially says that
every algebraic curve defined over a number field can be obtained as a
covering of the projective line ramified only over the points 0, 1 and ∞.
This result seems to have remained more or less unobserved. Yet it appears
to me to have considerable importance. To me, its essential message is
that there is a profound identity between the combinatorics of finite maps
on the one hand, and the geometry of algebraic curves defined over number fields on the other. This deep result, together with the algebraic-
geometric interpretation of maps, opens the door onto a new, unexplored
world – within reach of all, who pass by without seeing it.

It was only close to three years later, seeing that decidedly the vast horizons opening here caused nothing to quiver in any of my students, nor even
in any of the three or four high-flying colleagues to whom I had occasion
to talk about it in a detailed way, that I made a first scouting voyage into
this "new world", from January to June 1981. This first foray materialised into a packet of some 1300 handwritten pages, baptised "The Long
March through Galois theory". It is first and foremost an attempt at understanding the relations between "arithmetic" Galois groups and profinite
"geometric" fundamental groups. Quite quickly it became oriented towards
a work of computational formulation of the action of $\mathrm{Gal}(\overline{\mathbb{Q}}/\mathbb{Q})$ on $\widehat{\pi}_{0,3}$,
and at a later stage, on the somewhat larger group $\widehat{\mathrm{Sl}(2,\mathbb{Z})}$, which gives
rise to a more elegant and efficient formalism. Also during the course of
this work (but developed in a different set of notes) appeared the central
theme of anabelian algebraic geometry, which is to reconstitute certain so-
called "anabelian" varieties X over an absolute field K from their mixed
fundamental group, the extension of $\mathrm{Gal}(\overline{K}/K)$ by $\pi_1(X_{\overline{K}})$; this is when I
discovered the "fundamental conjecture of anabelian algebraic geometry",
close to the conjectures of Mordell and Tate recently proved by Faltings
([3]). This period also saw the appearance of the first reflection on the Teichmüller groups, and the first intuitions on the many-faceted structure of the

"Teichmüller tower" – the open modular multiplicities $M_{g,\nu}$ also appearing as the first important examples in dimension > 1, of varieties (or rather, multiplicities) seeming to deserve the appellation of "anabelian". Towards the end of this period of reflection, it appeared to me as a fundamental reflection on a theory still completely up in the air, for which the name "Galois-Teichmüller theory" seems to me more appropriate than the name "Galois Theory" which I had at first given to my notes. Here is not the place to give a more detailed description of this set of questions, intuitions, ideas – which even includes some tangible results. The most important thing seems to me to be the one pointed out in par. 2, namely the faithfulness of the outer action of $\Gamma = \mathrm{Gal}(\overline{\mathbb{Q}}/\mathbb{Q})$ (and of its open subgroups) on $\hat{\pi}_{0,3}$, and more generally (if I remember rightly) on the fundamental group of any "anabelian" algebraic curve (i.e. whose genus g and "number of holes" ν satisfy the equality $2g + \nu \geq 3$, i.e. such that $\chi(X) < 0$) defined over a finite extension of $\mathbb{Q}$. This result can be considered to be essentially equivalent to Bielyi's theorem – it is the first concrete manifestation, via a precise mathematical statement, of the "message" which was discussed above.

I would like to conclude this rapid outline with a few words of commentary on the truly unimaginable richness of a typical anabelian group such as $\mathrm{Sl}(2, \mathbb{Z})$ – doubtless the most remarkable discrete infinite group ever encountered, which appears in a multiplicity of avatars (of which certain have been briefly touched on in the present report), and which from the point of view of Galois-Teichmüller theory can be considered as the fundamental "building block" of the "Teichmüller tower". The element of the structure of $\mathrm{Sl}(2, \mathbb{Z})$ which fascinates me above all is of course the outer action of Γ on its profinite compactification. By Bielyi's theorem, taking the profinite compactifications of subgroups of finite index of $\mathrm{Sl}(2, \mathbb{Z})$, and the induced outer action (up to also passing to an open subgroup of Γ), we essentially find the fundamental groups of all algebraic curves (not necessarily compact) defined over number fields K, and the outer action of $\mathrm{Gal}(\overline{K}/K)$ on them – at least it is true that every such fundamental group appears as a quotient of one of the first groups (*). Taking the "anabelian yoga" (which remains conjectural) into account, which says that an anabelian algebraic curve over a number field K (finite extension of $\mathbb{Q}$) is known up to isomorphism when we know its mixed fundamental group (or what comes to the same thing, the outer action of $\mathrm{Gal}(\overline{K}/K)$ on its profinite geometric fundamental group), we can thus say that all algebraic curves defined over number fields are "contained" in the profinite compactification

(*) In fact, we are considering quotients of a particularly trivial nature, by abelian subgroups which are products of "Tate modules" $\hat{\mathbb{Z}}(1)$, corresponding to "loop-groups" around points at infinity.

$\widehat{\text{Sl}(2,\mathbb{Z})}$, and in the knowledge of a certain subgroup $\mathbb{T}$ of its group of outer automorphisms! Passing to the abelianisations of the preceding fundamental groups, we see in particular that all the abelian ℓ-adic representations dear to Tate and his circle, defined by Jacobians and generalised Jacobians of algebraic curves defined over number fields, are contained in this single action of $\mathbb{T}$ on the anabelian profinite group $\widehat{\text{Sl}(2,\mathbb{Z})}$! ([4])

17
18

There are people who, faced with this, are content to shrug their shoulders with a disillusioned air and to bet that all this will give rise to nothing, except dreams. They forget, or ignore, that our science, and every science, would amount to little if since its very origins it were not nourished with the dreams and visions of those who devoted themselves to it.

4. From the very start of my reflection on 2-dimensional maps, I was most particularly interested by the "regular" maps, those whose automorphism group acts transitively (and consequently, simply transitively) on the set of flags. In the oriented case and in terms of the algebraic-geometric interpretation given in the preceding paragraph, it is these maps which correspond to Galois coverings of the projective line. Very quickly also, and even before the appearance of the link with algebraic geometry, it appears necessary not to exclude the infinite maps, which in particular occur in a natural way as universal coverings of finite maps. It appears (as an immediate consequence of the "dictionary" of maps, extended to the case of maps which are not necessarily finite) that for every pair of natural integers p, $q \geq 1$, there exists up to non-unique isomorphism one and only one 1-connected map of type (p,q), i.e. all of whose vertices are of order p and whose faces are of order q, and this map is a regular map. It is pinned down by the choice of a flag, and its automorphism group is then canonically isomorphic to the quotient of the cartographic group (resp. of the oriented cartographic group, in the oriented case) by the additional relations

$$\rho_s^p = \rho_f^q = 1.$$

18
9

The case where this group is finite is the "Pythagorean" case of regular spherical maps, the case where it is infinite gives the regular tilings of the Euclidean plane or of the hyperbolic plane (*). The link between combinatorial theory and the "conformal" theory of regular tilings of the hyperbolic plane was foreshadowed, before the appearance of the link between finite maps and finite coverings of the projective line. Once this link is understood,

---

(*) In these statements, we must not exclude the case where p, q can take the value $+\infty$, which is encountered in particular in a very natural way as tilings associated to certain regular infinite polyhedra, cf. below.

it becomes obvious that it should also extend to infinite maps (regular or
not): <u>every map, finite or not, can be canonically realised on a conformal
surface</u> (compact if and only if the map is finite), <u>as a ramified covering of
the complex projective line, ramified only over the points</u> 0, 1 and ∞. The
only difficulty here was to develop the dictionary between topological maps
and sets with operators, which gave rise to some conceptual problems in
the infinite case, starting with the very notion of a "topological map". It
appears necessary in particular, both for reasons of internal coherence of the
dictionary and not to let certain interesting cases of infinite maps escape,
to avoid excluding vertices and faces of infinite order. This foundational
work was also done by J. Malgoire and C. Voisin, in the wake of their first
work on finite maps, and their theory indeed gives everything that we could
rightly expect (and even more...)

In 1977 and 1978, in parallel with two C4 courses on the geometry of the
cube and that of the icosahedron, I began to become interested in regular
polyhedra, which then appeared to me as particularly concrete "geometric
realizations" of combinatorial maps, the vertices, edges and faces being
realised as points, lines and planes respectively in a suitable 3-dimensional
affine space, and respecting incidence relations. This notion of a geometric
realisation of a combinatorial map keeps its meaning over an arbitrary
base field, and even over an arbitrary base ring. It also keeps its meaning
for regular polyhedra in any dimension, if the cartographic group $\underline{C_2}$ is
replaced by a suitable n-dimensional analogue $\underline{C_n}$. The case $n = 1$, i.e.
the theory of regular polygons in any characteristic, was the subject of a
DEA course in 1977/78, and already sparks the appearance of some new
phenomena, as well as demonstrating the usefulness of working not in an
ambient affine space (here the affine plane), but in a <u>projective</u> space. This
is in particular due to the fact that in certain characteristics (in particular
in characteristic 2) the centre of a regular polyhedron is sent off to infinity.
Moreover, the projective context, contrarily to the affine context, enables us
to easily develop a duality formalism for regular polyhedra, corresponding
to the duality formalism of combinatorial or topological maps (where the
roles of the vertices and the faces, in the case $n = 2$ say, are exchanged). We
find that for every projective regular polyhedron, we can define a canonical
associated hyperplane, which plays the role of a canonical hyperplane at
infinity, and allows us to consider the given polyhedron as an affine regular
polyhedron.

The extension of the theory of regular polyhedra (and more generally,
of all sorts of geometrico-combinatorial configurations, including root systems...) of the base field $\mathbb{R}$ or $\mathbb{C}$ to a general base ring, seems to me to
have an importance comparable, in this part of geometry, to the analogous

extension which has taken place since the beginning of the century in algebraic geometry, or over the last twenty years in topology (*), with the introduction of the language of schemes and of topoi.   My sporadic reflection on this question, over some years, was limited to discovering some simple basic principles, concentrating my attention first and foremost on the case of <u>pinned</u> regular polyhedra, which reduces to a minimum the necessary conceptual baggage, and practically eliminates the rather delicate questions of rationality. For such a polyhedron, we find a canonical basis (or flag) of the ambient affine or projective space, such that the operations of the cartographic group $\underline{C}_n$, generated by the fundamental reflections σ_i ($0 \leq i \leq n$), are written in that basis by universal formulae, in terms of the n parameters $\alpha_1, \ldots, \alpha_n$, which can be geometrically interpreted as the doubles of the cosines of the "fundamental angles" of the polyhedron. The polyhedron can be reconstituted from this action, and from the affine or projective flag associated to the chosen basis, by transforming this flag by all the elements of the group generated by the fundamental reflections. Thus the "universal" pinned n-polyhedron is canonically defined over the ring of polynomials with n indeterminates

$$\mathbb{Z}[\underline{\alpha_1}, \ldots, \underline{\alpha_n}],$$

its specialisations to arbitrary base fields k (via values $\alpha_i \in k$ given to the indeterminates $\underline{\alpha_i}$) giving regular polyhedra corresponding to various combinatorial types. In this game, there is no question of limiting oneself to finite regular polyhedra, nor even to regular polyhedra whose facets are of finite order, i.e. for which the parameters α_i are roots of suitable "semicyclotomic" equations, expressing the fact that the "fundamental angles" (in the case where the base field is $\mathbb{R}$) are commensurable with 2π. Already when $n = 1$, perhaps the most interesting regular polygon (morally the regular polygon with only one side!) is the one corresponding to $\alpha = 2$, giving rise to a parabolic circumscribed conic, i.e. tangent to the line at infinity. The finite case is the one where the group generated by the fundamental reflections, which is also the automorphism group of the regular polyhedron considered, is finite. In the case where the base field is $\mathbb{R}$ (or $\mathbb{C}$, which comes to the same thing), and for $n = 2$, the finite cases have been well-known since antiquity – which does not exclude that the schematic point of view unveils new charms; we can however say that when specialising the icosahedron (for example) to finite base fields of arbitrary characteristic, it remains

---

(*) In writing this, I am aware that rare are the topologists, even today, who realise the existence of this conceptual and technical generalisation of topology, and the resources it offers.

an icosahedron, with its own personal combinatorics and the same simple
group of automorphisms of order 60. The same remark applies to finite regular polyhedra in higher dimension, which were systematically studied in
two beautiful books by Coxeter. The situation is entirely different if we start
from an infinite regular polyhedron, over a field such as $\mathbb{Q}$, for instance, and
"specialise" it to the prime fields $\mathbb{F}_p$ (a well-defined operation for all p except
a finite number of primes). It is clear that every regular polyhedron over a
finite field is finite – we thus find an infinity of finite regular polyhedra as
p varies, whose combinatorial type, or equivalently, whose automorphism
group varies "arithmetically" with p. This situation is particularly intriguing in the case where $n = 2$, where we can use the relation made explicit
in the preceding paragraph between combinatorial 2-maps and algebraic
curves defined over number fields. In this case, an infinite regular polyhedron defined over any infinite field (and therefore, over a sub-$\mathbb{Z}$-algebra of it
with two generators) thus gives rise to an infinity of algebraic curves defined
over number fields, which are Galois coverings ramified only over 0, 1 and
∞ of the standard projective line. The optimal case is of course the one
deduced by passage to the field of fractions $\mathbb{Q}(\alpha_1, \alpha_2)$ of its base ring. This
raises a host of new questions, both vague and precise, none of which I have
up till now had leisure to examine closely – I will cite only this one: exactly
which are the finite regular 2-maps, or equivalently, the finite quotients of
the 2-cartographic group, which come from regular 2-polyhedra over finite
fields (*)? Do we obtain them all, and if yes: how?

These reflections shed a special light on the fact, which to me was completely unexpected, that the theory of finite regular polyhedra, already in
the case of dimension $n = 2$, is infinitely richer, and in particular gives
infinitely many more different combinatorial forms, in the case where we
admit base fields of non-zero characteristic, than in the case considered up
to now, where the base fields were always restricted to $\mathbb{R}$, or at best $\mathbb{C}$ (in
the case of what Coxeter calls "complex regular polyhedra", and which I
prefer to call "regular pseudo-polyhedra defined over $\mathbb{C}$") (**). Moreover,
it seems that this extension of the point of view should also shed new light

(*) These are actually the same as those coming from regular polyhedra defined over arbitrary fields, or algebraically closed fields, as can be seen using
standard specialisation arguments.

(**) The pinned pseudo-polyhedra are described in the same way as the pinned
polyhedra, with the only difference that the fundamental reflections σ_i ($0 \leq
i \leq n$) are here replaced by pseudo-reflections (which Coxeter assumes of
finite order, since he restricts himself to finite combinatorial structures).
This simply leads to the introduction for each of the σ_i of an additional
numerical invariant β_i, such that the universal n-pseudo-polyhedron can

on the already known cases. Thus, examining the Pythagorean polyhedra
one after the other, I saw that the same small miracle was repeated each
time, which I called the combinatorial paradigm of the polyhedra under
consideration. Roughly speaking, it can be described by saying that when
we consider the specialisation of the polyhedra in the or one of the most
singular characteristic(s) (namely characteristics 2 and 5 for the icosahedron, characteristic 2 for the octahedron), we read off from the geometric
regular polyhedron over the finite field ($\mathbb{F}_4$ and $\mathbb{F}_5$ for the icosahedron, $\mathbb{F}_2$
for the octahedron) a particularly elegant (and unexpected) description of
the combinatorics of the polyhedron. It seems to me that I perceived there
a principle of great generality, which I believed I found again for example in
a later reflection on the combinatorics of the system of 27 lines on a cubic
surface, and its relations with the root system E_7. Whether it happens that
such a principle really exists, and even that we succeed in uncovering it from
its cloak of fog, or that it recedes as we pursue it and ends up vanishing like
a Fata Morgana, I find in it for my part a force of motivation, a rare fascination, perhaps similar to that of dreams. No doubt that following such an
unformulated call, the unformulated seeking form, from an elusive glimpse
which seems to take pleasure in simultaneously hiding and revealing itself
– can only lead far, although no one could predict where...

However, occupied by other interests and tasks, I have not up to now
followed this call, nor met any other person willing to hear it, much less to
follow it. Apart from some digressions towards other types of geometrico-
combinatorial structures, my work on the question has been limited to a
first effort of refining and housekeeping, which it is useless for me to describe further here ([5]). The only point which perhaps still deserves to be
mentioned is the existence and uniqueness of a hyperquadric circumscrib-
ing a given regular n-polyhedron, whose equation can be given explicitly
by simple formulae in terms of the fundamental parameters α_i (*). The
case which interests me most is when $n = 2$, and the moment seems ripe
to rewrite a new version, in modern style, of Klein's classic book on the
icosahedron and the other Pythagorean polyhedra. Writing such an exposé
on regular 2-polyhedra would be a magnificent opportunity for a young researcher to familiarise himself with the geometry of polyhedra as well as

---

also be defined over a ring of polynomials with integral coefficients, in the
$n + (n + 1)$ variables $\underline{\alpha_i}$ $(1 \le i \le n)$ and $\underline{\beta_j}$ $(0 \le i \le n)$.

(*) An analogous result is valid for pseudo-polyhedra. It would seem that
the "exceptional characteristics" we discussed above, for specialisations of
a given polyhedron, are those for which the circumscribed hyperquadric is
either degenerate or tangent to the hyperplane at infinity.

their connections with spherical, Euclidean and hyperbolic geometry and
with algebraic curves, and with the language and the basic techniques of
modern algebraic geometry. Will there be found one, some day, who will
seize this opportunity?

5. I would like to say a few words now about some topological consider-
ations which have made me understand the necessity of new foundations
for "geometric" topology, in a direction quite different from the notion of
topos, and actually independent of the needs of so-called "abstract" algebraic geometry (over general base fields and rings). The problem I started
from, which already began to intrigue me some fifteen years ago, was that of
defining a theory of "dévissage" for stratified structures, in order to rebuild
them, via a canonical process, out of "building blocks" canonically deduced
from the original structure. Probably the main example which had led me
to that question was that of the canonical stratification of a singular algebraic variety (or a complex or real singular space) through the decreasing
sequence of its successive singular loci. But I probably had the premonition
of the ubiquity of stratified structures in practically all domains of geometry
(which surely others had seen clearly a long time before). Since then, I have
seen such structures appear, in particular, in any situation where "moduli"
are involved for geometric objects which may undergo not only continuous
variations, but also "degeneration" (or "specialisation") phenomena – the
strata corresponding then to the various "levels of singularity" (or to the
associated combinatorial types) for the objects in question.   The compactified modular multiplicities $\widehat{M}_{g,\nu}$ of Mumford-Deligne for the stable
algebraic curves of type (g, ν) provide a typical and particularly inspiring
example, which played an important motivating role when I returned to
my reflection about stratified structures, from December 1981 to January
1982. Two-dimensional geometry provides many other examples of such
modular stratified structures, which all (if not using rigidification) appear
as "multiplicities" rather than as spaces or manifolds in the usual sense
(as the points of these multiplicities may have non-trivial automorphism
groups). Among the objects of two-dimensional geometry which give rise to
such modular stratified structures in arbitrary dimensions, or even infinite
dimension, I would list polygons (Euclidean, spherical or hyperbolic), systems of straight lines in a plane (say projective), systems of "pseudo straight
lines" in a projective topological plane, or more general immersed curves
with normal crossings, in a given (say compact) surface.

The simplest non-trivial example of a stratified structure is obtained by
considering a pair (X, Y) of a space X and a closed subspace Y, with a
suitable assumption of equisingularity of X along Y, and assuming moreover

(to fix ideas) that both strata Y and $X \setminus Y$ are topological __manifolds__. The naive idea, in such a situation, is to consider "the" tubular neighbourhood T of Y in X, whose boundary ∂T should also be a smooth manifold, fibred with compact smooth fibres over Y, whereas T itself can be identified with the conical fibration associated to the above one. Setting

$$U = X \setminus \mathrm{Int}(T),$$

one finds a manifold with boundary, whose boundary is canonically isomorphic to the boundary of T. This being said, the "building blocks" are the manifold with boundary U (compact if X is compact, and which replaces and refines the "open" stratum $X \setminus Y$) and the manifold (without boundary) Y, together with, as an additional structure which connects them, the "glueing" map

$$f : \partial U \longrightarrow Y$$

which is a proper and smooth fibration. The original situation (X, Y) can be recovered from $(U, Y, f : \partial U \to Y)$ via the formula

$$X \cong U \coprod_{\partial U} Y$$

(amalgamated sum over ∂U, mapping into U and Y by inclusion resp. the glueing map).

This naive vision immediately encounters various difficulties. The first is the somewhat vague nature of the very notion of tubular neighbourhood, which acquires a tolerably precise meaning only in the presence of structures which are much more rigid than the mere topological structure, such as "piecewise linear" or Riemannian (or more generally, space with a distance function) structure; the trouble here is that in the examples which naturally come to mind, one does not have such structures at one's disposal – at best an equivalence class of such structures, which makes it possible to rigidify the situation somewhat. If on the other hand one assumes that one might find an expedient in order to produce a tubular neighbourhood having the desired properties, which moreover would be unique modulo an automorphism (say a topological one) of the situation – an automorphism which moreover respects the fibred structure provided by the glueing map, there still remains the difficulty arising from the lack of canonicity of the choices involved, as the said automorphism is obviously not unique, whatever may be done in order to "normalise" it. The idea here, in order to make canonical something which is not, is to work systematically in the framework of the "__isotopic categories__" associated to the categories of a topological nature which are naturally present in such questions (such as the category of

admissible pairs (X, Y) and homeomorphisms of such pairs etc.), retaining
the same objects, but defining as "morphisms" the isotopy classes (in a
sense which is dictated unambiguously by the context) of isomorphisms (or
even morphisms more general than isomorphisms). I used this idea, which
is taken up successfully in the thesis of Yves Ladegaillerie (see beginning
of par. 3), in a systematic way in all my later reflections on combinatorial
topology, when it came to a precise formulation of translation theorems of
topological situations in terms of combinatorial situations. In the present
situation, my hope was to be able to formulate (and prove!) a theorem
of equivalence between two suitable isotopic categories, one being the category of "admissible pairs" (X, Y), and the other the category of "admissible
triples" (U, Y, f), where Y is a manifold, U a manifold with boundary, and
$f : \partial U \to Y$ a smooth and proper fibration. Moreover, I hoped that such a
statement could be naturally extended, modulo some essentially algebraic
work, to a more general statement, which would apply to general stratified
structures.

It soon appeared that there could be no question of getting such an ambitious statement in the framework of topological spaces, because of the
sempiternal "wild" phenomena. Already when X itself is a manifold and
Y is reduced to a point, one is confronted with the difficulty that the cone
over a compact space Z can be a manifold at its vertex, even if Z is not
homeomorphic to a sphere, nor even a manifold. It was also clear that the
contexts of the most rigid structures which existed then, such as the "piecewise linear" context were equally inadequate – one common disadvantage
consisting in the fact that they do not make it possible, given a pair (U, S)
of a "space" U and a closed subspace S, and a glueing map $f : S \to T$,
to build the corresponding amalgamated sum. Some years later, I was told
of Hironaka's theory of what he calls, I believe, (real) "semi-analytic" sets
which satisfy certain essential stability conditions (actually probably all of
them), which are necessary to develop a usable framework of "tame topology". This triggered a renewal of the reflection on the foundations of such
a topology, whose necessity appears more and more clearly to me.

After some ten years, I would now say, with hindsight, that "general
topology" was developed (during the thirties and forties)by analysts and in
order to meet the needs of analysis, not for topology per se, i.e. the study
of the topological properties of the various geometrical shapes. That the
foundations of topology are inadequate is manifest from the very beginning,
in the form of "false problems" (at least from the point of view of the topological intuition of shapes) such as the "invariance of domains", even if the
solution to this problem by Brouwer led him to introduce new geometrical
ideas. Even now, just as in the heroic times when one anxiously witnessed

for the first time curves cheerfully filling squares and cubes, when one tries
to do topological geometry in the technical context of topological spaces, one
is confronted at each step with spurious difficulties related to wild phenomena. For instance, it is not really possible, except in very low dimensions,
to study for a given space X (say a compact manifold), the homotopy type
of (say) the automorphism group of X, or of the space of embeddings, or
immersions etc. of X into some other space Y – whereas one feels that these
invariants should be part of the toolbox of the essential invariants attached
to X, or to the pair (X, Y), etc. just as the function space $\underline{\mathrm{Hom}}(X, Y)$ which
is familiar in homotopical algebra. Topologists elude the difficulty, without
tackling it, moving to contexts which are close to the topological one and
less subject to wildness, such as differentiable manifolds, PL spaces (piecewise linear) etc., of which it is clear that none is "good", i.e. stable under
the most obvious topological operations, such as contraction-glueing operations (not to mention operations like $X \to \mathrm{Aut}(X)$ which oblige one to leave
the paradise of finite dimensional "spaces"). This is a way of beating about
the bush! This situation, like so often already in the history of our science,
simply reveals the almost insurmountable inertia of the mind, burdened by
a heavy weight of conditioning, which makes it difficult to take a real look
at a foundational question, thus at the context in which we live, breathe,
work – accepting it, rather, as immutable data. It is certainly this inertia
which explains why it tooks millenia before such childish ideas as that of
zero, of a group, of a topological shape found their place in mathematics.
It is this again which explains why the rigid framework of general topology
is patiently dragged along by generation after generation of topologists for
whom "wildness" is a fatal necessity, rooted in the nature of things.

My approach toward possible foundations for a tame topology has been
an axiomatic one. Rather than declaring (which would indeed be a perfectly
sensible thing to do) that the desired "tame spaces" are no other than (say)
Hironaka's semianalytic spaces, and then developing in this context the
toolbox of constructions and notions which are familiar from topology, sup-
plemented with those which had not been developed up to now, for that very
reason, I preferred to work on extracting which exactly, among the geometrical properties of the semianalytic sets in a space $\mathbb{R}^n$, make it possible to
use these as local "models" for a notion of "tame space" (here semianalytic),
and what (hopefully!) makes this notion flexible enough to use it effectively
as the fundamental notion for a "tame topology" which would express with
ease the topological intuition of shapes. Thus, once this necessary foundational work has been completed, there will appear not one "tame theory",
but a vast infinity, ranging from the strictest of all, the one which deals
with "piecewise $\overline{\mathbb{Q}}_r$-algebraic spaces" (with $\overline{\mathbb{Q}}_r = \overline{\mathbb{Q}} \cap \mathbb{R}$), to the one which

appears (whether rightly or not) to be likely to be the vastest of all, namely using "piecewise real analytic spaces" (or semianalytic using Hironaka's terminology). Among the foundational theorems which I envision in my programme, there is a <u>comparison theorem</u> which, to put it vaguely, would say that <u>one will essentially find the same isotopic categories</u> (or even ∞-isotopic) whatever the tame theory one is working with. In a more precise way, the question is to put one's finger on a system of axioms which is rich enough to imply (among many other things) that if one has two tame theories T, T' with T finer than T' (in the obvious sense), and if X, Y are two T-tame spaces, which thus also define corresponding T'-tame spaces, the canonical map

$$\underline{\mathrm{Isom}}_T(X,Y) \to \underline{\mathrm{Isom}}_{T'}(X,Y)$$

induces a bijection on the set of connected components (which will imply that the isotopic category of the T-spaces is equivalent to the T'-spaces), and is even a homotopy equivalence (which means that one even has an equivalence for the "∞-isotopic" categories, which are finer than the isotopic categories in which one retains only the π_0 of the spaces of isomorphisms). Here the <u>Isom</u> may be defined in an obvious way, for instance as semisimplicial sets, in order to give a precise meaning to the above statement. Analogous statements should be true, if one replaces the "spaces" <u>Isom</u> with other spaces of maps, subject to standard geometric conditions, such as those of being embeddings, immersions, smooth, etale, fibrations etc. One also expects analogous statements where X, Y are replaced by systems of tame spaces, such as those which occur in a theory of dévissage of stratified structures – so that in a precise technical sense, this dévissage theory will also be essentially independent of the tame theory chosen to express it.

$\frac{31}{32}$

The first decisive test for a good system of axioms defining the notion of a "tame subset of $\mathbb{R}^n$" seems to me to consist in the possibility of proving such comparison theorems. I have settled for the time being for extracting a temporary system of plausible axioms, without any assurance that other axioms will not have to be added, which only working on specific examples will cause to appear. The strongest among the axioms I have introduced, whose validity is (or will be) most likely the most delicate to check in specific situations, is a <u>triangulability axiom</u> (in a tame sense, it goes without saying) of a tame part of $\mathbb{R}^n$. I did not try to prove the comparison theorem in terms of these axioms only, however I had the impression (right or wrong again!) that this proof, whether or not it necessitates the introduction of some additional axiom, will not present serious technical difficulties. It may well be that the technical difficulties in the development of satisfactory foundations for tame topology, including a theory of dévissage for tame stratified structures are actually already essentially concentrated in the axioms, and consequently already essentially overcome by triangulability theorems à la Lojasiewicz and Hironaka. What is again lacking is not the technical virtuosity of the mathematicians, which is sometimes impressive, but the audacity (or simply innocence...) to free oneself from a familiar context accepted by a flawless consensus...

The advantages of an axiomatic approach towards the foundations of tame topology seem to me to be obvious enough. Thus, in order to consider a complex algebraic variety, or the set of real points of an algebraic variety defined over $\mathbb{R}$, as a tame space, it seems preferable to work with the "piecewise $\mathbb{R}$-algebraic" theory , maybe even the $\overline{\mathbb{Q}}_r$-algebraic theory (with $\overline{\mathbb{Q}}_r = \overline{\mathbb{Q}} \cap \mathbb{R}$) when dealing with varieties defined over number fields, etc. The introduction of a subfield $K \subset \mathbb{R}$ associated to the theory $\mathcal{T}$ (consisting in the points of $\mathbb{R}$ which are $\mathcal{T}$-tame, i.e. such that the corresponding one-point set is $\mathcal{T}$-tame) make it possible to introduce for any point x of a tame space X, a residue field $k(x)$, which is an algebraically closed subextension of $\mathbb{R}/K$, of finite transcendence degree over K (bounded by the topological dimension of X). When the transcendence degree of $\mathbb{R}$ over K is infinite, we find a notion of transcendence degree (or "dimension") of a point of a tame space, close to the familiar notion in algebraic geometry. Such notions are absent from the "semianalytic" tame topology, which however appears as the natural topological context for the inclusion of real and complex analytic spaces.

Among the first theorems one expects in a framework of tame topology as I perceive it, aside from the comparison theorems, are the statements which establish, in a suitable sense, the existence and uniqueness of "the" tubular neighbourhood of closed tame subspace in a tame space (say compact to make things simpler), together with concrete ways of building it (starting for instance from any tame map $X \to \mathbb{R}^+$ having Y as its zero set), the description of its "boundary" (although generally it is in no way a manifold with boundary!) ∂T, which has in T a neighbourhood which is isomorphic to the product of T with a segment, etc. Granted some suitable equisingularity hypotheses, one expects that T will be endowed, in an essentially unique way, with the structure of a locally trivial fibration over Y, with ∂T as a subfibration. This is one of the least clear points in my temporary intuition of the situation, whereas the homotopy class of the predicted structure map $T \to Y$ has an obvious meaning, independent of any equisingularity hypothesis, as the homotopic inverse of the inclusion map $Y \to T$, which must be a homotopism. One way to a posteriori obtain such a structure would be via the hypothetical equivalence of isotopic categories which was considered at the beginning, taking into account the fact that the functor

$(U, Y, f) \longmapsto (X, Y)$ is well-defined in an obvious way, independently of any theory of tubular neighbourhoods.

It will perhaps be said, not without reason, that all this may be only dreams, which will vanish in smoke as soon as one sets to work on specific examples, or even before, taking into account some known or obvious facts which have escaped me. Indeed, only working out specific examples will make it possible to sift the right from the wrong and to reach the true substance. The only thing in all this which I have no doubt about, is the very necessity of such a foundational work, in other words, the artificiality of the present foundations of topology, and the difficulties which they cause at each step. It may be however that the formulation I give of a theory of dévissage of stratified structures in terms of an equivalence theorem of suitable isotopic (or even ∞-isotopic) categories is actually too optimistic. But I should add that I have no real doubts about the fact that the theory of these dévissages which I developed two years ago, although it remains in part heuristic, does indeed express some very tangible reality. In some part of my work, for want of a ready-to-use "tame" context, and in order to have precise and provable statements, I was led to postulate some very plausible additional structures on the stratified structure I started with, especially concerning the local retraction data, which do make it possible to construct a canonical system of spaces, parametrised by the ordered set of flags Drap(I) of the ordered set I indexing the strata; these spaces play the role of the spaces (U, Y) above, and they are connected by embedding and proper fibration maps, which make it possible to reconstitute in an equally canonical way the original stratified structure, including these "additional structures" ([7]). The only trouble here, is that these appear as an additional artificial element of structure, which is no way part of the data in the usual geometric situations, as for example the compact moduli space $\widehat{M}_{g,\nu}$ with its canonical "stratification at infinity", defined by the Mumford-Deligne divisor with normal crossings. Another, probably less serious difficulty, is that this so-called moduli "space" is in fact a multiplicity – which can be technically expressed by the necessity of replacing the index set I for the strata with an (essentially finite) category of indices, here the "MD graphs" which "parametrise" the possible "combinatorial structures" of a stable curve of type (g, ν). This said, I can assert that the general theory of dévissage, which has been developed especially to meet the needs of this example, has indeed proved to be a precious guide, leading to a progressive understanding, with flawless coherence, of some essential aspects of the Teichmüller tower (that is, essentially the "structure at infinity" of the ordinary Teichmüller groups). It is this approach which finally led me, within some months, to the principle of a purely combinatorial construction of the tower of Teichmüller groupoids, in the spirit sketched above (cf. par. 2).

Another satisfying test of coherence comes from the "topossic" viewpoint. Indeed, as my interest for the multiplicities of moduli was first prompted by their algebrico-geometric and arithmetic meaning, I was first and foremost interested by the modular algebraic multiplicities, over the absolute basefield $\mathbb{Q}$, and by a "dévissage" at infinity of their geometric fundamental groups (i.e. of the profinite Teichmüller groups) which would be compatible with the natural operations of $\Gamma = \mathrm{Gal}(\overline{\mathbb{Q}}/\mathbb{Q})$. This requirement seemed to exclude from the start the possibility of a reference to a hypothetical theory of dévissage of stratified structures in a context of "tame topology" (or even, at worst, of ordinary topology), beyond a purely heuristic guiding thread. Thus the question arose of translating, in the context of the topoi (here etale topoi) which were present in the situation, the theory of dévissage I had arrived at in a completely different context – with the additional task, in the sequel, of extracting a general comparison theorem, patterned after well-known theorems, in order to compare the invariants (in particular the homotopy types of various tubular neighbourhoods) obtained in the transcendent and schematic frameworks. I have been able to convince myself that such a formalism of dévissage indeed had some meaning in the (so-called "abstract"!) context of general topoi, or at least noetherian topoi (like those occurring in this situation), via a suitable notion of canonical tubular neighbourhood of a subtopos in an ambient topos. Once this notion is acquired, together with some simple formal properties, the description of the "dévissage" of a stratified topos is even considerably simpler in that framework than in the (tame) topological one. True, there is foundational work to be done here too, especially around the very notion of the tubular neighbourhood of a subtopos – and it is actually surprising that this work (as far as I know) has still never been done, i.e. that no one (since the context of etale topology appeared, more than twenty years ago) apparently ever felt the need for it; surely a sign that the understanding of the topological structure of schemes has not made much progress since the work of Artin-Mazur...

Once I had accomplished this (more or less heuristic) double work of refining the notion of dévissage of a stratified space or topos, which was a crucial step in my understanding of the modular multiplicities, it actually appeared that, as far as these are concerned, one can actually take a short cut for at least a large part of the theory, via direct geometric arguments. Nonetheless, the formalism of dévissage which I reached has proved its usefulness and its coherence to me, independently of any question about the most adequate foundations which make it completely meaningful.

6. One of the most interesting foundational theorems of (tame) topology which should be developed would be a theorem of "dévissage" (again!) of a proper tame map of tame spaces

$$f : X \to Y,$$

via a decreasing filtration of Y by closed tame subspaces Y^i, such that above the "open strata" $Y^i \backslash Y^{i-1}$ of this filtration, f induces a locally trivial fibration (from the tame point of view, it goes without saying). It should be possible to generalise such a statement even further and to make it precise in various ways, in particular by requiring the existence of an analogous simultaneous dévissage for X and for a given finite family of (tame) closed subspaces of X. Also the very notion of locally trivial fibration in the tame sense can be made considerably stronger, taking into account the fact that the open strata U_i are better than spaces whose tame structure is purely local, because they are obtained as differences of two tame spaces, compact if Y is compact. Between the notion of a compact tame space (which is realised as one of the starting "models" in an $\mathbb{R}^n$) and that of a "locally tame" (locally compact) space which can be deduced from it in a relatively obvious way, there is a somewhat more delicate notion of a "globally tame" space X, obtained as the difference $\hat{X} \backslash Y$ of two compact tame spaces, it being understood that we do not distinguish between the space defined by a pair $(\hat{X}, Y)$ and that defined by a pair $(\hat{X}', Y')$ deduced from it by a (necessarily proper) tame map

$$g : \hat{X}' \to \hat{X}$$

inducing a bijection $g^{-1}(X) \to X$, taking $Y' = g^{-1}(Y)$. Perhaps the most interesting natural example is the one where we start from a separated
scheme of finite type over $\mathbb{C}$ or $\mathbb{R}$, taking for X the set of its real or complex points, which inherits a global tame structure with the help of schematic compactifications (which exist according to Nagata) of the scheme we started with. This notion of a globally tame space is associated to a notion of a globally tame map, which in turn allows us to strengthen the notion of a locally trivial fibration, in stating a theorem of dévissage for a map $f : X \to Y$ (now not necessarily proper) in the context of globally tame spaces.

I was informed last summer by Zoghman Mebkhout that a theorem of dévissage in this spirit has been recently obtained in the context of real and/or complex analytic spaces, with Y^i which here are analytic subspaces of Y. This result makes it plausible that we already have at our disposal techniques which are powerful enough to also prove a dévissage theorem in the tame context, apparently more general, but probably less arduous.

The context of tame topology should also, it seems to me, make it possible to formulate with precision a certain very general principle which I frequently use in a great variety of geometric situations, which I call the "principle of anodine choices" – as useful as vague in appearance! It says that when for the needs of some construction of a geometric object in terms of others, we are led to make a certain number of arbitrary choices along the way, so that the final object appears to depend on these choices, and is thus stained with a defect of canonicity, that this defect is indeed serious (and to be removed requires a more careful analysis of the situation, the notions used, the data introduced etc.) whenever at least one of these choices is made in a space which is not "contractible", i.e. whose π_0 or one of whose higher invariants π_i is non-trivial, and that this defect is on the contrary merely apparent, and the construction itself is "essentially" canonical and will not bring along any troubles, whenever the choices made are all "anodine", i.e. made in <u>contractible</u> spaces. When we try in actual examples to really understand this principle, it seems that each time we stumble onto the same notion of "∞-isotopic categories" expressing a given situation, and finer than the more naive isotopic (= 0-isotopic) categories obtained by considering only the π_0 of the spaces of isomorphisms introduced in the situation, while the ∞-isotopic point of view considers all of their homotopy type. For example, the naive isotopic point of view for compact surfaces with boundary of type (g, ν) is "good" (without any hidden boomerangs!) exactly in the cases which I call "anabelian" (and which Thurston calls "hyperbolic"), i.e. distinct from $(0,0)$, $(0,1)$ $(0,2)$, $(1,0)$ – which are also exactly the cases where the connected component of the identity of the automorphism group of the surface is <u>contractible</u>. In the other cases, except for the case $(0,0)$ of the sphere without holes, it suffices to work with 1-isotopic categories to express in a satisfying way via algebra the essential geometrico-topological facts, since the said connected component is then a $K(\pi, 1)$. Working in a 1-isotopic category actually comes down to working in a bicategory, i.e. with $\underline{\mathrm{Hom}}(X, Y)$ which are (no longer discrete sets as in the 0-isotopic point of view, but) groupoids (whose π_0 are exactly the 0-isotopic Hom). This is the description in purely algebraic terms of this bicategory which is given in the last part of the thesis of Yves Ladegaillerie (cf. par. 3).

If I allowed myself to dwell here at some length on the theme of the foundations of tame topology, which is not one of those to which I intend to devote myself prioritarily in the coming years, it is doubtless because I feel that it is yet another cause which needs to be pleaded, or rather: a work of great current importance which needs hands! Just as years ago for the new foundations of algebraic geometry, it is not pleadings which will surmount the inertia of acquired habits, but tenacious, meticulous long-term work, which will from day to day bring eloquent harvests.

I would like to say some last words on an older reflection (end of the sixties?), very close to the one I just discussed, inspired by ideas of Nash which I found very striking. Instead of axiomatically defining a notion of "tame theory" via a notion of a "tame part of $\mathbb{R}^n$" satisfying certain conditions (mainly of stability), I was interested by an axiomatisation of the notion of "non-singular variety" via, for each natural integer n, a subring $\mathcal{A}_n$ of the ring of germs of real functions at the origin in $\mathbb{R}^n$. These are the functions which will be admitted to express the "change of chart" for the corresponding notion of $\mathcal{A}_n$-variety, and I was first concerned with uncovering a system of axioms on the system $\mathcal{A} = (\mathcal{A}_n)_{n \in \mathbb{N}}$ which ensures for this notion of variety a suppleness comparable to that of a C^∞ variety, or a real analytic one (or a Nash one). According to the familiar type of construction which one wants to be able to do in the context of $\mathcal{A}$-varieties, the relevant system of axioms is more or less reduced or rich. One doesn't need much if one only wants to develop the differential formalism, with the construction of jet bundles, De Rham complexes etc. If we want a statement of the type "quasi-finite implies finite" (for a map in the neighbourhood of a point), which appeared as a key statement in the local theory of analytic spaces, we need a more delicate stability axiom, in Weierstrass' "Vorbereitungssatz" (*). In other questions, a stability axiom by analytic continuation (in $\mathbb{C}^n$) appears necessary. The most Draconian axiom which I was led to introduce, also a stability axiom, concerns the integration of Pfaff systems, ensuring that certain (even all) Lie groups are $\mathcal{A}$-varieties. In all this, I took care not to suppose that the $\mathcal{A}_n$ are $\mathbb{R}$-algebras, so a constant function on a $\mathcal{A}$-variety is "admissible" only if its value belongs to a certain subfield K of $\mathbb{R}$ (which is, if one likes, $\mathcal{A}_0$). This subfield can very well be $\mathbb{Q}$, or its algebraic closure $\overline{\mathbb{Q}}_r$ in $\mathbb{R}$, or any other subextension of $\mathbb{R}/\mathbb{Q}$, preferably even of finite or at least countable transcendence degree over $\mathbb{Q}$. This makes it possible, for example, as before for tame spaces, to have every point x of a variety (of type $\mathcal{A}$) correspond to a residue field $k(x)$, which is a subextension of $\mathbb{R}/K$. A fact which appears important to me here, is that even in its strongest form, the system of axioms does not imply that we must have $K = \mathbb{R}$. More precisely, because all the axioms are stability axioms, it follows that for a given set S of germs of real analytic functions at the origin (in various spaces $\mathbb{R}^n$), there exists a smaller theory

---

(*) It could seem simpler to say that the (local) rings $\mathcal{A}_n$ are Henselian, which is equivalent. But it is not at all clear a priori in this latter form that the condition in question is in the nature of a stability condition, and this is an important circumstance as will appear in the following reflections.

$\mathcal{A}$ for which these germs are admissible, and that it is "countable", i.e. the $\mathcal{A}_n$ are countable, whenever S is. A fortiori, K is then countable, i.e. of countable transcendence degree over $\mathbb{Q}$.

The idea here is to introduce, via this axiomatic system, a notion of an "elementary" (real analytic) function, or rather, a whole hierarchy of such notions. For a function of 0 variables i.e. a constant, this notion gives that of an "elementary constant", including in particular (in the case of the strongest axiomatic system) constants such as π, e and many others, taking values of admissible functions (such as exponentials, logarithms etc.) for systems of "admissible" values of the argument. One feels that the relation between the system $\mathcal{A} = (\mathcal{A}_n)_{n \in \mathbb{N}}$ and the corresponding rationality field K must be very tight, at least for $\mathcal{A}$ which can be generated by a finite "system of generators" S – but one must fear that even the least of the interesting questions one could ask about this situation still remains out of reach (^{1}).

These old reflections have taken on some current interest for me due to my more recent reflection on tame theories. Indeed, it seems to me that it is possible to associate in a natural way to a "differentiable theory" $\mathcal{A}$ a tame theory $\mathcal{T}$ (doubtless having the same field of constants), in such a way that every $\mathcal{A}$-variety is automatically equipped with a $\mathcal{T}$-tame structure and conversely for every $\mathcal{T}$-tame compact space X, we can find a rare tame closed subset Y in X, such that $X \backslash Y$ comes from an $\mathcal{A}$-variety, and moreover such that this $\mathcal{A}$-variety structure is unique at least in the following sense: two such structures coincide in the complement of a rare tame subset $Y' \supset Y$ of X. The theory of dévissage of stratified tame structures (which was discussed in the preceding par.), in the case of smooth strata, should moreover raise much more precise questions of comparison of tame structures with structures of differentiable (or rather, $\mathbb{R}$-analytic) type. I suspect that the type of axiomatisation proposed here for the notion of "differentiable theory" would give a natural framework for the formulation of such questions with all desirable precision and generality.

7. Since the month of March last year, so nearly a year ago, the greater part of my energy has been devoted to a work of reflection on the foundations of non-commutative (co)homological algebra, or what is the same, after all, of homotopic algebra. These reflections have taken the concrete form of a voluminous stack of typed notes, destined to form the first volume (now being finished) of a work in two volumes to be published by Hermann, under the overall title "Pursuing Stacks". I now foresee (after successive extensions of the initial project) that the manuscript of the whole of the two volumes, which I hope to finish definitively in the course of this year, will be about

1500 typed pages in length. These two volumes are moreover for me the first in a vaster series, under the overall title "Mathematical Reflections", in which I intend to develop some of the themes sketched in the present report.

Since I am speaking here of work which is actually now being written up and is even almost finished, the first volume of which will doubtless appear this year and will contain a detailed introduction, it is undoubtedly less interesting for me to develop this theme of reflection here, and I will content myself with speaking of it only very briefly. This work seems to me to be somewhat marginal with respect to the themes I sketched before, and does not (it seems to me) represent a real renewal of viewpoint or approach with respect to my interests and my mathematical vision of before 1970. If I suddenly resolved to do it, it is almost out of desperation, for nearly twenty years have gone by since certain visibly fundamental questions, which were ripe to be thoroughly investigated, without anyone seeing them or taking the trouble to fathom them. Still today, the basic structures which occur in the homotopic point of view in topology are not understood, and to my knowledge, after the work of Verdier, Giraud and Illusie on this theme (which are so many beginnings still waiting for continuations...) there has been no effort in this direction. I should probably make an exception for the axiomatisation work done by Quillen on the notion of a category of models, at the end of the sixties, and taken up in various forms by various authors. At that time, and still now, this work seduced me and taught me

43
44

a great deal, even while going in quite a different direction from the one which was and still is close to my heart. Certainly, it introduces derived categories in various non-commutative contexts, but without entering into the question of the essential internal structures of such a category, also left open in the commutative case by Verdier, and after him by Illusie. Similarly, the question of putting one's finger on the natural "coefficients" for a non-commutative cohomological formalism, beyond the stacks (which should be called 1-stacks) studied in the book by Giraud, remained open – or rather, the rich and precise intuitions concerning it, taken from the numerous examples coming in particular from algebraic geometry, are still waiting for a precise and supple language to give them form.

I returned to certain aspects of these foundational questions in 1975, on the occasion (I seem to remember) of a correspondence with Larry Breen (two letters from this correspondence will be reproduced as an appendix to Chap. I of volume 1, "History of Models", of Pursuing Stacks). At that moment the intuition appeared that ∞-groupoids should constitute particularly adequate models for homotopy types, the n-groupoids corresponding to truncated homotopy types (with $\pi_i = 0$ pour $i > n$). This same intuition, via very different routes, was discovered by Ronnie Brown and some of his students in Bangor, but using a rather restrictive notion of ∞-groupoid (which, among the 1-connected homotopy types, model only products of Eilenberg-Mac Lane spaces). Stimulated by a rather haphazard correspondence with Ronnie Brown, I finally began this reflection, starting with an attempt to define a wider notion of ∞-groupoid (later rebaptised stack in ∞-groupoids or simply "stack", the implication being: over the 1-point topos), and which, from one thing to another, led me to Pursuing Stacks. The volume "History of Models" is actually a completely unintended digression with respect to the initial project (the famous stacks being temporarily forgotten, and supposed to reappear only around page 1000...).

This work is not completely isolated with respect to my more recent interests. For example, my reflection on the modular multiplicities $\widehat{M}_{g,\nu}$ and their stratified structure renewed the reflection on a theorem of van Kampen in dimension > 1 (also one of the preferred themes of the group in Bangor), and perhaps also contributed to preparing the ground for the more important work of the following year. This also links up from time to time with a reflection dating from the same year 1975 (or the following year) on a "De Rham complex with divided powers", which was the subject of my last public lecture, at the IHES in 1976; I lent the manuscript of it to I don't remember whom after the talk, and it is now lost. It was at the moment of this reflection that the intuition of a "schematisation" of homotopy types germinated, and seven years later I am trying to make it precise in a (particularly hypothetical) chapter of the History of Models.

The work of reflection undertaken in Pursuing Stacks is a little like a debt which I am paying towards a scientific past where, for about fifteen years (from 1955 to 1970), the development of cohomological tools was the constant Leitmotiv in my foundational work on algebraic geometry. If in this renewal of my interest in this theme, it has taken on unexpected dimensions, it is however not out of pity for a past, but because of the numerous unexpected phenomena which ceaselessly appear and unceremoniously shatter the previously laid plans and projects – rather like in the thousand and one nights, where one awaits with bated breath through twenty other tales the final end of the first.

8. Up to now I have spoken very little of the more down-to-earth reflections on two-dimensional topological geometry, directly associated to my activities of teaching and "directing research". Several times, I saw opening before me vast and rich fields ripe for the harvest, without ever succeeding in communicating this vision, and the spark which accompanies it, to one of my students, and having it open out into a more or less long-term com-

mon exploration. Each time up through today, after a few days or weeks of
investigating where I, as scout, discovered riches at first unsuspected, the
voyage suddenly stopped, upon its becoming clear that I would be pursuing it alone. Stronger interests then took precedence over a voyage which
at that point appeared more as a digression or even a dispersion, than a
common adventure.

One of these themes was that of planar polygons, centred around the
modular varieties which can be associated to them. One of the surprises
here was the irruption of algebraic geometry in a context which had seemed
to me quite distant. This kind of surprise, linked to the omnipresence of
algebraic geometry in plain geometry, occurred several times.

Another theme was that of curves (in particular circles) immersed in a
surface, with particular attention devoted to the "stable" case where the singular points are ordinary double points (and also the more general theme
where the different branches at a point mutually cross), often with the additional hypothesis that the immersion is "cellular", i.e. gives rise to a map.
A variation on the situations of this type is that of immersions of a surface with non-empty boundary, and first of all a disk (which was pointed
out to me by A'Campo around ten years ago). Beyond the question of
the various combinatorial formulations of such situations, which really represent no more than an exercise of syntax, I was mainly interested in a
dynamical vision of the possible configurations, with the passage from one
to another via continuous deformations, which can be decomposed into com-
positions of two types of elementary operations and their inverses, namely
the "sweeping" of a branch of a curve over a double point, and the erasing
or the creation of a bigon. (The first of these operations also plays a key
role in the "dynamical" theory of systems of pseudo-lines in a real projective
plane.) One of the first questions to be asked here is that of determining
the different classes of immersions of a circle or a disk (say) modulo these
elementary operations; another, that of seeing which are the immersions of
the boundary of the disk which come from an immersion of the disk, and
to what extent the first determine the second. Here also, it seems to me
that it is a systematic study of the relevant modular varieties (of infinite
dimension here, unless a purely combinatorial description of them can be
given) which should give the most efficient "focus", forcing us in some sense
to ask ourselves the most relevant questions. Unfortunately, the reflection
on even the most obvious and down-to-earth questions has remained in an
embryonic state. As the only tangible result, I can cite a theory of canonical
"dévissage" of a stable cellular immersion of a circle in a surface into "unde-
composable" immersions, by "telescoping" such immersions. Unfortunately
I did not succeed in transforming my lights on the question into a DEA

46
47

thesis, nor other lights (on a complete theoretical description, in terms of
fundamental groups of topological 1-complexes, of the immersions of a surface with boundary which extend a given immersion of its boundary) into
the beginnings of a doctoral thesis...

A third theme, pursued simultaneously over the last three years at different levels of teaching (from the option for first year students to the
three third-cycle theses now being written on this theme) deals with the
topologico-combinatorial classification of systems of lines or pseudo-lines.
Altogether, the participation of my students here has been less disappoint-
ing than elsewhere, and I have had the pleasure of occasionally learning
interesting things from them which I would not have thought of. Things
being what they are, however, our common reflection was limited to a
very elementary level. Lately, I finally devoted a month of intensive reflection to the development of a purely combinatorial construction of a sort of
"modular surface" associated to a system of n pseudo-lines, which classifies
the different possible "relative positions" (stable or not) of an $(n + 1)$-st
pseudo-line with respect to the given system, in other words: the different
possible "affinisations" of this system, by the different possible choices of a
"pseudo-line at infinity". I have the impression of having put my finger on
a remarkable object, causing an unexpected order to appear in questions
of classification which up to now appeared fairly chaotic! But the present
report is not the place to dwell further on this subject.

Since 1977, in all the questions (such as the two last themes evoked above)
where two-dimensional maps occur, the possibility of realising them canonically on a conformal surface, so on a complex algebraic curve in the compact
oriented case, remains constantly in filigree throughout my reflection. In
practically every case (in fact, in all cases except that of certain spherical
maps with "few automorphisms") such a conformal realisation implies in
fact a <u>canonical Riemannian metric</u>, or at least, canonical up to a multiplicative constant. These new elements of structure (without even taking
into account the arithmetic element which was considered in par. 3) are of
a nature to deeply transform the initial aspect of the questions considered,
and the methods of approaching them. A beginning of familiarisation with
the beautiful ideas of Thurston on the construction of Teichmüller space,
in terms of a very simple game of hyperbolic Riemannian surgery, confirms
me in this presentiment. Unfortunately, the very modest level of culture
of almost all the students who have worked with me over these last ten
years does not allow me to investigate these possibilities with them even by
allusion, since the assimilation of even a minimal combinatorial language already frequently encounters considerable psychical obstacles. This is why,
in some respect and more and more in these last years, my teaching activity

$\frac{47}{48}$

$\frac{48}{49}$

has often acted like a weight, rather than a stimulus for the unfolding of a
somewhat advanced or even merely delicate geometric reflection.

9. The occasion appears to be auspicious for a brief assessment of my teaching activity since 1970, that is, since it has taken place in a university. This
contact with a very different reality taught me many things, of a completely
different order than simply pedagogic or scientific. Here is not the place to
dwell on this subject. I also mentioned at the beginning of this report the
role which this change of professional milieu played in the renewal of my
approach to mathematics, and that of my centres of interest in mathematics. If I pursue this assessment of my teaching activity on the research level,
I come to the conclusion of a clear and solid failure. In the more than ten
years that this activity has taken place, year after year in the same university, I was never at any moment able to suscitate a place where "something
happened" – where something "passed", even among the smallest group of
people, linked together by a common adventure. Twice, it is true, around
the years 1974 to 1976, I had the pleasure and the privilege of awakening
a student to a work of some consequence, pursued with enthusiasm: Yves
Ladegaillerie in the work mentioned earlier (par. 3) on questions of isotopy
in dimension 2, and Carlos Contou-Carrère (whose mathematical passion
did not await a meeting with myself to blossom) an unpublished work on
the local and global Jacobians over general base schemes (of which one part
was announced in a note in the CR). Apart from these two cases, my role
has been limited throughout these ten years to somehow or other conveying
the rudiments of the mathematician's trade to about twenty students on
the research level, or at least to those among them who persevered with
me, reputed to be more demanding than others, long enough to arrive at a
first acceptable work written black on white (and even, sometimes, at something better than acceptable and more than just one, done with pleasure
and worked out through to the end). Given the circumstances, among the
rare people who persevered, even rarer are those who will have the chance
of carrying on the trade, and thus, while earning their bread, learning it
ever more deeply.

10. Since last year, I feel that as regards my teaching activity at the university, I have learned everything I have to learn and taught everything I
can teach there, and that it has ceased to be really useful, to myself and to
others. To insist on continuing it under these circumstances would appear
to me to be a waste both of human resources and of public funds. This
is why I have applied for a position in the CNRS (which I left in 1959 as
freshly named director of research, to enter the IHES). I know moreover

that the employment situation is tight in the CNRS as everywhere else, that the result of my request is doubtful, and that if a position were to be attributed to me, it would be at the expense of a younger researcher who would remain without a position. But it is also true that it would free my position at the USTL to the benefit of someone else. This is why I do not scruple to make this request, and to renew it if is not accepted this year.

In any case, this application will have been the occasion for me to write this sketch of a programme, which otherwise would probably never have seen the light of day. I have tried to be brief without being sybilline and also, afterwards, to make it easier reading by the addition of a summary. If in spite of this it still appears rather long for the circumstances, I beg to be excused. It seems short to me for its content, knowing that ten years of work would not be too much to explore even the least of the themes sketched here through to the end (assuming that there is an "end"...), and one hundred years would be little for the richest among them!

Behind the apparent disparity of the themes evoked here, an attentive reader will perceive as I do a profound unity. This manifests itself particularly by a common source of inspiration, namely the geometry of surfaces, present in all of these themes, and most often front and centre. This source, with respect to my mathematical "past", represents a renewal, but certainly not a rupture. Rather, it indicates the path to a new approach to the still mysterious reality of "motives", which fascinated me more than any other in the last years of this past (*). This fascination has certainly not vanished, rather it is a part of the fascination with the most burning of all the themes evoked above. But today I am no longer, as I used to be, the voluntary prisoner of interminable tasks, which so often prevented me from springing into the unknown, mathematical or not. The time of tasks is over for me. If age has brought me something, it is lightness.

Janvier 1984

(*) On this subject, see my commentaries in the "Thematic Sketch" of 1972 attached to the present report, in the last section "motivic digressions", (loc. cit. pages 17-18).

51
52

(^{1}) The expression "out of reach" here (and also later for a completely different question), appears to me to be decidedly hasty and unfounded. I have noted myself on other occasions that when oracles (here myself!) declare with an air of deep understanding (or doubt) that such and such a problem is "out of reach", it is actually an entirely subjective affirmation. It simply means, apart from the fact that the problem is supposed to be not yet solved, that the person speaking has no ideas on the question, or probably more precisely, that he has no feelings and no motivation with regard to it, that it "does nothing to him" and that he has no desire to do anything with it – which is frequently a sufficient reason to want to discourage others. As in the remark of M. de la Palisse, this did not stop the beautiful and regretted conjectures of Mordell, Tate, and Shafarevitch from succumbing although they were all reputed to be "out of reach", poor things ! – Besides, in the very days which followed the writing up of the present report, which put me into contact with questions from which I had distanced myself during the last year, I noticed a new and remarkable property of the outer action of an absolute Galois group on the fundamental group of an algebraic curve, which had escaped me until now and which undoubtedly constitutes at least a new step towards the formulation of an algebraic characterisation of $\mathrm{Gal}(\overline{\mathbb{Q}}/\mathbb{Q})$. This, with the "fundamental conjecture" (mentioned in par. 3 below) appears at present as the principal open question for the foundation of an "anabelian algebraic geometry", which starting a few years ago, has represented (by far) my strongest centre of interest in mathematics.

52
53

(^{2}) With the exception of another "fact", at the time when, around the age of twelve, I was interned in the concentration camp of Rieucros (near Mende). It is there that I learnt, from another prisoner, Maria, who gave me free private lessons, the definition of the circle. It impressed me by its simplicity and its evidence, whereas the property of "perfect rotundity" of the circle previously had appeared to me as a reality mysterious beyond words. It is at that moment, I believe, that I glimpsed for the first time (without of course formulating it to myself in these terms) the creative power of a "good" mathematical definition, of a formulation which describes the essence. Still today, it seems that the fascination which this power exercised on me has lost nothing of its force.

(^{3}) More generally, beyond the so-called "anabelian" varieties, over fields of finite type, anabelian algebraic geometry (as it revealed itself some years ago) leads to a description, uniquely in terms of profinite groups, of the category of schemes of finite type over the absolute base $\mathbb{Q}$ (or even $\mathbb{Z}$), and from there, in principle, of the category of all schemes (by suitable passages to limits). It is thus a construction which "pretends" to ignore the rings (such as $\mathbb{Q}$, algebras of finite type over $\mathbb{Q}$, etc.) and the algebraic equations which traditionally serve to describe schemes, while working directly with their etale topoi, which can be expressed in terms of systems of profinite groups. A grain of salt nevertheless: to be able to hope to reconstitute a scheme (of finite type over $\mathbb{Q}$ say) from its etale topos, which is a purely topological invariant, we must place ourselves not in the category of schemes (here of finite type over $\mathbb{Q}$), but in the one which is deduced from it by "localisation", by making the morphisms which are "universal homeomorphisms", i.e. finite, radicial and surjective, be invertible. The development of such a translation of a "geometric world" (namely that of schemes, schematic multiplicities etc.) in terms of an "algebraic world" (that of profinite groups and systems of profinite groups describing suitable topoi (called "etale") can be considered as the ultimate goal of Galois theory, doubtless even in the very spirit of Galois. The sempiternal question "and why all this?" seems to me to have neither more nor less meaning in the case of the anabelian geometry now in the process of birth, than in the case of Galois theory in the time of Galois (or even today, when the question is asked by an overwhelmed student...); the same goes for the commentary which usually accompanies it, namely "all this is very general indeed!".

$\frac{53}{54}$

(^{4}) We thus easily conceive that a group like $\mathrm{Sl}(2,\mathbb{Z})$, with its "arithmetic" structure, is positively a machine for constructing "motivic" representations of $\mathrm{Gal}(\overline{\mathbb{Q}}/\mathbb{Q})$ and its open subgroups, and that we thus obtain, at least in principle, all the motivic representations which are of weight 1, or contained in a tensor product of such representations (which already makes quite a packet!) In 1981 I began to experiment with this machine in a few specific cases, obtaining various remarkable representations of Γ in groups $G(\hat{\mathbb{Z}})$, where G is a (not necessarily reductive) group scheme over $\mathbb{Z}$, starting from suitable homomorphisms

$$\mathrm{Sl}(2,\mathbb{Z}) \to G_0(\mathbb{Z}),$$

where G_0 is a group scheme over $\mathbb{Z}$, and G is constructed as an extension of G_0 by a suitable group scheme. In the "tautological" case $G_0 = \mathrm{Sl}(2)_{\mathbb{Z}}$, we find for G a remarkable extension of $\mathrm{Gl}(2)_{\mathbb{Z}}$ by a torus of dimension 2, with a motivic representation which "covers" those associated to the class fields of the extensions $\mathbb{Q}(i)$ and $\mathbb{Q}(j)$ (as if by chance, the "fields of complex multiplication" of the two "anharmonic" elliptic curves). There is here a principle of construction which seemed to me very general and very efficient, but I didn't have (or take) the leisure to unravel it and follow it through to the end – this is one of the numerous "hot points" in the foundational programme of anabelian algebraic geometry (or "Galois theory", extended

version) which I propose to develop. At this time, and in an order of priority
which is probably very temporary, these points are:

a) Combinatorial construction of the Teichmüller tower.

b) Description of the automorphism group of the profinite compactifica-
tion of this tower, and reflection on a characterisation of $\Gamma = \mathrm{Gal}(\overline{\mathbb{Q}}/\mathbb{Q})$ as
a subgroup of the latter.

c) The "motive machine" $\mathrm{Sl}(2, \mathbb{Z})$ and its variations.

d) The anabelian dictionary, and the fundamental conjecture (which is
perhaps not so "out of reach" as all that!). Among the crucial points of
this dictionary, I foresee the "profinite paradigm" for the fields $\mathbb{Q}$ (cf. b)),
$\mathbb{R}$ and $\mathbb{C}$, for which a plausible formalism remains to be uncovered, as well
as a description of the inertia subgroups of Γ, via which the passage from
characteristic zero to characteristic $p > 0$ begins, and to the absolute ring
$\mathbb{Z}$.

e) Fermat's problem.

(^{5}) I would like to point out, however, a more delicate task (apart from the
task pointed out in passing on cubic complexes), on the combinatorial interpretation of regular maps associated to congruence subgroups of $\mathrm{Sl}(2, \mathbb{Z})$.
This work was developed with a view to expressing the "arithmetic" oper-
ation of $\Gamma = \mathrm{Gal}(\overline{\mathbb{Q}}/\mathbb{Q})$ on these "congruence maps", which is essentially
done via the intermediary of the cyclotomic character of Γ. A point of
departure was the combinatorial theory of the "bi-icosahedron" developed
in a C4 course starting from purely geometric motivations, and which (it
afterwards proved) gives rise to a very convenient expression for the action
of Γ on the category of icosahedral maps (i.e. congruence maps of index 5).

(^{6}) Let us note in relation to this that the isomorphism classes of compact tame spaces are the same as in the "piecewise linear" theory (which
is not, I recall, a tame theory). This is in some sense a rehabilitation of
the "Hauptvermutung", which is "false" only because for historical reasons
which it would undoubtedly be interesting to determine more precisely,
the foundations of topology used to formulate it did not exclude wild phenomena. It need (I hope) not be said that the necessity of developing new
foundations for "geometric" topology does not at all exclude the fact that
the phenomena in question, like everything else under the sun, have their
own reason for being and their own beauty. More adequate foundations
would not suppress these phenomena, but would allow us to situate them
in a suitable place, like "limiting cases" of phenomena of "true" topology.

(^{7}) In fact, to reconstruct the system of spaces

$$(i_0,\ldots,i_n) \mapsto X_{i_0,\ldots,i_n}$$

contravariant on $\mathrm{Drap}(I)$ (for the inclusion of flags), it suffices to know the X_i (or "<u>unfolded strata</u>") and the X_{ij} (or "<u>joining tubes</u>") for $i,j \in I$, $i < j$, and the morphisms $X_{ij} \to X_j$ (which are "bounding" inclusions) and $X_{ij} \to X_i$ (which are proper fibrations, whose fibres F_{ij} are called "<u>joining fibres</u>" for the strata of index i and j). In the case of a tame multiplicity, however, we must also know the "<u>junction spaces</u>" X_{ijk} ($i < j < k$) and their morphisms in X_{ij}, X_{jk} and above all $X_{i,k}$, included in the following hexagonal commutative diagram, where the two squares on the right are Cartesian, the arrows $\hookrightarrow$ are immersions (not necessarily embeddings here), and the other arrows are proper fibrations:

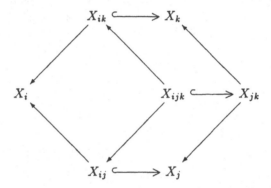

(N.B. This diagram defines X_{ijk} in terms of X_{ij} and X_{jk} over X_j, but not the arrow $X_{ijk} \to X_{ik}$, since $X_{ik} \to X_k$ is <u>not</u> necessarily an embedding.)

In the case of actual stratified tame spaces (which are not, strictly speaking, multiplicities) we can conveniently express the unfolding of this structure, i.e. the system of spaces $X_{i_0,\ldots,i_n}$ in terms of the tame space X_* sum of the X_i, which is equipped with a <u>structure of an ordered object</u> (in the category of tame spaces) having as graph X_{**} of the order relation the sum of the X_{ij} and the X_i (the latter being on the diagonal). Among the essential properties of this ordered structure, let us only note here that $\mathrm{pr}_1 : X_{**} \to X_*$ is a (locally trivial) proper fibration, and $\mathrm{pr}_2 : X_{**} \to X_*$ is a "bounding" embedding. We have an analogous interpretation of the unfolding of a stratified tame multiplicity, in terms of a <u>category structure</u> (replacing a simple ordered structure) "in the sense of tame multiplicities", such that the composition map is given by the morphisms $X_{ijk} \to X_{ik}$ above.

Dear Mr. Faltings,

Many thanks for your quick answer and for sending me your reprints! Your comments on the so-called "Theory of Motives" are of the usual kind, and for a large part can be traced to a tradition which is deeply rooted in mathematics. Namely that research (possibly long and exacting) and attention is devoted only to mathematical situations and relations for which one entertains not merely the hope of coming to a provisional, possibly in part conjectural understanding of a hitherto mysterious region – as it has indeed been and should be the case in the natural sciences – but also at the same time the prospect of a possibility of permanently supporting the newly gained insights by means of conclusive arguments. This attitude now appears to me as an extraordinarily strong psychological obstacle to the development of the visionary power in mathematics, and therefore also to the progress of mathematical insight in the usual sense, namely the insight which is sufficiently penetrating or comprehending to finally lead to a "proof". What my experience of mathematical work has taught me again and again, is that the proof always springs from the insight, and not the other way round – and that the insight itself has its source, first and foremost, in a delicate and obstinate feeling of the relevant entities and concepts and their mutual relations. The guiding thread is the inner coherence of the image which gradually emerges from the mist, as well as its consonance with what is known or foreshadowed from other sources – and it guides all the more surely as the "exigence" of coherence is stronger and more delicate.

To return to Motives, there exists to my knowledge no "theory" of motives, for the simple reason that nobody has taken the trouble to work out such a theory. There is an impressive wealth of available material both of known facts and anticipated connections – incomparably more, it seems to me, than ever presented itself for working out a physical theory! There exists at this time a kind of "yoga des motifs", which is familiar to a handful of initiates, and in some situations provides a firm support for guessing precise relations, which can then sometimes be actually proved in one way or another (somewhat as, in your last work, the statement on the Galois action on the Tate module of abelian varieties). It has the status, it seems to me, of some sort of secret science – Deligne seems to me to be the person who is most fluent in it. His first [published] work, about the degeneration of the Leray spectral sequence for a smooth proper map between algebraic varieties over $\mathbb{C}$, sprang from a simple reflection on "weights" of cohomology groups, which at that time was purely heuristic, but now (since the proof

$\frac{1}{2}$

of the Weil conjectures) can be realised over an arbitrary base scheme. It is also clear to me that Deligne's generalisation of Hodge theory finds for a large part its source in the unwritten "Yoga" of motives – namely in the effort of establishing, in the framework of transcendent Hodge structures, certain "facts" from this Yoga, in particular the existence of a filtration of the cohomology by "weights", and also the semisimplicity of certain actions of fundamental groups.

Now, some words about the "Yoga" of anabelian geometry. It has to do with "absolute" alg. geometry, that is over (arbitrary) ground fields which are finitely generated over the prime fields. A general fundamental idea is that for certain, so-called "anabelian", schemes X (of finite type) over K, the geometry of X is completely determined by the (profinite) fundamental group $\pi_1(X, \xi)$ (where ξ is a "geometric point" of X, with value in a prescribed algebraic closure $\overline{K}$ of K), together with the extra structure given by the homomorphism:

$$(1) \qquad \pi_1(X, \xi) \to \pi_1(K, \xi) = \text{Gal}(\overline{K}/K).$$

The kernel of this homomorphism is the "geometric fundamental group"

$$(2) \qquad \pi_1(\overline{X}, \xi) \qquad (\overline{X} = X \otimes_K \overline{K}),$$

which is also the profinite compactification of the transcendent fundamental group, when $\overline{K}$ is given as a subfield of the field $\mathbb{C}$ of the complex numbers. The image of (1) is an open subgroup of the profinite Galois group, which is of index 1 exactly when $\overline{X}$ is connected.

The first question is to determine which schemes X can be regarded as "anabelian". On this matter, I will in any case restrict myself to the case of non-singular X. And I have obtained a completely clear picture only when $\dim X = 1$. In any case, being anabelian is a purely geometric property, that is, one which depends only on $\overline{X}$, defined over the algebraic closure $\overline{K}$ (or the corresponding scheme over an arbitrary algebraically closed extension of $\overline{K}$, such as $\mathbb{C}$). Moreover, $\overline{X}$ should be anabelian if and only if its connected components are. Finally (in the one dimensional case), a (non-singular connected) curve over $\overline{K}$ is anabelian when its Euler-Poincaré characteristic is < 0, in other words, when its fundamental group is not abelian; this latter formulation is valid at least in the characteristic zero case, or in the case of a proper ("compact") curve – otherwise, one should consider the "prime-to-p" fundamental group. Other equivalent formulations: the group scheme of the automorphisms should be of dimension zero, or still the automorphism group should be finite. For a curve of type (g, ν), where g is the genus, and ν the number of "holes" or "points at infinity", then the anabelian curves

$\frac{2}{3}$

are exactly those whose type is not one of

$$(0,0), (0,1), (0,2) \text{ and } (1,0)$$

in other words

$$2g + \nu > 2 \quad (\text{i.e. } -\chi = 2g - 2 + \nu > 0).$$

When the ground field is $\mathbb{C}$, the anabelian curves are exactly those whose (transcendent) universal cover is "hyperbolic", namely isomorphic to the Poincaré upper half plane – that is, exactly those which are "hyperbolic" in the sense of Thurston.

In any case, I regard a variety as "anabelian" (I could say "elementary anabelian"), when it can be constructed by successive smooth fibrations from anabelian curves. Consequently (following a remark of M.Artin), any point of a smooth variety X/K has a fundamental system of (affine) anabelian neighbourhoods.

Finally, my attention has been lately more and more strongly attracted by the moduli varieties (or better modular <u>multiplicities</u>) $M_{g,\nu}$ of algebraic curves. I am rather convinced that these also may be approached as "anabelian", namely that their relation with the fundamental group is just as tight as in the case of anabelian curves. I would assume that the same should hold for the multiplicities of moduli of polarized abelian varieties.

A large part of my reflections of two years ago were restricted to the case of char. zero, an assumption which, as a precaution, I will now make. As I have not occupied myself with this complex of questions for more than a year, I will rely on my memory, which at least is more easily accessible than a pile of notes – I hope I will not weave too many errors into what follows! A point of departure –among others– was the known fact that for varieties X, Y over an algebraically closed field K, when Y can be embedded into a [quasi-]abelian variety A, a map $X \to Y$ is determined, up to a translation of A, by the corresponding map on H_1 (ℓ-adic). From this, it follows in many situations (as when Y is "elementary anabelian"), that for a dominant morphism f (i.e. $f(X)$ dense in Y), f is known exactly when $H_1(f)$ is. Yet the case of a constant map cannot obviously be included. But precisely the case when X is reduced to a point is of particular interest, if one is aiming at a "characterization" of the points of Y.

Going now to the case of a field K of finite type, and replacing H_1 (namely the "abelianised" fundamental group) with the full fundamental group, one obtains, in the case of an "elementary anabelian" Y, that f is known when $\pi_1(f)$ is known "up to inner automorphism". If I understand correctly, one may work here with the quotients of the fundamental group which are obtained by replacing (2) with the corresponding abelianised group $H_1(\overline{X}, \hat{\mathbb{Z}})$,

$\frac{3}{4}$

instead of with the full fundamental group. The proof follows rather easily
from the Mordell-Weil theorem stating that the group $A(K)$ is a finitely
generated $\mathbb{Z}$-module, where A is the "jacobienne généralisée" of Y, corresponding to the "universal" embedding of Y into a torsor under a quasi-
abelian variety. Here the crux of the matter is the fact that a point of
A over K, i.e. a "section" of A over K, is completely determined by the
corresponding splitting of the exact sequence

$$(3) \qquad\qquad 1 \to H_1(\overline{A}) \to \pi_1(A) \to \pi_1(K) \to 1$$

(up to inner automorphism); in other words by the corresponding cohomology class in

$$H^1(K, \pi_1(\overline{A})),$$

where $\pi_1(\overline{A})$ can be replaced by the ℓ-adic component, namely the Tate
module $T_\ell(\overline{A})$.

From this result, the following easily follows, which rather amazed me
two and a half years ago: let K and L be two fields of finite type (called
"absolute fields" for short), then a homomorphism

$$K \to L$$

is completely determined when one knows the corresponding map

$$(4) \qquad\qquad \pi_1(K) \to \pi_1(L)$$

of the corresponding "outer fundamental groups" (namely when this map
is known up to inner automorphism). This strongly recalls the topological
intuition of $K(\pi, 1)$ spaces and their fundamental groups – namely the homotopy classes of the maps between the spaces are in one-to-one correspondence with the maps between the outer groups. However, in the framework
of absolute alg. geometry (namely over "absolute" fields), the homotopy
class of a map already determines it. The reason for this seems to me to lie
in the extraordinary rigidity of the full fundamental group, which in turn
springs from the fact that the (outer) action of the "arithmetic part" of this
group, namely $\pi_1(K) = \mathrm{Gal}(\overline{K}/L)$, is extraordinarily strong (which is also
reflected in particular in the Weil-Deligne statements).

The last statement ("The reason for this...") came quickly into the typewriter – I now remember that for the above statement on field homomorphisms, it is in no way necessary that they be "absolute" – it is enough
that they should be of finite type over a common ground field k, as long as
one restricts oneself to k-homomorphisms. Besides, it is obviously enough
to restrict attention to the case when k is algebraically closed. On the

other hand, the aforementioned "rigidity" plays a decisive role when we turn to the problem of characterizing those maps (4) which correspond to a homomorphism $K \to L$. In this perspective, it is easy to conjecture the following: when the ground field k is "absolute", then the "geometric" outer homomorphisms are exactly those which commute with the "augmentation homomorphism" into $\pi_1(K)$. [see the correction in the PS: the image must be of finite index] Concerning this statement, one can obviously restrict oneself to the case when k is the prime field, i.e. $\mathbb{Q}$ (in char. zero). The "Grundobjekt" of anabelian alg. geometry in char. zero, for which the prime field is $\mathbb{Q}$, is therefore the group

$$(5) \qquad\qquad \Gamma = \pi_1(\mathbb{Q}) = \mathrm{Gal}(\overline{\mathbb{Q}}/\mathbb{Q}),$$

where $\overline{\mathbb{Q}}$ stands for the algebraic closure of $\mathbb{Q}$ in $\mathbb{C}$.

The above conjecture may be regarded as the main conjecture of "birational" anabelian alg. geometry – it asserts that the category of "absolute birational alg. varieties" in char. zero can be embedded into the category of Γ-augmented profinite groups. There remains the further task of obtaining a ("purely geometric") description of the group Γ, and also of understanding which Γ-augmented profinite groups are isomorphic to some $\pi_1(K)$. I will not go into these questions for now, but will rather formulate a related and considerably sharper conjecture for anabelian curves, from which the above follows. Indeed I see two apparently different but equivalent formulations:

1) Let X, Y be two (connected, assume once and for all) anabelian curves over the absolute field of char. zero, and consider the map

$$(6) \qquad\qquad \mathrm{Hom}_K(X, Y) \to \mathrm{Hom\ ext}_{\pi_1(K)}(\pi_1(X), \pi_1(Y)),$$

where Hom ext denotes the set of outer homomorphisms of the profinite groups, and the index $\pi_1(K)$ means the compatibility with augmentation into $\pi_1(K)$. From the above, one knows that this map is injective. I conjecture that it is bijective [see the correction in the P.S.]

2) This second form can be seen as a reformulation of 1) in the case of a constant map from X into Y. Let $\Gamma(X/K)$ be the set of all K-valued points (that is "sections") of X over K; one considers the map

$$(7) \qquad\qquad \Gamma(X/K) \to \mathrm{Hom\ ext}_{\pi_1(K)}(\pi_1(K), \pi_1(X)),$$

where the second set is thus the set of all the "splittings" of the group extension (3) (where A is replaced by $X - \pi_1(X) \to \pi_1(K)$ is actually surjective, at least if X has a K-valued point, so that X is also "geometrically connected"), or better the set of conjugacy classes of such splittings under the action of the group $\pi_1(\overline{X})$. It is known that (7) is injective, and the main conjecture asserts that it is bijective [see the correction below].

Formulation 1) follows from 2), with K replaced by the function field of X. Moreover, it is indifferent whether X is anabelian or not and, if I am not mistaken, assertion 1) follows even for arbitrary non-singular X (without the assumption $\dim X = 1$). Concerning Y, it follows from the conjecture that assertion 1) remains true, as far as Y is "elementary anabelian" [see the correction in the PS], and correspondingly of course for assertion 2). This in principle now gives the possibility, by applying Artin's remark, to obtain a complete description of the category of schemes of finite type over K "en termes de" $\Gamma(K)$ and systems of profinite groups. Here again I have typed something a little too quickly, as indeed the main conjecture should first be justified and completed with an assertion about which (up to isomorphism) complete $\Gamma(K)$-augmented profinite groups arise from anabelian curves over K. Concerning only an assertion of "pleine fidélité" as in formulations 1) and 2) above, it should be possible to deduce the following, without too much difficulty, from these assertions, or even already (if I am not mistaken) from the above considerably weaker birational variant. Namely, let X and Y be two schemes which are "essentially of finite type over $\mathbb{Q}$", e.g. each one is of finite type over an absolute field of char. zero. (remaining undetermined). X and Y need to be neither non-singular nor connected, let alone "normal" or the like – but they must be assumed to be reduced. I consider the etale topoi X_{et} and Y_{et}, and the map

$$(8) \qquad \mathrm{Hom}(X,Y) \to \mathrm{Iskl}\,\mathrm{Hom}_{\mathrm{top}}(X_{\mathrm{et}}, Y_{\mathrm{et}}),$$

where $\mathrm{Hom}_{\mathrm{top}}$ denotes the (set of) homomorphisms of the topos X_{et} in Y_{et}, and Iskl means that one passes to the (set of) isomorphism classes. (It should be noted moreover that the category $\underline{\mathrm{Hom}}_{\mathrm{top}}(X_{\mathrm{et}}, Y_{\mathrm{et}})$ is rigid, namely that there can be only one isomorphism between two homomorphisms $X_{\mathrm{et}} \to Y_{\mathrm{et}}$. When X and Y are multiplicities and not schemes, the assertion below should be replaced with a correspondingly finer one, namely one should state an equivalence of categories of $\underline{\mathrm{Hom}}(X,Y)$ with $\underline{\mathrm{Hom}}_{\mathrm{top}}(X_{\mathrm{et}}, Y_{\mathrm{et}})$.) It is essential here that X_{et} and Y_{et} are considered simply as topological spaces, that is without their structure sheaves, whereas the left-hand side of (8) can be interpreted as Iskl $\mathrm{Hom}_{\mathrm{top.ann.}}(X_{\mathrm{et}}, Y_{\mathrm{et}})$. Let us first notice that from the already "known" facts, it should follow without difficulty that (8) is injective. In fact I now realize that in the description of the right-hand side of (8), I forgot an important element of the structure, namely that X_{et} and Y_{et} must be considered as topoi over the absolute base $\mathbb{Q}_{\mathrm{et}}$, which is completely described by the profinite group $\Gamma = \pi_1(K)$ (5). So $\mathrm{Hom}_{\mathrm{top}}$ should be read $\mathrm{Hom}_{\mathrm{top}/\mathbb{Q}_{\mathrm{et}}}$. With this correction, we can now state the tantalizing conjecture that (8) should be bijective. This may not be [altogether] correct for the reason that there can exist radicial morphisms $Y \to X$ (so-called "universal homomorphisms"), which produce a topological equivalence $Y_{et} \overset{\sim}{\to} X_{et}$, without being an iso-morphisms, so that there does not exist an inverse map $X \to Y$, whereas it does exist for the etale topoi. If one now assumes that X is normal, then I conjecture that (8) is bijective. In the general case, it should be true that for any ϕ on the right-hand side, one can build a diagram

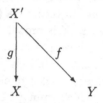

(where g is a "universal homomorphism"), from which ϕ arises in the obvious way. I even conjecture that the same assertion is still valid without the char. zero assertion, that is when $\mathbb{Q}$ is replaced by $\mathbb{Z}$ – which is connected with the fact that the "birational" main conjecture must be valid in arbitrary characteristic, as long as we replace the "absolute" fields with their "perfect closures" $K^{p^{-\infty}}$, which indeed have the same π_1.

I am afraid I have been led rather far afield by this digression about arbitrary schemes of finite type and their etale topoi – you may be more interested by a third formulation of the main conjecture, which sharpens it a little and has a peculiarly "geometric" ring. It is also the formulation I told Deligne about some two years ago, and of which he told me that it would imply Mordell's conjecture. Again let X be an anabelian geometrically connected curve over the absolute field K of char. zero, $\widetilde{X}$ its universal cover, considered as a scheme (but not of finite type) over $\overline{K}$, namely as the universal cover of the "geometric" curve $\overline{X}$. It stands here as a kind of algebraic analogue for the transcendental construction, in which the universal cover is isomorphic to the Poincaré upper half-plane. I also consider the completion $X^{\wedge}$ of X (which is thus a projective curve, not necessarily anabelian, as X can be of genus 0 or 1), together with its normalisation $\widetilde{X}^{\wedge}$ with respect to $\widetilde{X}$, which represents a kind of compactification of $\widetilde{X}$. (If you prefer, you can assume from the start that X is proper, so that $X = X^{\wedge}$ and $\widetilde{X} = \widetilde{X}^{\wedge}$.) The group $\pi_1(X)$ can be regarded as the group of the X-automorphisms of $\widetilde{X}$, and it acts also on the "compactification" $\widetilde{X}^{\wedge}$. This action commutes with the action on $\overline{K}$ via $\pi_1(K)$. I am now interested in the corresponding action

$$\text{Action of } \pi_1(X) \text{ on } \widetilde{X}^{\wedge}(\overline{K}),$$

(the [set of] $\overline{K}$-valued points, or what comes to the same thing, the points of $X^{\wedge}$ distinct from the generic point), and in particular, for a given section

$\frac{7}{8}$

of (3)
$$\pi_1(K) \to \pi_1(X),$$

I consider the corresponding action of the Galois group $\pi_1(K)$. The conjecture is now that <u>the latter action has (exactly) one fixed point</u>.

That it can have at most one fixed point follows from the injectivity in (7), or in any case can be proved along the same lines, using the Mordell-Weil theorem. What remains unproved is the <u>existence</u> of the fixed point, which is more or less equivalent to the surjectivity of (7). It now occurs to me that the formulation of the main conjecture via (7), which I gave a while ago, is correct only in the case where X is proper – and in that case, it is in effect equivalent to the third (just given) formulation. In the contrary case where X is not proper, so has "points infinitely far away", each of these points clearly furnishes a considerable packet of classes of sections (which has the power of the continuum), which <u>cannot</u> be obtained via points lying at a finite distance. These correspond to the case of a fixed point in $\widetilde{X}^\wedge$ which does not lie in $\widetilde{X}$. The uniqueness of the fixed point means among other things, besides the injectivity of (7), that the "packets" which correspond to <u>different</u> points at infinity have empty intersection; and thus any class of sections which does not come from a finite point can be assigned to a uniquely defined point at infinity.

The third formulation of the main conjecture was stimulated by certain transcendental reflections on the action of <u>finite</u> groups on complex algebraic curves and their (transcendentally defined) universal covers, which have played a decisive role in my reflections (during the first half of 1981, that is some two years ago) on the action of Γ on certain profinite anabelian fundamental groups (in particular that of $\mathbb{P}^1 - (0, 1, \infty)$). (This role was mainly that of a guiding thread into a previously completely unknown region, as the corresponding assertions in char $p > 0$ remained unproved, and still do today.) To come back to the action of the Galois groups as $\pi_1(K)$, these appear in several respects as analogous to the action of finite groups, something which for instance is expressed in the above conjecture in a particularly striking and precise way.

I took up the anabelian reflections again between December 81 and April 82, that time with a different emphasis – namely in an effort toward understanding the many-faceted structure of the [Teichmüller] fundamental groups $T_{g,\nu}$ (or better, the fundamental groupoids) of the multiplicities of moduli $\overline{M}_{g,\nu}$, and the action of Γ on their profinite completions. (I would like to return to this investigation next fall, if I manage to extricate myself this summer from the writing up of quite unrelated reflections on the foundations of cohomological resp. homotopical algebra, which has occupied me for four months already.) I appeal to your indulgence for the somewhat chaotic presentation of a circle of ideas which intensively held my attention for six months, but with which I have had for the past two years only very fleeting contacts, if any. If these ideas were to interest you, and if you happened at some point to be in the south of France, it would be a pleasure for me to meet with you and to go into more details of these or other aspects of the "anabelian Yoga". It would also surely be possible to invite you to Montpellier University for some period of time at your convenience; only I am afraid that under the present circumstances, the procedure might be a little long, as the university itself does not at present have funds for such invitations, so that the invitation would have to be decided resp. approved in Paris – which may well mean that the corresponding proposal would have to be made roughly one year in advance.

On this cheering note, I will put an end to this letter, which has somehow grown out of all proportion, and just wish you very pleasant holidays!
Best regards

Your Alexander Grothendieck

PS Upon rereading this letter, I realize that, like the second formulation of the main conjecture, also the generalisation to "elementary anabelian" varieties must be corrected, sorry! Besides I now see that the first formulation must be corrected in the same way – namely in the case where Y is not proper, it is necessary to restrict oneself, on the left-hand side of (6), to nonconstant homomorphisms, and on the right-hand side to homomorphisms $\pi_1(X) \to \pi_1(Y)$, whose images are of finite index (i.e. open). In the case where Y is replaced by an elementary anabelian variety, the bijectivity of (6) is valid, as long as one restricts oneself to <u>dominant</u> homomorphisms on the left-hand side, keeping the same restriction (finite index image) on the right-hand side. The "birational" formulation should be corrected analogously – namely one must restrict oneself to homomorphisms (4) with finite index image.

Returning now to the map (7) in the case of an anabelian curve, one can specify explicitly which classes of sections on the right-hand side do not correspond to a "finite" point, thus do not come from an element on the left-hand side; and if I remember correctly, such a simple characterization of the image of (7) can be extended to the more general situation of an "elementary anabelian" X. As far as I now remember, this characterization (which is of course just as conjectural, and indeed in both directions, "necessary" and "sufficient") goes as follows. Let

$$\pi_1(K)^o = \text{Kernel of } \pi_1(K) \to \widehat{\mathbb{Z}}^* \text{ (the cyclotomic character).}$$

Given a section $\pi_1(K) \to \pi_1(X)$, $\pi_1(K)$ and therefore also $\pi_1(K)^o$ operates on $\pi_1(\overline{X})$, the geometric fundamental group. The condition is now that the subgroup fixed under this action be reduced to 1!

Printed in the United States
By Bookmasters